FERRI'S
BEST TEST

FERRI'S BEST TEST

A Practical Guide to
Clinical Laboratory Medicine
and Diagnostic Imaging

Fourth Edition

FRED F. FERRI, MD, FACP
Clinical Professor
Warren Alpert Medical School
Brown University
Providence, Rhode Island

ELSEVIER

ELSEVIER

1600 John F. Kennedy Blvd.
Ste 1800
Philadelphia, PA 19103-2899

FERRI'S BEST TEST: A PRACTICAL GUIDE TO CLINICAL LABORATORY MEDICINE
AND DIAGNOSTIC IMAGING   ISBN: 978-0-323-51140-7

Notices

Knowledge and best practice in this field are constantly changing. As new research and experience broaden our understanding, changes in research methods, professional practices, or medical treatment may become necessary.

Practitioners and researchers must always rely on their own experience and knowledge in
evaluating and using any information, methods, compounds, or experiments described herein.
In using such information or methods they should be mindful of their own safety and the safety
of others, including parties for whom they have a professional responsibility.

With respect to any drug or pharmaceutical products identified, readers are advised to check
the most current information provided (i) on procedures featured or (ii) by the manufacturer
of each product to be administered, to verify the recommended dose or formula, the method and
duration of administration, and contraindications. It is the responsibility of practitioners, relying
on their own experience and knowledge of their patients, to make diagnoses, to determine dosag-
es and the best treatment for each individual patient, and to take all appropriate safety precautions.

To the fullest extent of the law, neither the Publisher nor the authors, contributors, or editors, assume any liability for any injury and/or damage to persons or property as a matter of
products liability, negligence or otherwise, or from any use or operation of any methods, products, instructions, or ideas contained in the material herein.

Previous editions copyright © 2015, 2010, 2004.

Library of Congress Cataloging-in-Publication Data

Names: Ferri, Fred F., author.
Title: Ferri's best test : a practical guide to clinical laboratory medicine
    and diagnostic imaging / Fred F. Ferri.
Other titles: Best test
Description: Fourth edition. | Philadelphia, PA : Elsevier, [2019] | Includes
    bibliographical references and index.
Identifiers: LCCN 2017039958 | ISBN 9780323511407 (pbk.)
Subjects: | MESH: Clinical Laboratory Techniques | Diagnostic Imaging |
    Reference Values | Handbooks
Classification: LCC RT48.5 | NLM QY 39 | DDC 616.07/5--dc23 LC record available at
https://lccn.loc.gov/2017039958

Executive Content Strategist: James Merritt
Content Development Specialist: Janice Gaillard
Publishing Services Manager: Catherine Jackson
Project Manager: Kate Mannix
Design Direction: Paula Catalano
Illustrations Manager: Nichole Beard

Printed in China

Last digit is the print number: 9   8   7   6   5   4   3   2

Working together
to grow libraries in
developing countries

www.elsevier.com • www.bookaid.org

Preface

This book is intended to be a practical and concise guide to clinical laboratory medicine and diagnostic imaging. It is designed for use by medical students, interns, residents, practicing physicians, and other health care personnel who deal with laboratory testing and diagnostic imaging in their daily work.

As technology evolves, physicians and other health care professionals are faced with a constantly changing armamentarium of diagnostic imaging and laboratory tests to supplement their clinical skills in arriving at a correct diagnosis. In addition, with the advent of managed care, it is increasingly important for physicians to practice cost-effective medicine.

The aim of this book is to be a practical reference for ordering tests, whether they are laboratory tests or diagnostic imaging studies. As such, it is unique in medical publishing. This manual is divided into three main sections: clinical laboratory testing, diagnostic imaging, and diagnostic algorithms. Appendices have been added to the fourth edition, providing general guidelines about when to use contrast agents in ordering CT and MRI scans.

Section I deals with common diagnostic imaging tests. Each test is approached with the following format: Indications, Strengths, Weaknesses, and Comments. The approximate cost of each test is also indicated. For the fourth edition, we have added several new diagnostic modalities such as transient elastography (Fibroscan), CT enterography and CT enteroclysis, comparison tables, and 25 new illustrations.

Section II describes nearly 300 laboratory tests. The fourth edition features several new tables and illustrations. Each test in this section is approached with the following format:

- Laboratory test
- Normal range in adult patients
- Common abnormalities (e.g., positive test, increased or decreased value)
- Causes of abnormal result

Section III includes the diagnostic modalities (imaging and laboratory tests) and algorithms of common diseases and disorders. New algorithms on evaluation of immunodeficiency and hematochezia and several new illustrations have been added to the fourth edition.

Three new appendices provide general guidelines when ordering CT scans and MRIs.

I hope that this unique approach will simplify the diagnostic testing labyrinth and will lead the readers of this manual to choose the best test to complement their clinical skills. However, it is important to remember that laboratory tests and x-rays do not make diagnoses. Doctors and health care professionals do. As such, any laboratory and radiographic results should be integrated with the complete clinical picture to arrive at a diagnosis.

Fred F. Ferri, MD, FACP

Acknowledgments

I extend a special thank you to the authors and contributors of the following texts, who have lent multiple images, illustrations, and text material to this edition and prior editions:

Bennett JE, Dolin R, Blaser MJ. Mandell, Douglas, and Bennett's principles and practice of infectious diseases. 8th ed. Philadelphia: Saunders; 2015.

Boxt LM, Abbara S. The requisites: cardiac imaging. 4th ed. Philadelphia: Elsevier; 2016.

Broder JS. Diagnostic imaging for the emergency physician. Philadelphia: Saunders; 2011.

Feldman M, Friedman LS, Brandt LJ. Sleisenger and Fortran's gastrointestinal and liver disease. 10th ed. Philadelphia: Elsevier; 2016.

Grainger RG, Allison D. Grainger & Allison's diagnostic radiology: a textbook of medical imaging. 4th ed. Philadelphia: Churchill Livingstone; 2001.

Hoffman R, et al. Hematology: basic principles and practice. 5th ed. Philadelphia: Saunders; 2009.

Marx JA, et al. Rosen's emergency medicine. 8th ed. Philadelphia: Saunders; 2014.

McPherson RA, Pincus MR. Henry's clinical diagnosis and management by laboratory methods. 23rd ed. Philadelphia: Elsevier; 2017.

Melmed S, et al. Williams textbook of endocrinology. 12th ed. Philadelphia: Saunders; 2011.

Mettler FA. Primary care radiology. Philadelphia: Saunders; 2000.

Pagana KD, Pagana TJ. Mosby's diagnostic and laboratory test reference. 8th ed. St. Louis: Mosby; 2007.

Parrillo JE, Dellinger RP. Critical care medicine: principles of diagnosis and management in the adult. 4th ed. Philadelphia: Elsevier; 2014.

Pope TL, Bloem HL, Beltran J, Morrison WB, Wilson DJ. Musculoskeletal imaging. 2nd ed. Philadelphia: Saunders; 2014.

Rumack CM, Wilson SR, Charboneau JW, Levine D. Diagnostic ultrasound. 4th ed. Philadelphia: Elsevier; 2011.

Sellke FW, del Nido PJ, Swanson SJ. Sabiston & Spencer surgery of the chest. 9th ed. Philadelphia: Elsevier; 2016.

Skorecki K, Chertow GM, Marsden PA, Taal MW, Yu ASL, Wasser WG. Brenner & Rector's the didney. 10th ed. Philadelphia: Elsevier; 2016.

Soto JA, Lucey BC. Emergency radiology: the requisites. 2nd ed. Philadelphia: Elsevier; 2017.

Talley NJ, Martin CJ. Clinical gastroenterology. 2nd ed. Sidney: Churchill Livingstone; 2006.

Vincent JL, Abraham E, Moore FA, Kochanek PM, Fink MP. Textbook of critical care. 6th ed. Philadelphia: Saunders; 2011.

Webb WR, Brant WE, Major NM. Fundamentals of body CT. 4th ed. Philadelphia: Saunders; 2015.

Weissleder R, Wittenberg J, Harisinghani MG, Chen JW. Primer of diagnostic imaging. 5th ed. St. Louis: Mosby; 2011.

Wu AHB. Tietz clinical guide to laboratory tests. Philadelphia: Saunders; 2006.

Fred F. Ferri, MD, FACP
Clinical Professor
Warren Alpert Medical School
Brown University
Providence, Rhode Island

Contents

Section II   Laboratory Values and Interpretation of Results   129

Section III   Diseases and Disorders   229

Diagnostic Imaging

This section deals with common diagnostic imaging tests. Each test is approached with the following format: Indications, Strengths, Weaknesses, and Comments. The comparative cost of each test is also indicated. Please note that there is considerable variation in the charges and reimbursement for each diagnostic imaging procedure based on individual insurance and geographic region. The costs described in this book are in most cases based on the Resource-Based Relative Value Scale (RBRVS) fee schedule provided by the Centers for Medicare and Medicaid Services for total component billing.

$ Relatively inexpensive–$$$$$ Very expensive

A. Abdominal and Gastrointestinal (GI) Imaging
    1. Abdominal film, plain (kidney, ureter, and bladder [KUB])
    2. Barium enema (BE)
    3. Barium swallow (esophagram)
    4. Upper GI (UGI) series
    5. Computed tomographic colonography (CTC, Virtual Colonoscopy)
    6. CT scan of abdomen and pelvis
    7. CT enterography
    8. CT enteroclysis
    9. Magnetic resonance enterography (MRE)
    10. Cholescintigraphy (hepatobiliary iminodiacetic acid [HIDA] scan, diisopropyliminodiacetic acid [DISIDA] scan)
    11. Endoscopic retrograde cholangiopancreatography (ERCP)
    12. Percutaneous biliary procedures
    13. Magnetic resonance cholangiopancreatography (MRCP)
    14. Meckel's scan (Tc-99m pertechnetate scintigraphy)
    15. MRI scan of abdomen
    16. Small-bowel series
    17. Tc-99m sulfur colloid (Tc-99m SC) scintigraphy for GI bleeding
    18. Tc-99m–labeled red blood cell (RBC) scintigraphy for GI bleeding
    19. Ultrasound of abdomen
    20. Ultrasound of appendix
    21. Ultrasound of gallbladder and bile ducts
    22. Ultrasound of liver
    23. Ultrasound of pancreas
    24. Endoscopic ultrasound (EUS)
    25. Video capsule endoscopy (VCE)
    26. Transient elastography (fibroscan)
B. Breast Imaging
    1. Mammogram
    2. Breast ultrasound
    3. MRI of the breast
C. Cardiac Imaging
    1. Stress echocardiography
    2. Cardiovascular radionuclide imaging (thallium, sestamibi, dipyridamole [persantine] scan), positron emission tomography (PET), single-photon emission computed tomography (SPECT)
    3. Cardiac MRI (CMR) scan
    4. Multidetector CT scan
    5. Transesophageal echocardiogram (TEE)
    6. Transthoracic echocardiography (TTE)
    7. Intravascular ultrasound (IVUS)
D. Chest Imaging
    1. Chest radiograph
    2. CT scan of chest
    3. MRI scan of chest
E. Endocrine Imaging
    1. Adrenal medullary scintigraphy (metaiodobenzylguanidine [MIBG] scan)

    2. Parathyroid (PTH) scan
    3. Thyroid scan (radioiodine uptake study)
    4. Thyroid ultrasound
F. Genitourinary Imaging
    1. Obstetric ultrasound
    2. Pelvic ultrasound
    3. Prostate ultrasound
    4. Renal ultrasound
    5. Scrotal ultrasound
    6. Transvaginal (endovaginal) ultrasound
    7. Urinary bladder ultrasound
    8. Hysterosalpingography (HSG)
    9. Intravenous pyelography (IVP) and intravenous retrograde
       pyelography
G. Musculoskeletal and Spinal Cord Imaging
    1. Plain x-ray films of skeletal system
    2. Bone densitometry (dual-energy x-ray absorptiometry [DEXA] scan)
    3. MRI scan of spine
    4. MRI scan of shoulder
    5. MRI scan of hip and extremities
    6. MRI scan of pelvis
    7. MRI scan of knee
    8. CT scan of spinal cord
    9. Arthrography
    10. CT myelography
    11. Nuclear imaging (bone scan, gallium scan, white blood cell [WBC]
        scan)
H. Neuroimaging of Brain
    1. CT scan of brain
    2. MRI scan of brain
I. Positron Emission Tomography (PET)
J. Single-Photon Emission Computed Tomography (SPECT)
K. Vascular Imaging
    1. Angiography
    2. Aorta ultrasound
    3. Arterial ultrasound
    4. Captopril Renal Scan (CRS)
    5. Carotid ultrasonography
    6. Computed tomographic angiography (CTA)
    7. Magnetic resonance angiography (MRA)
    8. Magnetic resonance direct thrombus imaging (MRDTI)
    9. Pulmonary angiography
    10. Transcranial Doppler
    11. Venography
    12. Compression ultrasonography and venous Doppler ultrasound
    13. Ventilation/perfusion (V/Q) lung scan
L. Oncology
    1. Whole-body integrated (dual-modality) PET-CT
    2. Whole-body MRI

A. Abdominal and Gastrointestinal (GI) Imaging

1. Abdominal Film, Plain (Kidney, Ureter, and Bladder [KUB])
Indications
- Abdominal pain
- Suspected intraperitoneal free air (pneumoperitoneum)
- Bowel distention

Strengths
- Low cost
- Readily available
- Low radiation

Weaknesses
- Low diagnostic yield
- Contraindicated in pregnancy
- Presence of barium from recent radiographs will interfere with interpretation
- Nonspecific test

Comments
- KUB is a coned plain radiograph of the abdomen that includes kidneys, ureters, and bladder.
- A typical abdominal series includes flat and upright radiographs.
- KUB is valuable as a preliminary study when investigating abdominal pain and pathologic findings (e.g., pneumoperitoneum, bowel obstruction [Fig. 1.1], calcifications). Fig. 1.2 describes normal gas pattern. Normal gas collections under the diaphragm can also be seen on chest radiographs (Fig. 1.3).
- This is the least expensive but also least sensitive method to assess bowel obstruction radiographically.
- Cost: $

2. Barium Enema (BE)

Indications
- Colorectal carcinoma
- Diverticular disease (Fig. 1.4)
- Inflammatory bowel disease (IBD)
- Lower GI bleeding
- Polyposis syndromes

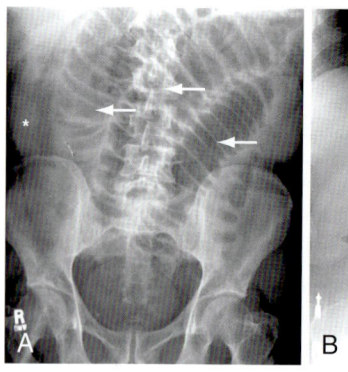
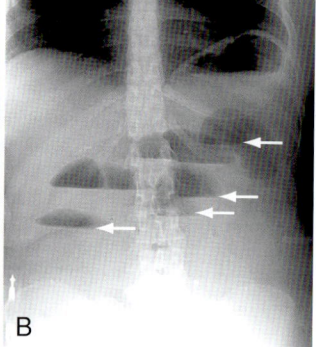

Fig. 1.1 Supine (A) and erect (B) abdominal plain films in 2 patients with adhesive small-bowel obstruction. A, Air-filled distended small-bowel loops (*arrows*) with collapsed colon (*asterisk*). Note that the small-bowel folds (valvulae conniventes) typically extend completely across the intestinal loops. B, Multiple air-fluid levels in dilated small-bowel loops (*arrows*) in the context of nondistended colon. (From Feldman M, Friedman LS, Brandt LJ: *Sleisenger and Fordtran's gastrointestinal and liver disease*, ed 10, Philadelphia, Elsevier, 2016.)

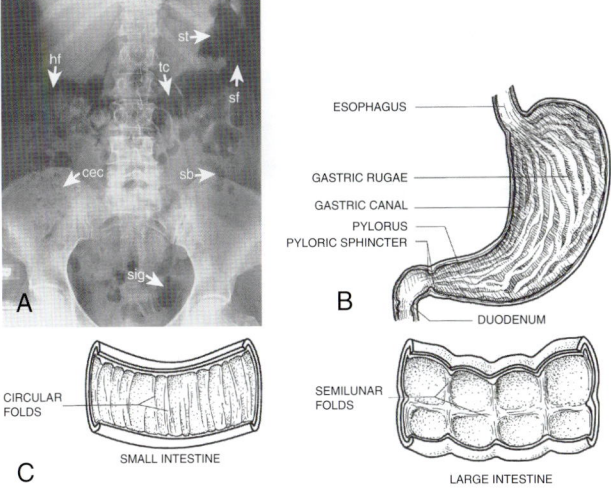

Fig. 1.2 A to C, Normal bowel gas pattern. Gas is normally swallowed and can be seen in the stomach (*st*). Small amounts of air normally can be seen in the small bowel (*sb*), usually in the left midabdomen or the central portion of the abdomen. In this patient, gas can be seen throughout the entire colon, including the cecum (*cec*). In the area where the air is mixed with feces, there is a mottled pattern. Cloverleaf-shaped collections of air are seen in the hepatic flexure (*hf*), transverse colon (*tc*), splenic flexure (*sf*), and sigmoid (*sig*). (From Mettler FA: *Primary care radiology*, Philadelphia, WB Saunders, 2000.)

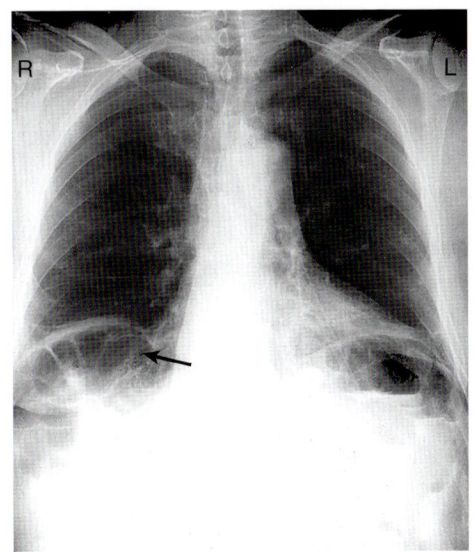

Fig. 1.3 Colonic interposition. This is a normal variant in which the hepatic flexure can be seen above the liver. This is seen as a gas collection under the right hemidiaphragm (*arrow*), but it is clearly identified as colon, owing to the transverse haustral markings. (From Mettler FA: *Essentials of radiology*, ed 3, Philadelphia, Elsevier, 2014.)

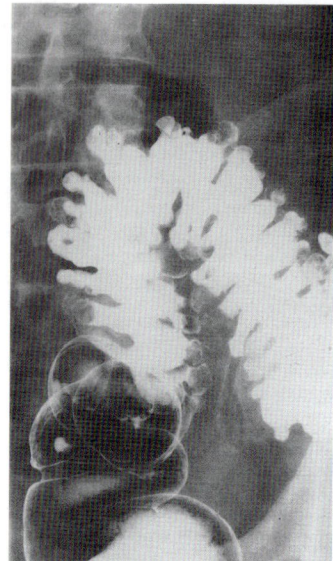

Fig. 1.4 Diverticular disease showing typical muscle changes in the sigmoid and diverticula arising from the apices of the clefts between interdigitating muscle folds. (From Grainger RG, Allison D: *Grainger and Allison's diagnostic radiology: a textbook of medical imaging,* ed 4, Sidney, Churchill Livingstone, 2001.)

- Constipation
- Evaluation for leak of postsurgical anastomotic site

Strengths
- Readily available
- Relatively inexpensive
- Good visualization of mucosal detail with double-contrast barium enema (DCBE)

Weaknesses
- Uncomfortable bowel preparation and procedure for most patients
- Risk of bowel perforation (incidence 1:5000)
- Contraindicated in pregnancy
- Can result in severe postprocedure constipation in older adult patients
- Poorly cleansed bowel will interfere with interpretation
- Poor visualization of rectosigmoid lesions

Comments
- BE is now rarely performed or indicated. Colonoscopy is more sensitive and specific for evaluation of suspected colorectal lesions.
- This test should not be performed in patients with suspected free perforation, fulminant colitis, severe pseudomembranous colitis, or toxic megacolon, or in a setting of acute diverticulitis.

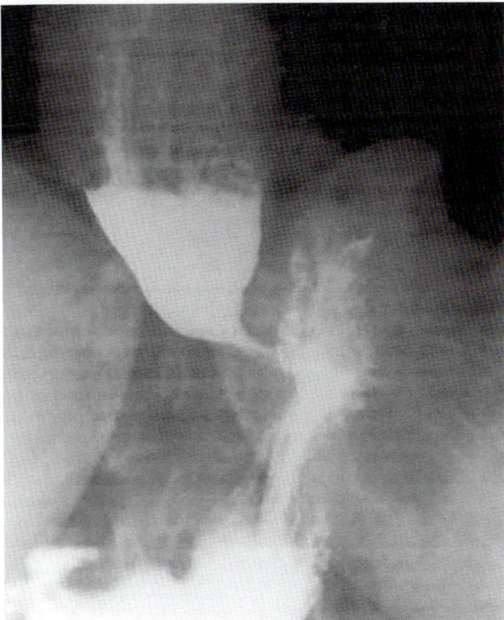

Fig. 1.5 "Bird's beak" appearance of lower esophagus during an upper gastrointestinal radiographic swallow study. (From Cameron JL, Cameron AC [eds]: Achalasic. *Current surgical therapy*, ed 10, 1269–1273, St. Louis, Saunders, 2011.)

- A single-contrast BE uses thin barium to fill the colon, whereas DCBE uses thick barium to coat the colon and air to distend the lumen. Single-contrast BE is generally used to rule out diverticulosis, whereas DCBE is preferable for evaluating colonic mucosa, detecting small lesions, and diagnosing IBD.
- Cost: $$

3. Barium Swallow (Esophagram)
Indications
- Achalasia
- Esophageal neoplasm (primary or metastatic) (Fig. 1.5)
- Esophageal diverticula (e.g., Zenker diverticulum), pseudodiverticula
- Suspected aspiration, evaluation for aspiration after stroke
- Suspected anastomotic leak
- Esophageal stenosis or obstruction
- Extrinsic esophageal compression
- Dysphagia
- Esophageal tear or perforation
- Fistula (aortoesophageal, tracheoesophageal)

- Esophagitis (infectious, chemical)
- Mucosal ring (e.g., Schatzki ring)
- Esophageal webs (e.g., Plummer-Vinson syndrome)

Strengths
- Low cost
- Readily available

Weaknesses
- Contraindicated in pregnancy
- Requires patient cooperation
- Radiation exposure

Comments
- In a barium swallow study, the radiologist observes the swallowing mechanism while films of the cervical and thoracic esophagus are obtained.
- Barium is generally used because it provides better anatomic detail than water-soluble contrast agents; however, diatrizoate (Hypaque) or Gastrografin should be used rather than barium sulfate in suspected perforation or anastomotic leak because free barium in the peritoneal cavity induces a granulomatous response that can result in adhesions and peritonitis; in the mediastinum, free barium can result in mediastinitis.
- Cost: $

4. Upper GI (UGI) Series
Indications
- Gastroesophageal reflux disease (GERD)
- Peptic ulcer disease
- Esophageal carcinoma
- Gastric carcinoma (Fig. 1.6)
- Gastric lymphoma
- Gastric polyps
- Gastritis (hypertrophic, erosive, infectious, granulomatous)
- Gastric outlet obstruction
- Gastroparesis

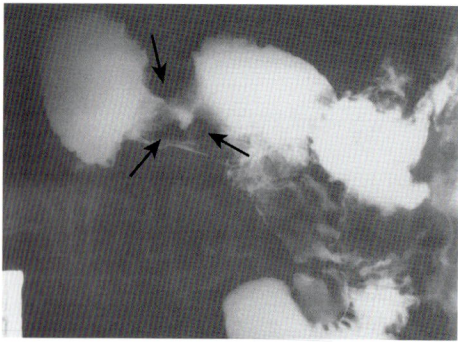

Fig. 1.6 Gastric adenocarcinoma of the stomach (*arrows*). (From Talley NJ, Martin CJ: *Clinical gastroenterology,* ed 2, Sidney, Churchill Livingstone, 2006.)

- Metastatic neoplasm (from colon, liver, pancreas, melanoma)
- Congenital abnormalities (e.g., hypertrophic pyloric stenosis, antral mucosal diaphragm)
- Evaluation for complications after gastric surgery

Strengths
- Inexpensive
- Readily available

Weaknesses
- Contraindicated in pregnancy
- Can result in significant postprocedure constipation in older adult patients
- Requires patient cooperation
- Radiation exposure

Comments
- Upper endoscopy is invasive and more expensive but is more sensitive and has replaced UGI series for evaluation of esophageal and gastric lesions.
- In a barium swallow examination, only films of the cervical and thoracic esophagus are obtained, whereas in a UGI series films are taken of the thoracic esophagus, stomach, and duodenal bulb.
- Barium provides better anatomic detail than water-soluble contrast agents; however, water-soluble contrast agents (Gastrografin, Hypaque) are preferred when perforation is suspected or postoperatively to assess anastomosis for leaks or obstruction because free barium in the peritoneal cavity can produce a granulomatous response that can result in adhesions.
- It is necessary to clean out the stomach with nasogastric (NG) suction before performing contrast examination when gastric outlet obstruction is suspected.
- Cost: $$

5. Computed Tomographic Colonography (CTC, Virtual Colonoscopy)

Indication
- Screening for colorectal carcinoma

Strengths
- May be more acceptable to patients than fiberoptic colonoscopy
- Does not require sedation
- Safer than fiberoptic colonoscopy
- Lower cost than fiberoptic colonoscopy
- Standard examination does not require intravenous (IV) contrast
- Also visualizes abdomen and lower thorax and can detect abnormalities there (e.g., aortic aneurysms, cancers of ovary, pancreas, lung, liver, kidney)

Weaknesses
- Failure to detect clinically important flat lesions, which do not protrude into the lumen of the colon
- Need for cathartic preparation; requires the same bowel preparation as colonoscopy
- Lack of therapeutic ability; nearly 10% of patients require follow-up traditional colonoscopies because of abnormalities detected by CTC
- Incidental findings detected on CTC can lead to additional and often unnecessary testing
- Radiation exposure

Comments
- CTC uses a computed tomographic (CT) scanner to take a series of radiographs of the colon and a computer to create a three-dimensional (3-D) view (Fig. 1.7). It can

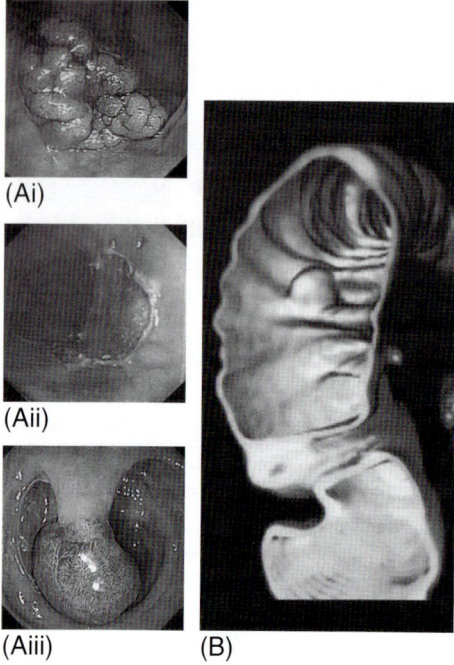

(Ai)

(Aii)

(Aiii)   (B)

Fig. 1.7 Colon polyps seen at (Ai–Aiii) colonoscopy and (B) computed tomography colonography. Aii is after endoscopic resection of the polyps in Ai. (From Ballinger A: *Kumar and Clark's essentials of clinical medicine,* ed 5, Edinburgh, Saunders, 2012.)

be uncomfortable because the patient isn't sedated and a small tube is inserted in the rectum to inflate the colon so that it can be more easily viewed.
- CTC uses a low-dose x-ray technique, typically 20% of the radiation used with standard diagnostic CT, and approximately 10% less than double-competent BE.
- Most insurance companies do not pay for CTC, but that could change if colon cancer screening guidelines endorse it.
- Sensitivity ranges from 85% to 94%, and specificity is approximately 96% for detecting large (≥1 cm) polyps.
- Cost: $$$

6. CT Scan of Abdomen and Pelvis
Indications
- Evaluation of abdominal mass, pelvic mass
- Suspected lymphoma
- Staging of neoplasm of abdominal and pelvic organs
- Splenomegaly
- Intraabdominal, pelvic, or retroperitoneal abscess

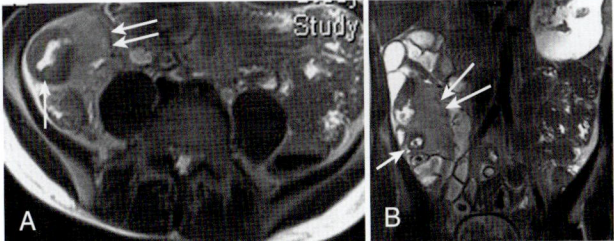

Fig. 1.8 Axial (A) and coronal (B) T2-weighted images from a magnetic resonance enterography showing thickening of the terminal ileum (*arrow*) with fat stranding (*double arrow*) in the surrounding mesentery in a patient with known Crohn's disease. (From Fielding JR et al: *Gynecologic imaging,* Philadelphia, Saunders, 2011.)

- Abdominal and pelvic trauma
- Jaundice
- Pancreatitis: contrast-enhanced CT is considered the gold standard for diagnosing pancreatic necrosis and peripancreatic collections, and for grading acute pancreatitis
- Suspected bowel obstruction
- Appendicitis
- IBD (Fig. 1.8)

Strengths
- Fast
- Noninvasive

Weaknesses
- Potential for significant contrast reaction
- Suboptimal sensitivity for traumatic injury of the pancreas, diaphragm, small bowel, and mesentery
- Retained barium from other studies will interfere with interpretation
- Expensive
- Relatively contraindicated in pregnancy
- Radiation exposure

Comments
- CT with contrast is the initial diagnostic imaging of choice in patients with left lower quadrant (LLQ) and right lower quadrant (RLQ) abdominal pain or mass in adults. Ultrasound is preferred as the initial imaging modality in children, young women, and in the evaluation of right upper quadrant (RUQ) and midabdominal pain or mass unless the patient is significantly obese (poor visualization).
- CT of abdomen and pelvis with contrast is the imaging procedure of choice for suspected abdominal abscess in adults.
- CT is 90% sensitive for small-bowel obstruction.
- The orientation of CT and magnetic resonance imaging (MRI) scans is described in Fig. 1.9. Fig. 1.10 illustrates the structures seen on a normal CT of the abdomen and pelvis. Fig. 1.11 illustrates the Hounsfield CT density scale and fat as a contrast agent. Fig. 1.11 illustrates the normal pancreas with IV and oral contrast.
- Cost: CT without contrast $$, CT with contrast $$$, CT with and without contrast $$$

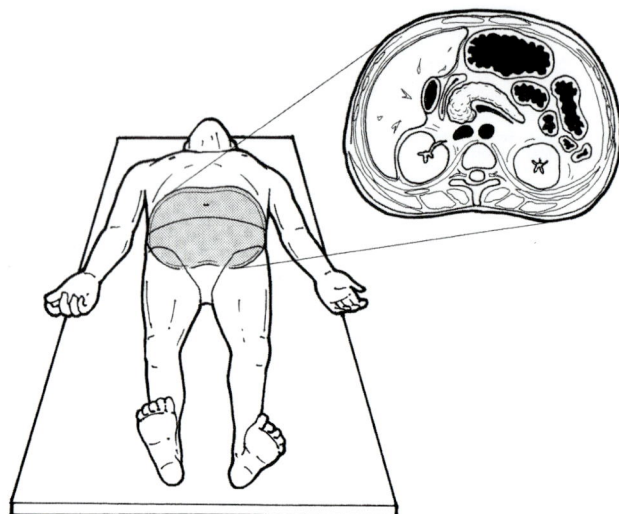

Fig. 1.9 Orientation of computed tomography (CT) and magnetic resonance (MR) images. CT and MR usually present images as transverse (axial) slices of the body. The orientation of most slices is the same as that of a patient viewed from the foot of the bed. (From Mettler FA: *Primary care radiology,* Philadelphia, Saunders, 2000.)

7. CT Enterography

Indications

- CT enterography using multidetector CT (MDCT), which improves image quality by limiting motion artifacts, is a first-line modality for examination of small-bowel disease.
- Indications for CT enterography include Crohn's disease and other suspected inflammatory bowel diseases, intermittent small-bowel obstruction, obscure GI bleeding, and suspected tumors of the small bowel.

Strengths

- Fast
- Noninvasive
- Advantages of CT enterography over traditional barium-based small-bowel follow-through examination include demonstration of the entire thickness of the bowel wall and disease in the mesentery, as well as display of bowel loops without superimposition.

Weaknesses

- Requires fasting for 4 to 6 hours before procedure
- Potential for significant contrast reaction
- Retained barium from other studies will interfere with interpretation
- Relatively contraindicated in pregnancy
- Radiation exposure

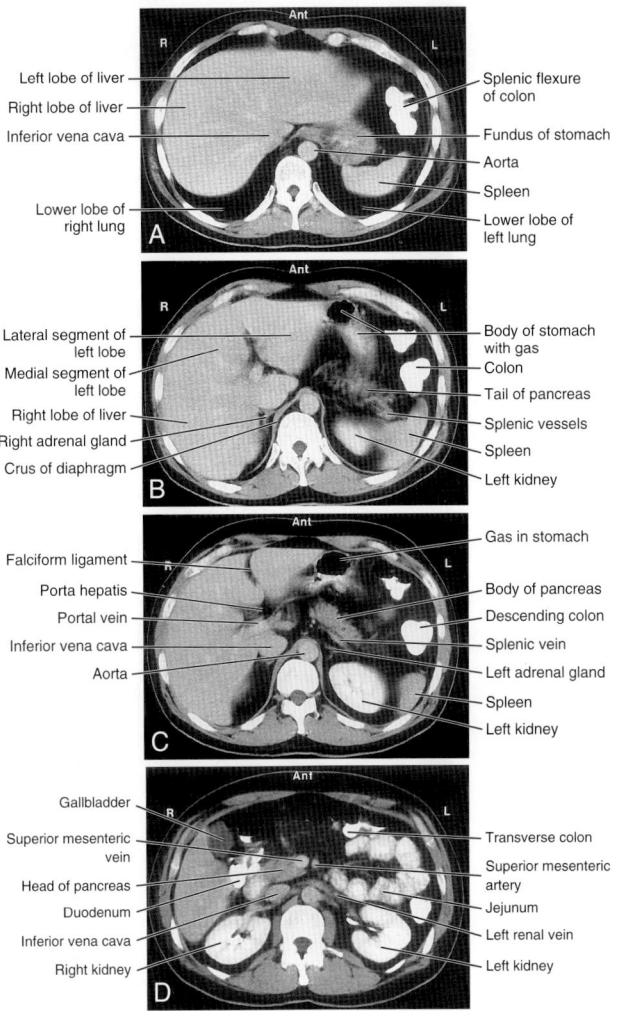

FIG. 1.10 Normal transverse computed tomography anatomy of the abdomen and pelvis. The patient has been given oral, rectal, and intravenous contrast media. (From Mettler FA: *Essentials of radiology*, ed 3, Philadelphia, Elsevier, 2014.)

Continued

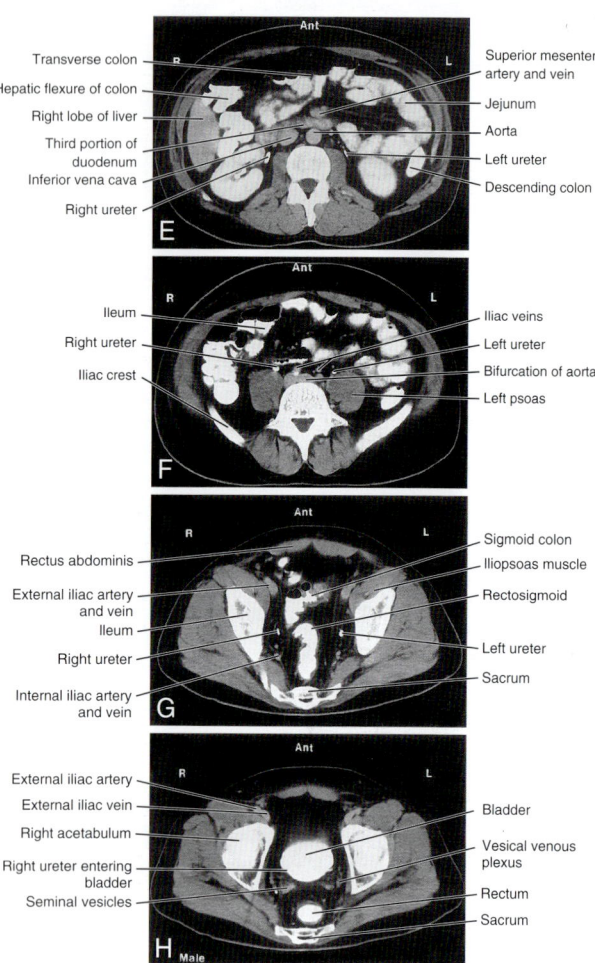

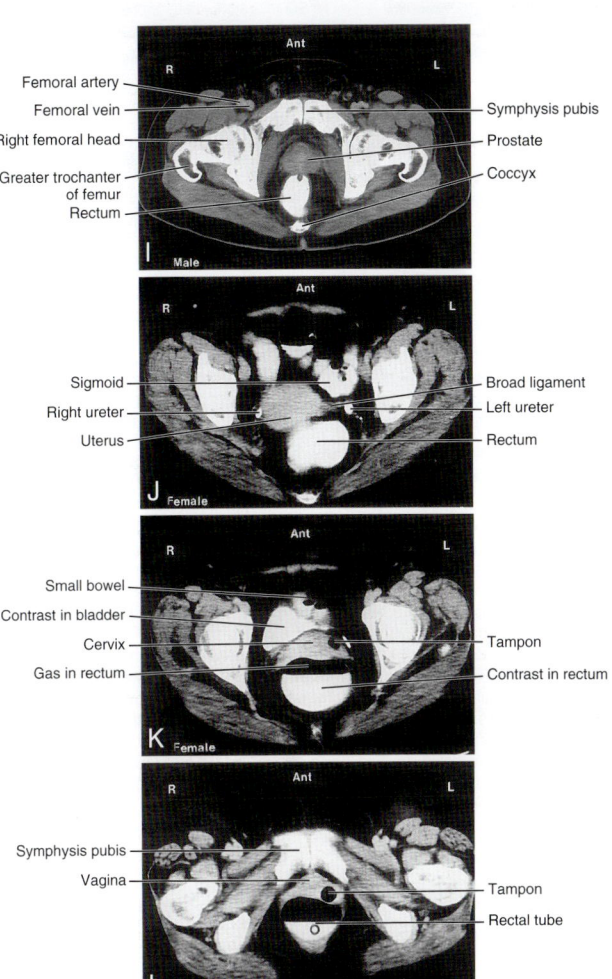

Fig. 1.10, cont'd

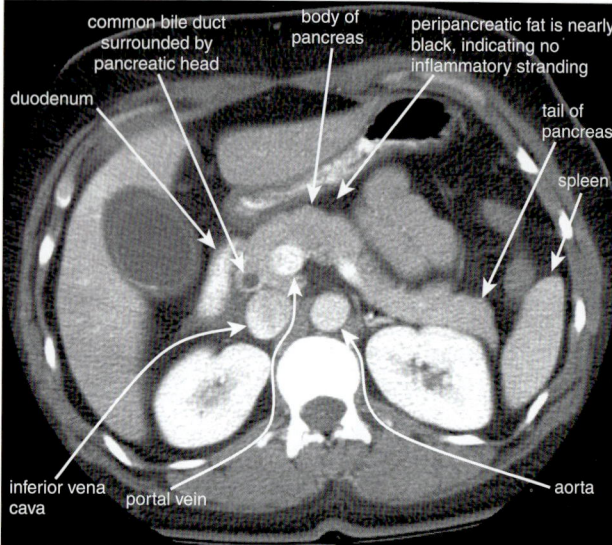

FIG. 1.11 Normal pancreas, computed tomography with intravenous (IV) and oral contrast, soft-tissue window. This scan shows a normal pancreas. In many patients, the pancreas is not so horizontally oriented and is therefore difficult to see in a single slice. Here the common course of the pancreas is seen. The pancreatic head is draped over the portal vein. The tail of the pancreas crosses the midline and then moves posteriorly. It crosses the left kidney and ends medial to the spleen. The duodenum is to the right of the pancreatic head, filled with oral contrast. The common bile duct is seen as a hypodense area within the pancreatic head because it is filled with bile. The contrast between the dark bile and the bright pancreatic tissue is increased by the administration of IV contrast, because the pancreas enhances as a result of high blood flow. The fat surrounding the pancreas is dark, which is normal and indicates the absence of inflammatory stranding; almost the entire pancreas is outlined in fat and has a distinct border. Incidentally, the patient has an abnormal dilated gallbladder with pericholecystic fluid. Given this finding, the prominent common bile duct should be inspected further for an obstructing stone. (From Broder JS: *Diagnostic imaging for the emergency physician*, Philadelphia, Saunders, 2011.)

Comments

- CT enterography differs from routine CT of the abdomen and pelvis in the use of large volumes of low-attenuation intraluminal contrast agent to optimally distend the bowel lumen, matched with thin-slice collimation and routine high-detail coronal and sagittal reformations. Low-attenuation intraluminal contrast paired with intravenous contrast administration optimally displays both the lumen and wall of the small bowel (Fig. 1.12).

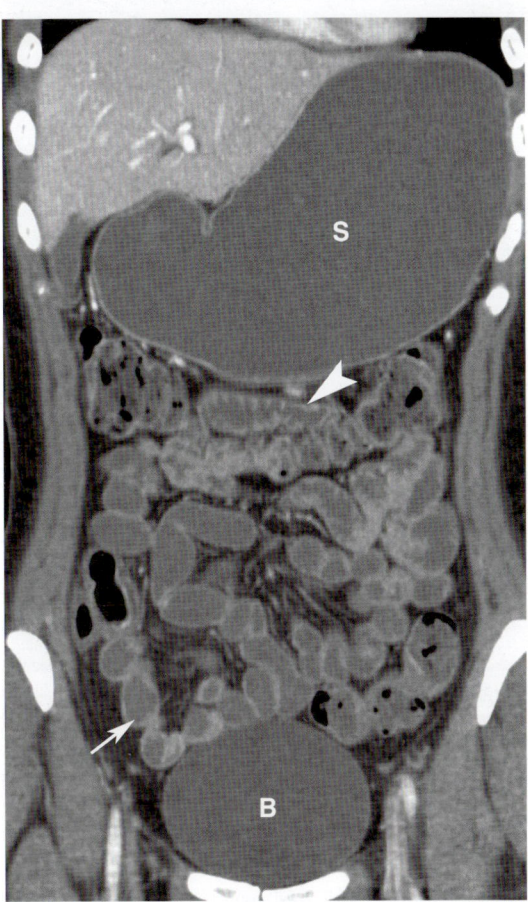

Fig. 1.12 Computed tomography (CT) enterography. Coronal image from a CT enterography examination using low-attenuation intraluminal contrast as well as intravenous contrast shows the normal fold pattern of the jejunum (*arrowhead*) and the normal absence of folds of the ileum (*arrow*). The stomach (*S*) is well distended with intraluminal contrast agent. The bladder (*B*) is also filled. (From Webb WR, Brant WE, Major NM: *Fundamentals of body CT*, ed 4, Philadelphia, Saunders, 2015.)

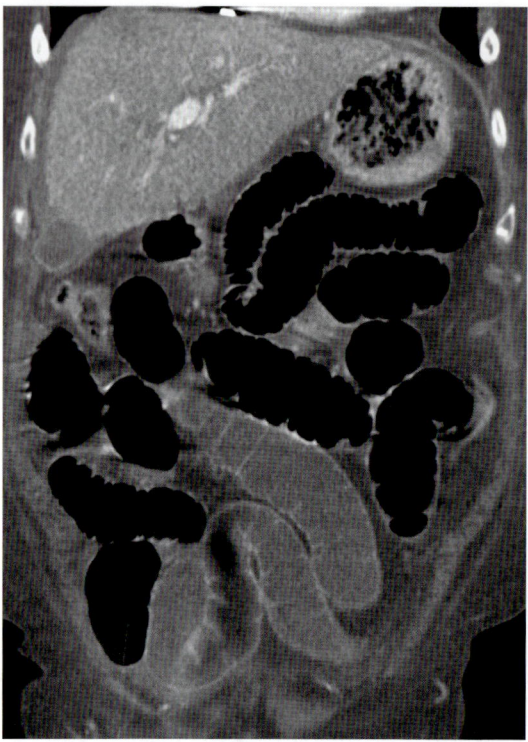

FIG. 1.13 Computed tomography (CT) enteroclysis: distal small-bowel obstruction. Coronal image from a CT enteroclysis examination shows diffuse dilatation of the small bowel with low-attenuation intraluminal contrast agent and air. An adhesion was the cause of this obstruction. (From Webb WR, Brant WE, Major NM: *Fundamentals of body CT,* ed 4, Philadelphia, Saunders, 2015.)

8. CT Enteroclysis

Indication
- Performed on patients with small-bowel obstruction to find the level and cause of obstruction (Fig. 1.13)

Weaknesses
- Potential for significant contrast reaction
- Relatively contraindicated in pregnancy
- Radiation exposure

Comments
- For CT enteroclysis, contrast medium for small-bowel distention is injected into the small bowel through a nasojejunal tube rather than given orally as for CT enterography. A nasojejunal tube is positioned at the duodenojejunal junction under

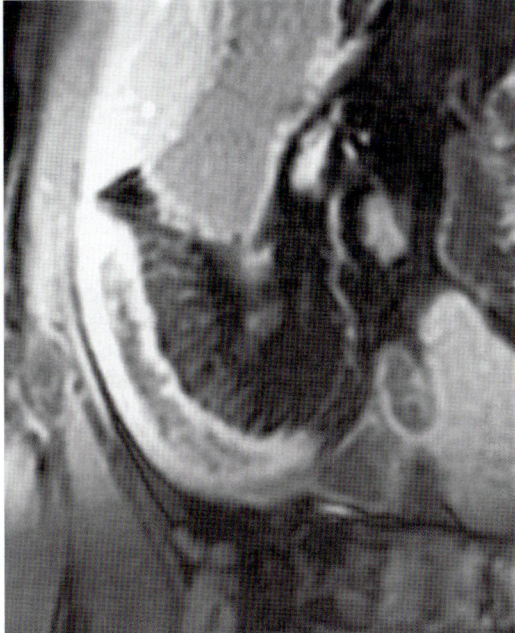

Fɪɢ. 1.14 MR enterography with gadolinium contrast in a patient with Crohn's disease. This coronal view shows mural hyperenhancement, mural thickening, and the comb sign (engorged perienteric vasculature) involving the terminal ileum. The vessels are seen medial to the inflamed loop and resemble the teeth of a comb. (Courtesy Dr. Edward Loftus and Dr. Jeffrey Fidler, Rochester, Minn.)

fluoroscopic guidance. A total volume of 1200 to 1600 mL of low-attenuation oral contrast medium is injected into the small bowel at 60 to 120 mL/minute. Glucagon or other antispasmodic medication is administered. Intravenous contrast is utilized, and MDCT scanning is performed using the same parameters as for CT enterography.

9. Magnetic Resonance Enterography (MRE)

Indication
• Evaluation of small and large bowel in patients with IBD

Strengths
• Depicts extraluminal abnormalities
• Useful to distinguish active from fibrotic strictures. Findings of intestinal wall thickening, submucosal edema, vasa recta engorgement, and lymphadenopathy are signs of active disease (Fig. 1.14)
• Better delineation of fistulas
• Not affected by overlying gas (unlike ultrasound)
• No ionizing radiation

Weaknesses
- Relatively long acquisition times
- Expensive
- May miss early mucosal lesions

Comments
- Requires fasting for 6 hours before procedure.
- Using dynamic fast imaging employing steady-state acquisition (FIESTA) images can add information regarding the functional status of fibrotic segments. Using these criteria, compared with the gold standard of ileocolonoscopy with biopsies, MRI images can yield a diagnostic accuracy of 90%.
- Cost: $$$

10. Cholescintigraphy (Hepatobiliary Iminodiacetic Acid [HIDA] Scan, Diisopropyliminodiacetic Acid [DISIDA] Scan)

Indications
- Acute cholecystitis (normal ultrasound but high clinical suspicion for acalculous cholecystitis)
- Chronic acalculous cholecystitis
- Bile leak
- Postcholecystectomy syndrome
- Obstruction of bile flow (normal ultrasound but high suspicion for cystic duct calculus)
- Biliary dyskinesia
- Biliary atresia
- Afferent loop syndrome
- Evaluation of focal liver lesions

Strengths
- Not operator dependent
- Can be performed on an emergency basis in a nonfasting patient
- High specificity for excluding acute cholecystitis
- Use of HIDA may be limited in jaundiced patients, but DISIDA may allow imaging of the biliary tree in patients with serum bilirubin values as high as 20 mg/dL

Weaknesses
- Severe hepatocellular dysfunction with bilirubin greater than 20 mg/dL will result in poor excretion and nondiagnostic study
- Recent or concomitant use of opiates or meperidine may interfere with bile flow
- False positives common
- Time-consuming (requires more than 1 hour of actual imaging time and patient preparation)

Comments
- In a normal scan, the radiopharmaceutical is cleared from the blood pool after 5 minutes, there is noticeable liver clearing after 30 minutes, and gallbladder and bowel activity is visualized after 60 minutes. Images are obtained every 5 minutes for 1 hour. Late images can be obtained for up to 4 hours after injection. Nonvisualization of the gallbladder is indicative of cholecystitis (Fig. 1.15).
- This test is most helpful when clinical suspicion for cholecystitis is high and ultrasound results are inconclusive. A positive scan result is defined as nonvisualization of the gallbladder with preserved excretion into the bile duct or small intestine. The accuracy of the test for detecting acute cholecystitis is 92%, superior to that of ultrasound.
- Food intake will interfere with test. Optimal fasting is 4 to 12 hours. Fasting longer than 24 hours will also lead to inconclusive examination.
- Cost: $$$

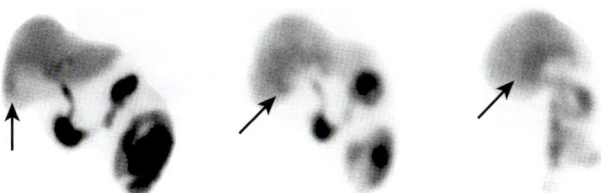

Fig. **1.15** Acute cholecystitis, hot rim sign (*arrows*), is suspicious for gangrenous gallbladder. Curvilinear area of relatively increased activity in liver adjacent to gallbladder (GB) persists in delayed images. Anterior, right anterior oblique, and right lateral views start at 40 minutes after injection. GB did not visualize at 4 hours (not shown). (From Specht N: *Practical guide to diagnostic imaging,* St. Louis, Mosby, 1998.)

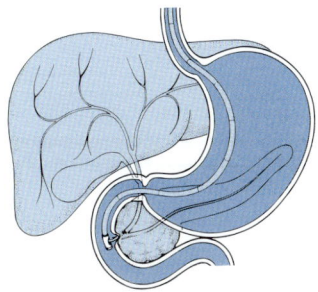

Fig. **1.16** Endoscopic retrograde cholangiopancreatography. The fiberoptic scope is passed into the duodenum. Note the small catheter being advanced into the biliary duct. (From Pagana KD, Pagana, TJ: *Mosby's diagnostic and laboratory test reference,* ed 8, St. Louis, Mosby, 2007.)

11. Endoscopic Retrograde Cholangiopancreatography (ERCP)
Indications
- Evaluation and treatment of diseases of the bile ducts and pancreas
- Treatment of choice for bile duct stones (Fig. 1.16) and for immediate relief of extrahepatic biliary obstruction in benign disease
- Other indications are biliary obstruction caused by cancer, acute and recurrent pancreatitis, pancreatic pseudocyst, suspected sphincter of Oddi dysfunction
- Can be used for diagnostic purposes when magnetic resonance cholangiography (MRCP) and other imaging studies are inconclusive or unreliable, such as in suspected cases of primary sclerosing cholangitis early in the disease, when the changes in duct morphologic characteristics are subtle, or in patients with nondilated bile duct and clinical signs and symptoms highly suggestive of gallstone or biliary sludge
- Preferred modality in patients with high pretest probability of sphincter dysfunction or ampullary stenosis

Strengths
- Preferred modality for treatment of bile duct stones (Fig. 1.17)
- Well suited to evaluate for and treat bile duct leaks and biliary tract injury after open or laparoscopic biliary surgery

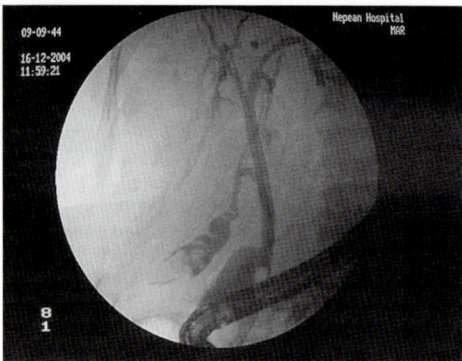

F<small>IG</small>. **1.17** Endoscopic retrograde cholangiopancreatography demonstrating gallstones within the gallbladder and common bile duct. (From Talley NJ, Martin CJ: *Clinical gastroenterology,* ed 2, Sidney, Churchill Livingstone, 2006.)

- ERCP in management of pancreatic and biliary cancer allows access to obstructed bile and pancreatic ducts for collecting tissue samples and placing stents to temporarily relieve obstruction

Weaknesses
- Invasive, technically difficult procedure
- Five percent to 7% risk of pancreatitis depending on patient, procedure, and operator expertise; other complications, such as bleeding, cholangitis, cholecystitis, cardiopulmonary events, perforation, and death, occur far less often

Comments
- In ERCP, contrast-agent injection is performed through the endoscope after cannulation of the common bile duct (CBD). Complications include pancreatitis, duodenal perforation, and GI bleeding.
- Although the complication rate of ERCP is acceptable compared with other invasive procedures such as biliary bypass surgery or open bile duct exploration, the rate is too high for patients with a low pretest probability of disease if the procedure is to be done purely diagnostically.
- Centers that perform a significant volume of ERCP have higher completion rates and lower complication rates.
- ERCP is not indicated for the management of mild pancreatitis or nonbiliary pancreatitis, and its overall use in patients with acute pancreatitis continues to be debated.
- Cost: $$$$

12. Percutaneous Biliary Procedures
Indications
- Transhepatic cholangiogram (Fig. 1.18): used for demonstration of biliary anatomy, first step before biliary drainage or stent placement
- Biliary drainage: used for biliary obstruction
- Biliary stent placement: used for malignant biliary stricture (Fig. 1.19), inability to place endoscopic stent

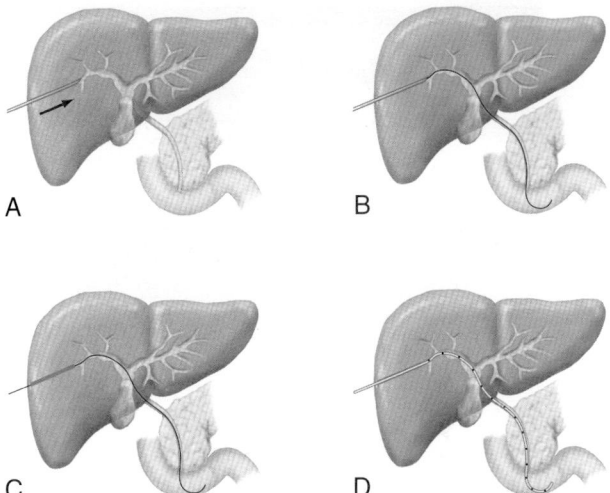

Fig. 1.18 Schematic showing percutaneous transhepatic cholangiography. A, A peripheral bile duct is identified and entered with a needle. B, A guidewire is passed through the needle across the obstructing lesion into the duodenum. C, The needle has been withdrawn. D, An internal-external catheter is inserted over the guidewire. (From Feldman M, Friedman LS, Brandt LJ: *Sleisenger and Fordtran's gastrointestinal and liver disease*, ed 10, Philadelphia, Elsevier, 2016.)

Weaknesses
- Invasive
- Operator dependent
- Cost: $$$$

13. Magnetic Resonance Cholangiopancreatography (MRCP)
Indications
- Suspected biliary or pancreatic disease
- Unsuccessful ERCP, contraindication to ERCP, and presence of biliary enteric anastomoses (e.g., choledochojejunostomy, Billroth II anastomosis)

Strengths
- Advantages over ERCP: noninvasive, less expensive, requires no radiation, less operator dependent, allows better visualization of ducts proximal to obstruction, and can allow detection of extraductal disease when combined with conventional T1W and T2W sequences
- Useful in patients who have biliary or pancreatic pain but no objective abnormalities in liver tests or routine imaging studies
- Can detect retained stone with sensitivity of 92% and specificity of 97% (Fig. 1.20)

Weaknesses
- Limitations of MRCP include artifacts caused by surgical clips, pneumobilia, or duodenal diverticula; and for use in patients with implantable devices or claustrophobia
- Accuracy diminished by stones 1 mm or less and normal bile duct diameter (<8 mm)

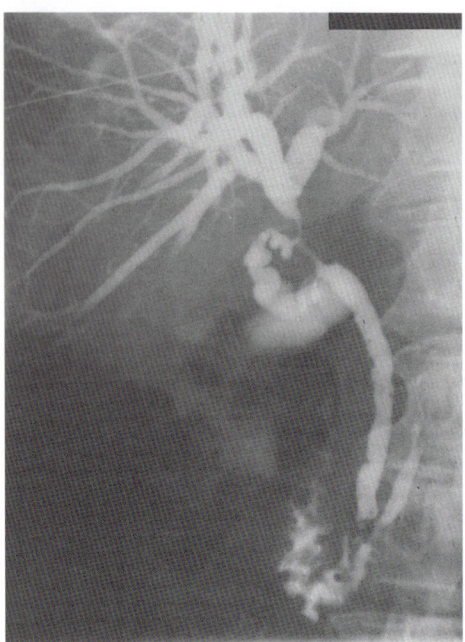

Fig. **1.19** Percutaneous transhepatic cholangiography in hilar tumor assessment. Relatively undistended ducts in a patient with a cholangiocarcinoma; a short stricture involves the junction of the common hepatic and common bile ducts. (From Grainger RG, Allison D: *Grainger and Allison's diagnostic radiology: a textbook of medical imaging*, ed 4, Sidney, Churchill Livingstone, 2001.)

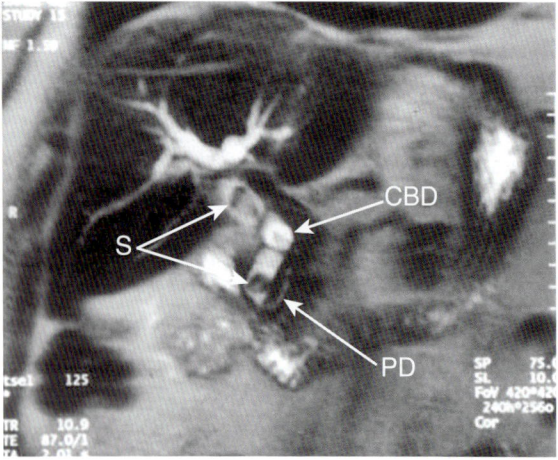

Fig. **1.20** Magnetic resonance cholangiopancreatogram of a dilated biliary tract. The common bile duct (CBD), pancreatic duct (PD), and two large common duct stones (S) are shown. (From Goldman L, Schafer AI: *Goldman's Cecil medicine*, ed 24, Philadelphia, Saunders, 2012.)

- Decreased spatial resolution makes MRCP less sensitive to abnormalities of the peripheral intrahepatic ducts (e.g., sclerosing cholangitis) and pancreatic ductal side branches (e.g., chronic pancreatitis)
- Cannot perform therapeutic endoscopic or percutaneous interventions for obstructing bile duct lesions; thus, in patients with high clinical suspicion for bile duct obstruction, ERCP should be initial imaging modality to provide timely intervention (e.g., sphincterotomy, dilatation, stent placement, stone removal) if necessary
- Pitfalls include pseudofilling defects, pseudodilations, and nonvisualization of ducts

Comments
- Overall sensitivity of MRCP for biliary obstruction is 95%. The procedure is less sensitive for stones (92%) and malignant conditions (92%) than for the presence of obstruction.
- Cost: $$$$

14. Meckel's Scan (Tc-99m Pertechnetate Scintigraphy)

Indication
- Identification of Meckel's diverticulum

Strengths
- In children, overall sensitivity for Meckel's diverticulum is 85%; specificity is 95%; sensitivity lower in adults (63%)

Weaknesses
- False-negative studies may occur because of lack of sufficient gastric mucosa (Meckel's diverticula that do not contain gastric mucosa are not detectable), poor technique, Meckel's diverticulum with impaired blood supply, or rapid washout of secreted pertechnetate
- False positives can be caused by several factors, including atrioventricular (AV) malformations, hemangiomas, peptic ulcer, IBD, neoplasms, intussusception, and hydronephrosis
- Barium in GI tract from prior studies may mask radionuclide concentration

Comments
- Meckel's diverticulum appears scintigraphically as a focal area of increased intraperitoneal activity, usually 5 to 10 minutes after tracer injection (Fig. 1.21).
- Full stomach or urinary bladder may obscure an adjacent Meckel's diverticulum; therefore fasting for 4 hours and voiding before, during, and after scan are important.
- Cost: $$

15. MRI Scan of Abdomen

Indications
- Suspected liver hemangioma (Fig. 1.22)
- Evaluation of adrenal mass
- Cervical cancer staging
- Endometrial cancer staging
- Evaluation of renal mass in patients allergic to iodine and in patients with diminished renal function
- Staging of renal cell carcinoma
- Evaluation of Müllerian duct anomalies when ultrasound is equivocal
- Characterization of pelvic mass indeterminate on ultrasound
- Evaluation of hepatic mass

Strengths
- Noninvasive
- Generally safe contrast agent (MRI uses gadolinium, an IV agent that is less nephrotoxic)

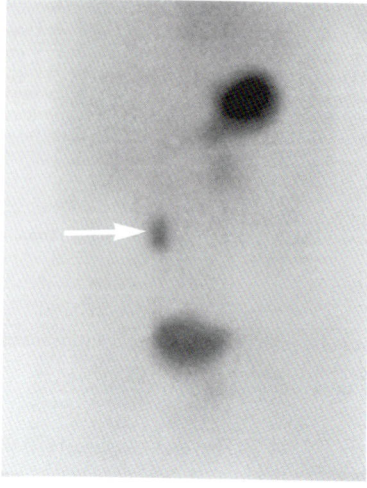

Fig. 1.21 Radionuclide image of Meckel's diverticulum. Increased radionuclide uptake by ectopic gastric mucosa (*arrow*) in the Meckel's diverticulum. The patient was an 11-month-old boy who presented with acute bleeding. (Courtesy Dr. Kieran McHugh and reproduced with permission from Nolan DJ: *Schweiz Med Wochenschr* 128:109–114, 1998.) (From Grainger RG, Allison DJ, Adam A, Dixon AK [eds]: *Grainger and Allison's diagnostic radiology*, ed 4, Philadelphia, Churchill Livingstone, 2001.)

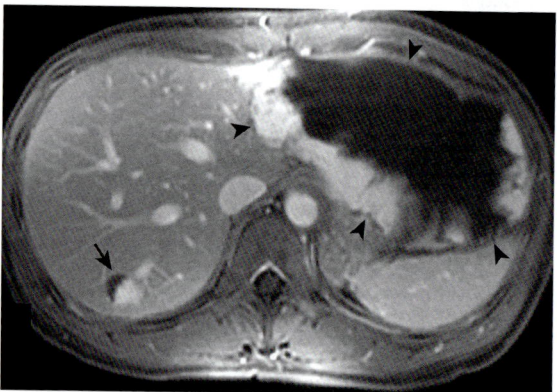

Fig. 1.22 Hepatic cavernous hemangiomas on magnetic resonance imaging (MRI). Contrast-enhanced fat-suppressed gradient-echo MRI scan shows characteristic findings of cavernous hemangiomas, including a giant left hepatic lobe lesion (*arrowheads*) and smaller right hepatic lobe lesion (*arrow*). Note the peripheral enhancement of the lesions that matches the signal intensity of the aortic blood pool. The findings are diagnostic, and tissue biopsy is unnecessary. (From Goldman L, Schafer AI: *Goldman's Cecil medicine*, ed 24, Philadelphia, Saunders, 2012.)

- No ionizing radiation
- Soft-tissue resolution
- Multiplanar

Weaknesses
- Expensive
- Needs cooperative patient
- Time-consuming
- Cannot be performed in patients with aneurysm clips that are not compatible with magnetic resonance or in patients with a pacemaker, cochlear implants, or metallic foreign body in the eyes; safe in women with intrauterine devices (IUDs), including copper ones, and in patients with surgical clips and staples.

Comments
- In patients with chronic liver disease, MRI is more sensitive (81% sensitivity) but less specific (85% specificity) than ultrasonography (sensitivity 61%, specificity 97%) or spiral CT (sensitivity 68%, specificity 93%) for diagnosis of hepatocellular carcinoma.
- Anxious patients (especially those with claustrophobia) should be premedicated with an anxiolytic agent, and imaging should be done with "open MRI" whenever possible.
- Cost: MRI with and without contrast $$$$$

16. Small-Bowel Series
Indications
- Small-bowel lymphoma and other small-bowel neoplasms
- Malabsorption (Fig. 1.23)
- Whipple's disease
- IBD
- Celiac sprue
- "Short-bowel" syndrome
- Pancreatic insufficiency
- Intestinal fistula
- GI bleeding
- Anemia (if other tests are negative or noncontributory)

Strengths
- Inexpensive
- Readily available
- Good visualization of mucosal detail

Weaknesses
- Contraindicated in pregnancy
- Requires cooperative patient
- Time-consuming
- Radiation exposure

Comments
- In a small-bowel series, sequential films are obtained at 15- to 30-minute intervals until the terminal ileum is visualized with fluoroscopy and spot films.
- Cost: $$

17. Tc-99m Sulfur Colloid (Tc-99m SC) Scintigraphy for GI Bleeding
Indication
- Localization of GI bleeding of undetermined source

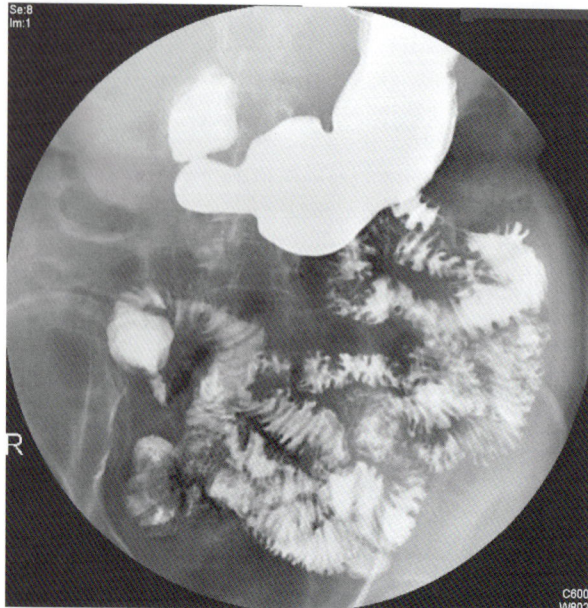

Fig. 1.23 Film from a barium follow-through study of the small bowel in a patient with tropical sprue showing loss of the normal feathery mucosal pattern in the small bowel, with dilation of jejunal loops and thickening of the mucosal folds. (From Feldman M, Friedman LS, Brandt LJ. *Sleisenger and Fortran's Gastrointestinal Liver Disease*, 10 ed. Philadelphia, Saunders, 2016.)

Strengths

- Fast: in patient who is actively bleeding, this study can be promptly performed and completed before angiography
- Active hemorrhage is most commonly detected within minutes of imaging
- In addition to detecting bleeding site, may also detect abnormalities such as vascular blushes of tumors, angiodysplasia, and arteriovenous malformations (AVMs)
- Useful in patients with heparin allergy

Weaknesses

- Main disadvantage is that bleeding must be active (bleeding rate >0.1 mL/min) at time of injection
- Window for detection of bleeding after injection approximately 1 hour
- Not useful for intermittent bleeding
- Inexact localization of bleeding site; because blood acts as an intestinal irritant, movement can often be rapid and bidirectional, making it difficult to localize site of bleeding
- Liver and spleen uptake may obscure splenic or hepatic flexure source
- Ectopic spleen and asymmetric bone marrow activity can interfere with detection of bleeding
- Presence of barium in GI tract may obscure bleeding site

Comments

- After injection of Tc-99m SC, radiotracer will extravasate at the bleeding site into the lumen with each recirculation of blood. The site of bleeding is seen as a focal area of radiotracer accumulation that increases in intensity and moves through the GI tract.
- Tc-99m SC is less sensitive than Tc-99 red blood cell (RBC) scan and is used less often for evaluation of GI hemorrhage.
- Cost: $$

18. Tc-99m–Labeled Red Blood Cell (RBC) Scintigraphy for GI Bleeding

Indication

- Localization of GI bleeding of undetermined source

Strengths

- Major advantage over Tc-99m SC is that a hemorrhagic site can be detected over a much longer period and can reimage if bleeding not seen immediately and patient rebleeds
- In addition to detecting active bleeding sites, may be able to detect vascular blushes of tumors, angiodysplasia, and AV malformations

Weaknesses

- False-positive results caused by misinterpretation of normal variants or poorly detailed delayed images
- Time-consuming; not indicated in patient actively bleeding and clinically unstable
- Inexact localization of bleeding site; because blood acts as an intestinal irritant, movement can often be rapid and bidirectional, making it difficult to localize site of bleeding
- Presence of barium in GI tract may obscure bleeding site
- Visualization requires a bleeding rate greater than 0.1 mL/min

Comments

- In an RBC scan, the patient's RBCs are collected, labeled with a radioisotope, and then returned to the patient's circulation.
- Criteria for positive Tc-RBC scintigraphy are as follows: abnormal radiotracer "hot" spot appears and conforms to bowel anatomy, normal activity persists or increases over time (Fig. 1.24), and there is noticeable movement of activity by peristalsis, retrograde, or anterograde.
- Cost: $$

19. Ultrasound of Abdomen

Indications

- Abdominal pain
- Jaundice
- Cholelithiasis
- Cholecystitis
- Elevated liver enzymes
- Splenomegaly
- Ascites
- Abdominal mass
- Pancreatitis
- Portal vein thrombosis (Fig. 1.25)

Strengths

- Fast
- Can be performed at bedside
- No ionizing radiation
- Widely available

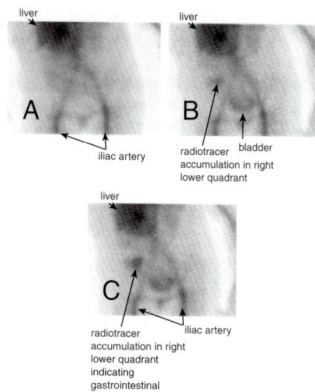

Fig. 1.24 Gastrointestinal hemorrhage, tagged red blood cell study. This 58-year-old presented with bright red blood per rectum. A Tc-99m tagged red blood cell study was performed. A, Acquired 10 minutes after injection of the labeled red cells. B, Acquired 45 minutes after injection. C, Acquired 55 minutes after injection. A focus of radiotracer activity is seen gradually accumulating in the right lower quadrant, consistent with hemorrhage within the cecum. Normal tracer is also seen in the region of the liver (because this is a vascular organ), in the iliac arteries, and in the urinary bladder. (From Broder JS: *Diagnostic imaging for the emergency physician,* Philadelphia, Saunders, 2011.)

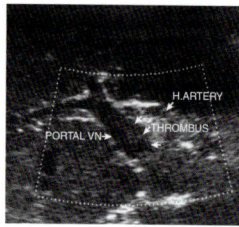

Fig. 1.25 Portal venous thrombosis. Partial portal venous thrombosis is visible on B mode as echoreflective material on one side of the vein (*arrows*). Doppler examination is always required to assess patency, as some thrombi are of reduced echoreflectivity and may not be visible on B mode. (From Grainger RG, Allison DJ, Adam A, Dixon AK [eds]: *Grainger and Allison's diagnostic radiology,* ed 4, Philadelphia, Churchill Livingstone, 2001.)

- Can provide Doppler and color flow information
- Lower cost than CT

Weaknesses
- Obscuring intestinal gas
- Only able to visualize 50% of bile duct stones
- Inferior anatomic detail compared with CT
- Affected by body habitus
- Cannot be used to definitely rule out abscess

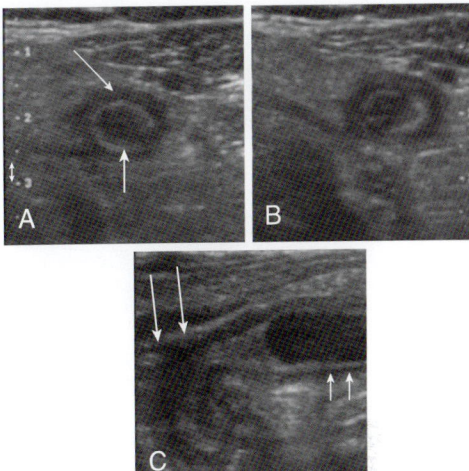

Fɪɢ. **1.26** Transabdominal ultrasonography in a 37-year-old woman with pelvic pain. A, Cross-sectional view of a dilated appendix (*large arrow*) with periappendiceal fluid (*small arrow*). B, Compression yields minimal change in appendiceal diameter and causes significant pain. C, A longitudinal view of the appendix (*small arrows*) and its origin at the cecum (*large arrows*). (From Adams JG et al: *Emergency medicine, clinical essentials,* ed 2, Philadelphia, Elsevier, 2013.)

Comments

- This is often the initial diagnostic procedure of choice in patients presenting with abdominal pain or mass in RUQ and midabdomen. CT of abdomen is preferred in LLQ and RLQ pain or mass and in significantly obese patients.
- The presence of a sonographic Murphy's sign (the ultrasonographer elicits focal gallbladder tenderness under the ultrasound transducer) has a positive predictive value >90% for acute cholecystitis if gallstones are present.
- Ultrasound should be considered as an initial test in all patients with pancreatitis, especially if gallstones are suspected.
- Cost: $$

20. Ultrasound of Appendix

Indication
- Suspected appendicitis (Fig. 1.26)

Strengths
- Fast
- Readily available
- Noninvasive
- No ionizing radiation

Weaknesses
- Can be affected by overlying bowel gas and body habitus (e.g., obese patient)
- Operator dependent; results may be affected by skill of technician

Comments
- This is the best initial study in suspected appendicitis in children and pregnant patients.
- Cost: $$

21. Ultrasound of Gallbladder and Bile Ducts
Indications
- Suspected cholelithiasis: Fig. 1.27 shows a normal gallbladder ultrasound in comparison with cholelithiasis seen on ultrasound (Fig. 1.28)

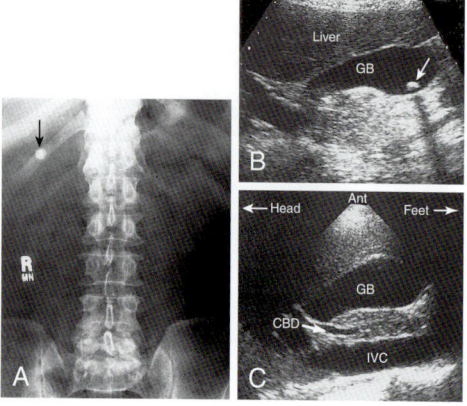

Fig. 1.27 Single gallstone. A, On the kidney, ureter, and bladder (plain radiograph of the abdomen), a single calcification is seen in the right upper quadrant (*arrow*). It is not possible to tell from this one picture whether this is a gallstone, kidney stone, or calcification in some other structure. B, A longitudinal ultrasound image in this patient clearly shows the liver, gallbladder (GB), and an echogenic focus (*arrow*) within the gallbladder lumen, representing the single gallstone. Also note the dark shadow behind the gallstone. C, Another longitudinal ultrasound image slightly more medial also shows the inferior vena cava (IVC) and the common bile duct (CBD), which can be measured. Here it is of normal diameter. (From Mettler FA: *Essentials of radiology*, ed 3, Philadelphia, Elsevier, 2014.)

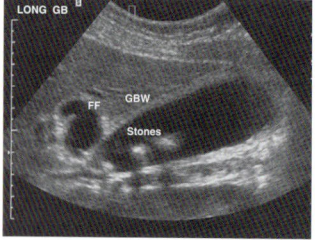

Fig. 1.28 Gallbladder with gallstones (Stones), thickened gallbladder wall (GBW), and pericholecystic fluid (FF). Together these findings constitute the sonographic signs of cholecystitis. (From Marx JA et al: *Rosen's emergency medicine*, ed 7, Philadelphia, Elsevier, 2010.)

- Cholecystitis
- Gallbladder polyps
- Gallbladder neoplasms
- Choledocholithiasis
- Biliary neoplasm
- Cholangitis
- Suspected congenital biliary abnormalities (e.g., biliary atresia, Caroli's disease, choledochal cyst)
- Biliary dyskinesia

Strengths
- Fast
- Readily available
- Can be performed at bedside
- Noninvasive
- No ionizing radiation

Weaknesses
- Is affected by overlying bowel gas and body habitus (e.g., obese patient)
- Operator dependent; results may be affected by skill of technician
- Stones smaller than 2 mm may be missed or confused with biliary sludge (layering echogenic material that does not cast shadows)

Comments
- This is the initial best test for suspected cholelithiasis and cholecystitis. Sensitivity for detection of gallstone in the gallbladder is better than 95% for stones larger than 2 mm. Specificity is >95% when stones produce acoustic shadowing.
- Patient must take nothing by mouth for 4 hours but should not fast longer than 24 hours (gallbladder may be contracted).
- Cost: $$

22. Ultrasound of Liver
Indications
- Elevated liver enzymes
- Hepatomegaly
- Liver mass (neoplasm, cystic disease, abscess)
- Jaundice
- Hepatic trauma
- Hepatic parenchymal disease (e.g., fatty infiltration, hemochromatosis, hepatitis, cirrhosis, portal hypertension)
- Ascites (Fig. 1.29)

Strengths
- Fast. Requires only an overnight or 8-hour fasting
- Widely available
- Portable (can be performed at bedside)
- Noninvasive
- No ionizing radiation
- Low cost

Weaknesses
- Can be affected by overlying bowel gas and body habitus
- Cannot be used to definitely rule out abscess
- Rib artifact may obscure images of the right lobe
- Rarely provides definitive diagnosis and usually requires confirmatory CT or MRI

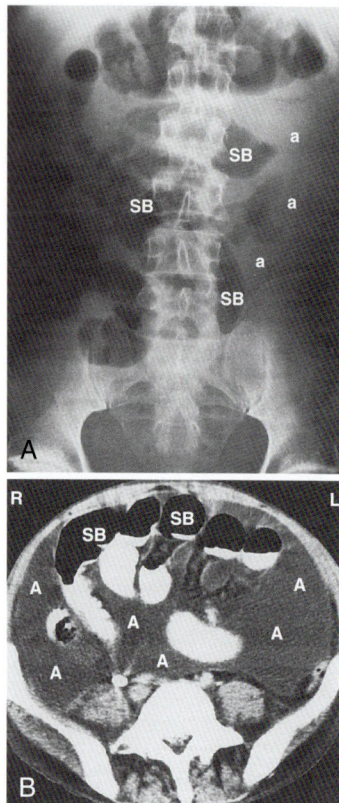

FIG. 1.29 Ascites. On a plain film of the abdomen (A), only gross amount of ascites (*a*) can be identified. This is usually seen, because the ascites have caused a rather gray appearance of the abdomen and pushed the gas-containing loops of small bowel (*SB*) toward the most nondependent and central portion of the abdomen. A transverse computed tomography scan (B) shows a cross-sectional view of the same appearance with the air- and contrast-filled small bowel floating in the ascitic fluid (*A*) (From Mettler FA, Guiberteau MJ, Voss CM, Urbina CE: *Primary care radiology*, Philadelphia, Elsevier, 2000.)

Comments

- Because of its widespread availability, noninvasive nature, and low cost, ultrasound is often performed as initial study in evaluation of suspected liver disease.
- Table 1.1 compares various imaging modalities of the biliary tract
- Cost: $$

TABLE 1.1	Imaging Studies of the Biliary Tract	
TECHNIQUE	**CONDITION TESTED FOR**	**FINDINGS/COMMENTS**
US	Cholelithiasis	Stones manifest as mobile, dependent echogenic foci within the gallbladder lumen with acoustic shadowing
		Sludge appears as layering echogenic material without shadows
		Sensitivity >95% for stones >2 mm
		Specificity >95% for stones with acoustic shadows
		Rarely, a stone-filled gallbladder may be contracted and difficult to see, with a "wall-echo-shadow" sign
		Best single test for stones in the gallbladder
	Choledocholithiasis	Stones are seen in the BD in only ≈50% of cases but can be inferred from the finding of a dilated BD (>6 mm diameter), with or without gallstones, in another ≈25% of cases
		Can confirm, but not exclude, BD stones
	Acute cholecystitis	Sonographic Murphy's sign (focal gallbladder tenderness under the transducer) has a positive predictive value of >90% in detecting acute cholecystitis when stones are seen
		Pericholecystic fluid (in the absence of ascites) and gallbladder wall thickening to >4 mm (in the absence of hypoalbuminemia) are nonspecific findings but are suggestive of acute cholecystitis
EUS	Choledocholithiasis	Highly accurate for excluding or confirming stones in the BD
		Concordance of EUS with the ERCP diagnosis ≈95%; many studies suggest slightly higher sensitivity rates for EUS than for ERCP
		Specificity ≈97%
		Positive predictive value ≈99%, negative predictive value ≈98%, accuracy ≈97%
		With experienced operators, EUS can be used in lieu of ERCP to exclude BD stones, particularly when the clinical suspicion is low or intermediate
		Considered for patients with low to moderate clinical probability of choledocholithiasis
Oral cholecystography	Cholelithiasis	Stones manifest as mobile filling defects in an opacified gallbladder
		Sensitivity and specificity exceed 90% when the gallbladder is opacified, but nonvisualization occurs in 25% of studies and can result from multiple causes other than stones
		Opacification of the gallbladder indicates cystic duct patency
		May be useful in the evaluation of acalculous gallbladder diseases such as cholesterolosis and adenomyomatosis

Continued

TABLE 1.1		Imaging Studies of the Biliary Tract—cont'd
TECHNIQUE	**CONDITION TESTED FOR**	**FINDINGS/COMMENTS**
Cholescintigraphy (hepatobiliary scintigraphy; hydroxyimino-diacetic acid or diisopropyl iminodiacetic acid scan)	Acute cholecystitis	Assesses patency of the cystic duct
		A normal scan shows radioactivity in the gallbladder, BD, and small bowel within 30-60 minutes
		A positive result is defined as nonvisualization of the gallbladder, with preserved hepatic excretion of radionuclide into the BD or small bowel
		Sensitivity is ≈95% and specificity is ≈90%, with false-positive results seen in fasted critically ill patients
		With cholecystokinin stimulation, the gallbladder "ejection fraction" can be determined and may help evaluate patients with acalculous biliary pain
		A normal scan result virtually excludes acute cholecystitis
ERCP	Choledocholithiasis	ERCP is the standard diagnostic test for stones in the BD, with sensitivity and specificity of ≈95%
		Use of ERCP to extract stones (or at least drain infected bile) is life-saving in severe cholangitis and reduces the need for BD exploration at the time of cholecystectomy
		Recommended for patients with a high clinical probability of choledocholithiasis
	Cholelithiasis	When contrast agent flows retrograde into the gallbladder, stones appear as filling defects and can be detected with a sensitivity rate of ≈80%, but US remains the mainstay for confirming cholelithiasis
MRCP	Choledocholithiasis	A rapid, noninvasive modality that provides detailed bile duct and pancreatic duct images equal to those of ERCP
		Sensitivity ≈93% and specificity ≈94%, comparable with those for ERCP
		Useful for examining nondilated ducts, particularly at the distal portion, which often is not well visualized by US
		Adjacent structures such as the liver and pancreas can be examined at the same time
		Recommended for patients with low to moderate clinical probability of choledocholithiasis
CT	Complications of gallstones	Not well suited for detecting uncomplicated stones but excellent for detecting complications such as abscess, perforation of gallbladder or BD, and pancreatitis
		Spiral CT may prove useful as a noninvasive means of excluding BD stones; some studies suggest improved diagnostic accuracy when CT is combined with an oral cholecystographic contrast agent

BD, bile duct; *ERCP,* endoscopic retrograde cholangiopancreatography; *EUS,* endoscopic ultrasound; *MRCP,* magnetic resonance cholangiopancreatography; *US,* ultrasound.
Feldman M, Friedman LS, Brandt LJ: *Sleisenger and Fordtran's gastrointestinal and liver disease,* ed 10, Philadelphia, Elsevier, 2016.

23. Ultrasound of Pancreas

Indications
- Pancreatitis
- Cystic fibrosis
- Pancreatic abscess
- Pancreatic pseudocyst
- Suspected neoplasm
- Trauma

Strengths
- Fast
- Noninvasive
- Can be performed at bedside
- No ionizing radiation

Weaknesses
- Is affected by overlying bowel gas and body habitus (e.g., in obese patient fat overlying the pancreas impedes visualization)
- Operator dependent; results may be affected by skill of technician
- Barium from recent radiographs will interfere with visualization
- Cannot be used to conclusively rule out abscess
- Difficult to evaluate tail of pancreas because of location

Comments
- Cost: $$

24. Endoscopic Ultrasound (EUS)

Indications
- Evaluation of choledocholithiasis; useful when CT and ultrasonography fail to show suspected CBD stones
- Preoperative staging of esophageal malignancies (Fig. 1.30)
- Detection of defects in internal and external sphincter in patients with fecal incontinence, detection of exophytic distal rectal tumors, fistula-in-ano, perianal abscess, rectal ulcer, and presacral cyst
- Localization of insulinomas and other pancreatic endocrine tumors
- Evaluation of submucosal lesions of the GI tract
- Guidance for fine-needle aspiration of pancreatic cysts
- Chronic pancreatitis: useful to delineate strictures and proximal dilation of CBD and intrahepatic biliary radicles
- Useful for selecting patients who might benefit from ERCP and early stone extraction

Strengths
- When used for evaluation of submucosal GI lesions, the sensitivity of EUS in determining the depth of tumor invasion is approximately 85% to 90%.
- In fecal incontinence, EUS-detected sphincter disruption correlates well with pressure measurements and operative findings.
- EUS is less invasive than MRCP or ERCP and has a sensitivity and specificity of 90% to 100% for evaluation of choledocholithiasis. The major benefit of EUS in patients with a clinical suspicion of choledocholithiasis is the ability to avoid unnecessary ERCP and sphincterectomy.

Weaknesses
- Can overestimate the extent of GI tumor invasion because of the presence of tissue inflammation and edema
- Operator dependent; results may be affected by skill of technician

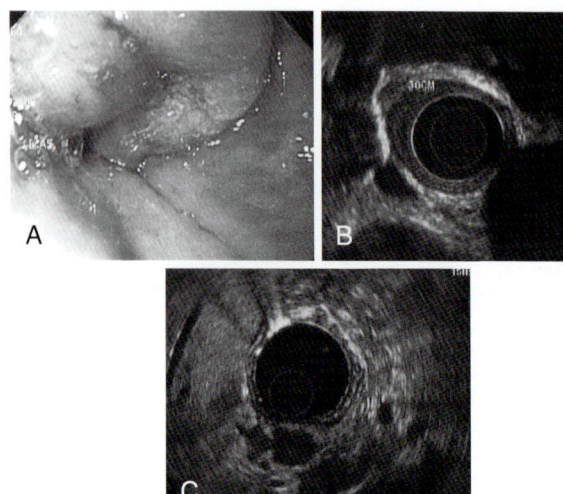

Fig. 1.30 A, Endoscopic picture of malignant esophageal stricture. B, Endoscopic ultrasound image showing T3 lesion. C, Malignant celiac lymphadenopathy. (Cameron JL, Cameron AM: Esophageal function tests, *Current surgical therapy*, ed 10, Philadelphia, Saunders, 2011.)

Comments
- EUS combines ultrasonography and endoscopic evaluation. It involves visualization of the GI tract via a high-frequency ultrasound transducer placed through an endoscope.
- Cost: $$$

25. Video Capsule Endoscopy (VCE)
Indications
- Determination of obscure source of GI bleeding
- Diagnosis of Crohn's disease in the small intestine
- Detection of tumors and polyps in the small bowel
- Detection of small intestinal ulcerations (Fig. 1.31)
- Diagnosis of Meckel's diverticulum
- Diagnosis of small-bowel varices in patients with portal hypertension and obscure GI bleeding

Strengths
- Noninvasive
- Ambulatory testing
- Minimal or no patient discomfort
- Able to visualize the entire small intestine
- Does not require sedation or analgesia

Weaknesses
- Cannot take biopsies
- Can result in capsule retention (<1%) requiring surgical intervention if there is an obstruction or stricture
- Labor-intensive for endoscopist (50–100 minutes to review images)

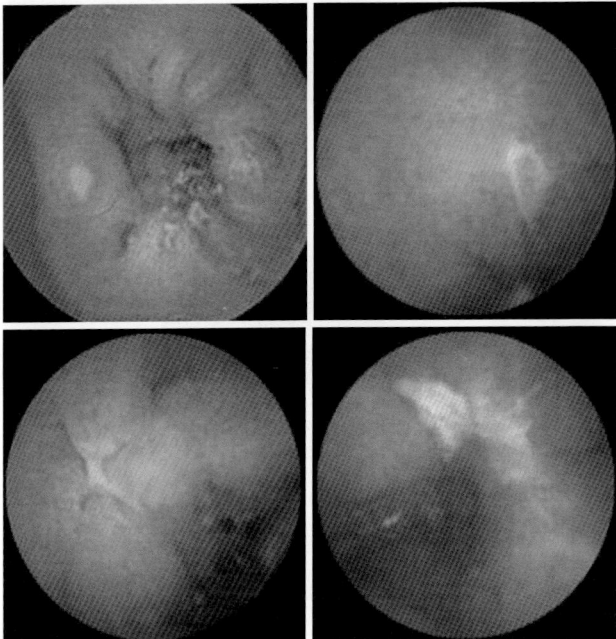

Fɪɢ. 1.31 Examples of small intestinal ulcerations found on capsule endoscopy. (Courtesy Dr. Laurel Fisher, University of Michigan.)

- Relatively contraindicated in patients with implanted pacemakers or defibrillators (possible interference)

Comments

- In VCE, the patient fasts for 12 hours, then swallows a miniature high-resolution camera that is propelled through the GI tract, allowing visualization of the small intestine that is inaccessible by conventional endoscopy. The capsule measures 11×23 mm and contains a color video camera and transmitters. The patient wears sensors and a data recorder. The capsule is propelled by peristalsis through the GI tract and acquires two or more video images per second. The capsule is used once and is not recovered. When the study is completed, the stored images are downloaded to a computer for viewing.
- Diagnostic yield for obscure GI bleeding is 50% to 70%.
- Cost: $$$

26. Transient Elastography (Fibroscan)

Indications

- Measurement of hepatic fibrosis
- Useful in patients with hepatitis C to quantify hepatic injury into discrete grades of inflammation and stages of fibrosis. It creates a composite score, adjusted for gender and age, derived from the serum levels of alpha-2-macroglobulin, haptoglobin, apolipoprotein A-1, GGTP, and total bilirubin.

Strengths
- Noninvasive
- Ambulatory testing
- No patient discomfort, does not require sedation or analgesia
- Used in many centers in place of liver biopsy before initiating therapy for hepatitis C

Weaknesses
- Less accurate than liver biopsy
- It accurately categorizes patients to stage 0 and 1 fibrosis and those with cirrhosis; however, it is less useful in patients with intermediate scores

Comments
- Table 1.2 compares the accuracy of noninvasive tests for predicting hepatic fibrosis in patients with hepatitis C
- Cost: $$

B. Breast Imaging

1. Mammogram
Indications
- Screening for breast cancer; American Cancer Society guidelines recommend baseline mammogram, age 35 to 40; yearly mammogram after age 40; younger than age 30, mammography generally not indicated unless positive family history of breast cancer at a very early age
- Evaluation of breast mass, tenderness

Strengths
- Inexpensive
- Readily available

Weaknesses
- Misses 15% to 20% of breast neoplasms
- Can be painful for patient
- Poor identification of nonpalpable intraductal papillomas
- Residue on breasts from powders, deodorants, or perfumes may interfere with diagnosis of lesions

Comments
- Digital mammography is the single best initial method for detecting breast cancer (Fig. 1.32) at a curable stage based on cost and availability.
- When ordering a mammogram, it is important to distinguish a screening mammogram from a diagnostic mammogram. Screening mammograms are indicated in healthy women (see guidelines earlier), whereas a diagnostic mammogram is indicated when patients present with signs or symptoms related to the breast or palpable abnormalities on breast examination.
- Mammography is available in both plain film and digital format. Digital mammography is often performed because it offers the following advantages over film mammography: significantly shorter examination times, 50% less radiation than traditional film radiography, and 27% more sensitivity for cancer in women younger than 50 and in women with dense breast tissue.
- The use of computer-aided detection in screening mammography is associated with reduced accuracy of interpretation of screening mammograms. The increased rate of biopsy with the use of computer-aided detection is not clearly associated with improved detection of invasive breast cancer.
- Cost: $

TEST	NUMBER OF PATIENTS STUDIED	FIBROSIS STAGING SYSTEM	HISTOLOGIC FIBROSIS (F) STAGES COMPARED	SENSITIVITY (%)*	SPECIFICITY (%)*	PPV FOR FIBROSIS-CIRRHOSIS (%)	TEST ACCURACY (%)†
APRI	270	Ishak	F0-2 vs. F3-6	41	95	88	70
			F0-4 vs. F5-6	89	75	57	77
FibroSure	339	METAVIR	F0-1 vs. F2-4	100	22	50	57
			F0-2 vs. F3-4	70	95	91	84
Transient elastography (Fibroscan)	327	METAVIR	F0-1 vs. F2-4	56	91	88	68
			F0-3 vs. F4	86	96	78	94

TABLE 1.2   Accuracy of Noninvasive Tests for Predicting Hepatic Fibrosis in Patients with Hepatitis C

*Sensitivity and specificity for distinguishing higher stages of fibrosis.

†Accuracy = (sensitivity)(prevalence) + (specificity) (1 − prevalence).

APRI, AST-to-platelets ratio index; *PPV*, positive predictive value.

From Wai CT, Greenson JK, Fontana RJ, et al: A simple noninvasive index can predict both significant fibrosis and cirrhosis in patients with chronic hepatitis C, *Hepatology* 38(2):518–526, 2003; Imbert-Bismut F, Ratziu V, Pieroni L, et al: Biochemical markers of liver fibrosis in patients with hepatitis C virus infection: A prospective study, *Lancet* 357(9262):1069–1075, 2001; and Ziol M, Handra-Luca A, Kettaneh A, et al: Noninvasive assessment of liver fibrosis by measurement of stiffness in patients with chronic hepatitis C, *Hepatology* 41(1):48–54, 2005; From Feldman M, Friedman LS, Brandt LJ: *Slesinger and Fordtran's gastrointestinal and liver disease*, ed 10, Philadelphia, Elsevier, 2016.

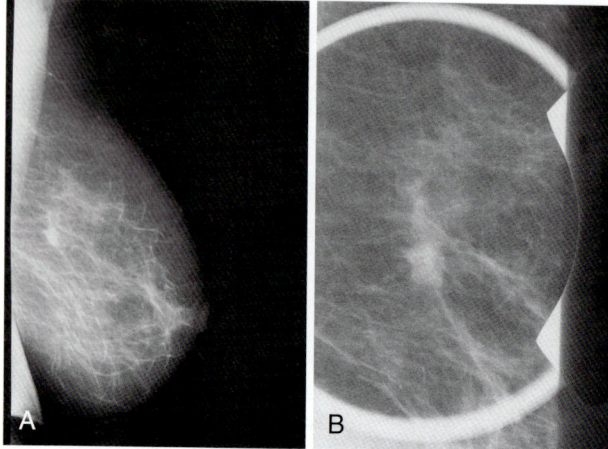

Fig. 1.32 Right mediolateral (A) and spot magnification views (B) from routine screening mammography demonstrate a small, ill-defined mass with minimal spiculation. This was nonpalpable, and biopsy demonstrated infiltrating ductal carcinoma. (From Specht N: *Practical guide to diagnostic imaging,* St. Louis, Mosby, 1998.)

2. Breast Ultrasound

Indications

- Characterization of breast mass or density as cystic or solid (Fig. 1.33). Simple cysts are anechoic and have enhanced sound transmission (Fig. 1.34)
- Sonography is the best diagnostic tool for assessing mammographic abnormalities that do not contain suspicious calcifications
- Guidance for interventional procedure, cyst aspiration, needle localization, fine-needle aspiration or core biopsy, prebiopsy localization
- Evaluation of palpable masses in young patients, those who are pregnant or lactating, or those with a palpable abnormality and negative mammogram
- Confirmation, identification, and characterization of masses or density seen on only one view on mammographic examination (Fig. 1.35)
- Evaluation of breast implant integrity
- Second look after abnormal MRI (Fig. 1.36)

Strengths

- Fast
- Noninvasive
- No ionizing radiation
- Readily available

Weaknesses

- Cannot detect microcalcifications
- Large masses can blend with background pattern, limiting their visibility as discrete entities on ultrasound
- Both benign and malignant solid tumors can have similar appearance

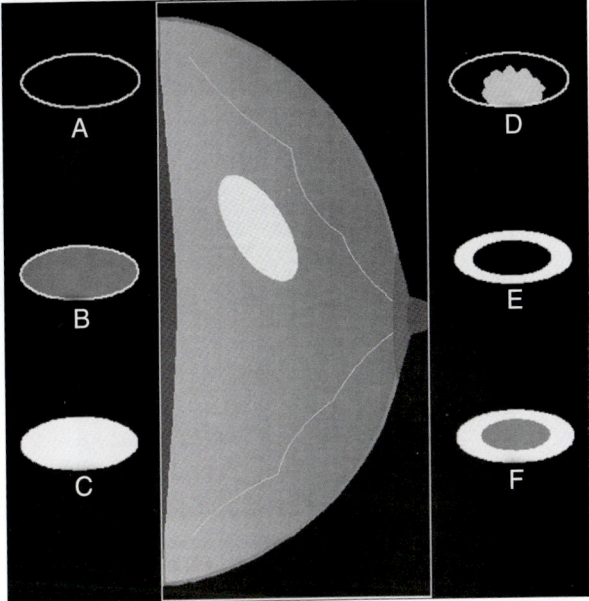

Fig. 1.33 Importance of mammographic-sonographic correlation. Everything that is water density could contribute to the size of the mammographic lesion. Thus a 3-cm, ovoid, circumscribed mammographic mass could represent A, a cyst, or B, a solid nodule, surrounded by a thin echogenic capsule; C, 3-cm collection of interlobular stromal fibrous tissue; D, 3-cm cyst containing a mural nodule; or E, smaller cyst, or F, solid nodule, surrounded by fibrous or glandular tissue. (From Rumack CM, Wilson SR, Charboneau JW, Levine D: *Diagnostic ultrasound*, ed 4, Philadelphia, Mosby, 2011.)

Comments
- Breast ultrasound is not indicated as a screening examination for breast disease or for evaluation of microcalcifications.
- Sensitivity for evaluation of breast implant rupture is 70%, specificity 70%.
- Cost: $

3. MRI of the Breast
Indications
- Staging of breast cancer for treatment planning (e.g., detection of chest wall involvement); preoperative MRI has been shown to detect unsuspected multifocal and multicentric disease in nearly 30% of patients and contralateral disease in up to 5%
- Breast augmentation: evaluation of silicone implant integrity and screening, including patients who have received silicone implants previously
- Malignant axillary adenopathy with occult primary (Fig. 1.37); useful in patients with positive axillary lymph node for cancer and negative mammogram and ultrasound

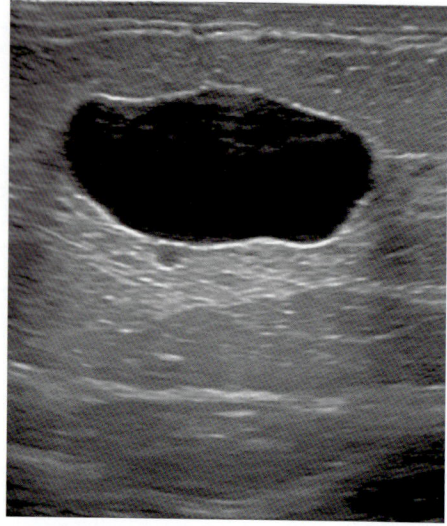

Fig. 1.34 Simple cysts. Simple cysts are anechoic and have enhanced sound transmission, well-circumscribed borders, thin-edge shadows, and thin, echogenic walls. They are benign (BIRADS 2) and require no aspiration or follow-up. (From Rumack CM, Wilson SR, Charboneau JW, Levine D: *Diagnostic ultrasound*, ed 4, Philadelphia, Mosby, 2011.)

- Screening for breast neoplasm in women at high risk (*BRCA* gene carriers, personal history of breast cancer, strong family history of breast cancer, prior radiation to chest, prior atypical ductal or lobular hyperplasia and lobular carcinoma in situ [LCIS])
- Additional evaluation of contradictory, inconclusive, or equivocal mammogram results
- Differentiation between scar tissue and recurrent breast cancer after lumpectomy

Strengths
- More sensitive than mammogram for detecting breast neoplasm; sensitivity 88% to 100%, specificity 30% to 90%
- Sensitivity for breast implant rupture is 94%, specificity 97%
- Allows evaluation of axillary lymph nodes
- Useful for evaluation of inverted nipples for cancer

Weaknesses
- High rate of false positives
- Contrast-enhanced MRI requires injection of contrast material

Comments
- Breast MRI has emerged as the most sensitive imaging modality for the detection of invasive breast carcinoma; however, it is much more expensive than mammography and is not currently a replacement for screening mammography.
- Scheduling guidelines: When used for additional evaluation of equivocal mammogram, patients should have recent (within 4 months) mammogram available for correlation.
- Cost: MRI with and without contrast $$$$

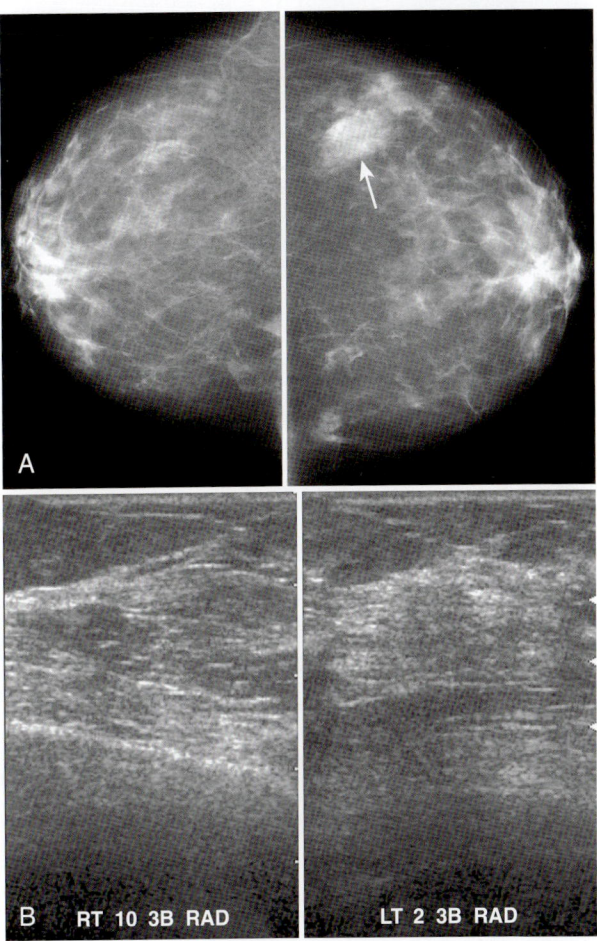

Fig. 1.35 Value of split-screen mirror ultrasound image. A, Mammography of both breasts showed a focal asymmetric density in the left breast, upper outer quadrant on the craniocaudal view (*arrow*). B, Split-screen mirror-image ultrasound images show focal fibrous tissue in the upper outer quadrant of the left breast that is markedly asymmetric with the thickness of tissue in the mirror-image upper outer quadrant location of the right breast. This collection of asymmetric fibrous tissue is the cause of the mammographic asymmetry. (From Rumack CM, Wilson SR, Charboneau JW, Levine D: *Diagnostic ultrasound*, ed 4, Philadelphia, Mosby, 2011.)

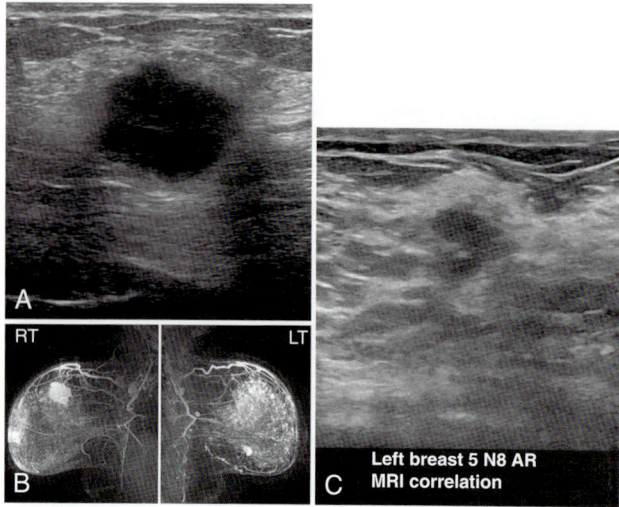

Fig. 1.36 Contralateral invasive carcinoma detected on magnetic resonance imaging and verified on "second look" ultrasound. (From Rumack CM, Wilson SR, Charboneau JW, Levine D: *Diagnostic ultrasound*, ed 4, Philadelphia, Mosby, 2011.)

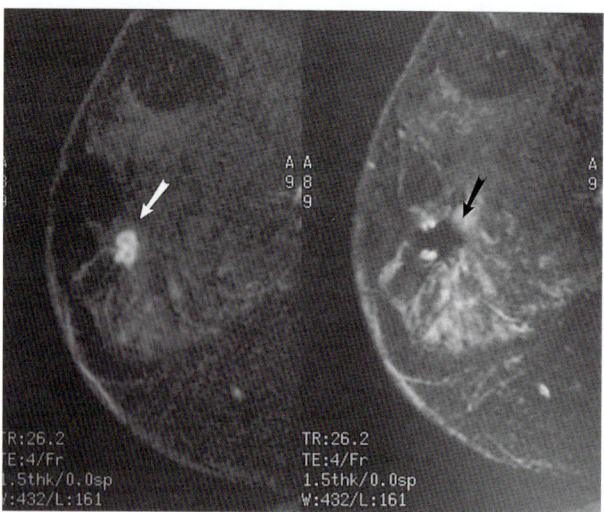

Fig. 1.37 Magnetic resonance imaging (MRI)–guided wire localization. Images of a patient with malignant axillary adenopathy and unknown primary. Sagittal, fat-suppressed, contrast-enhanced, three-dimensional FSPGR MRI reveals a peripherally enhancing lesion (*arrow in left image*) localized by an MRI-compatible needle (*arrow in right image*). Invasive ductal carcinoma was found at excisional biopsy. (From Grainger RG, Allison D: *Grainger and Allison's diagnostic radiology: a textbook of medical imaging*, ed 4, Sydney, Churchill Livingstone, 2001.)

C. Cardiac Imaging

1. Stress Echocardiography

Indications
- Suspected myocardial ischemia based on electrocardiogram (ECG) changes, history
- Post–myocardial infarction (MI), post–coronary artery bypass graft (CABG), postangioplasty risk stratification
- Evaluation of chest pain in patients with Wolff-Parkinson-White syndrome
- Evaluation of young women with chest pain (high rates of false-positive results with conventional stress test)
- Evaluation of adequacy of therapy while patient is on medication
- Evaluation of patients with significant abnormalities on resting ECG (e.g., left bundle branch block [LBBB] or paced rhythm, left ventricular hypertrophy [LVH] and baseline ST-segment or T-wave abnormalities, sloping ST segment secondary to digitalis administration)
- Preoperative risk assessment

Strengths
- Readily available at many institutions (e.g., can be used on weekends or evenings when nuclear testing may be difficult to arrange)
- Useful to detect regional wall abnormalities that occur during myocardial ischemia associated with coronary artery disease (CAD)
- Significantly higher sensitivity for diagnosing CAD than conventional treadmill exercise test
- Dobutamine echocardiography (Fig. 1.38) is preferable to dipyridamole or adenosine scintigraphy in patients with moderate or severe bronchospastic disease

Weaknesses
- More expensive than conventional treadmill exercise test

Comments
- In stress echocardiography, decrements in contractile function are directly related to decreases in regional subendocardial blood flow (Fig. 1.39).

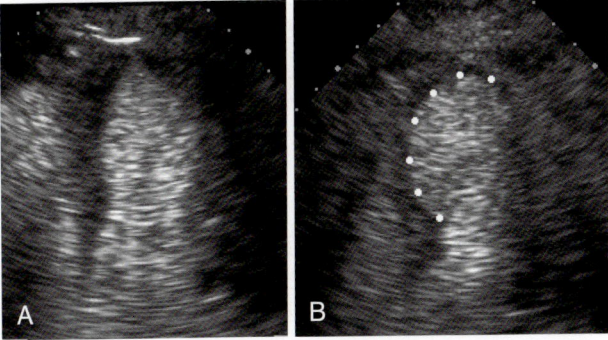

Fɪɢ. **1.38** Abnormal stress echocardiogram with images of the apical four-chamber view taken at rest (A) and after administration of dobutamine (B). Transpulmonary microbubble contrast is used to opacify the left ventricular chamber. After dobutamine administration, there is a wall motion abnormality, with hypokinesis and contour change in the apex and distal septum of the left ventricle (*dots*). (From Sellke FW, del Nido PJ, Swanson SJ: *Sabiston & Spencer surgery of the chest*, ed 9, Philadelphia, Elsevier, 2016.)

Resting   Stress

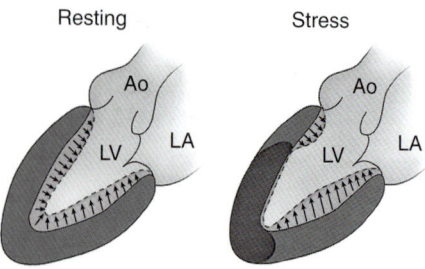

Fig. 1.39 The concept of stress echocardiography in a patient with 70% stenosis in the proximal third of the left anterior descending (LAD) coronary artery. At rest (*left*), endocardial motion and wall thickening are normal. After stress (*right*), either exercise or pharmacologic, the middle and apical segments of the anterior wall become ischemic, showing reduced endocardial wall motion and wall thickening. If the LAD extends around the apex, the apical segment of the posterior wall also will be affected, as shown here. The normal segment of the posterior wall shows compensatory hyperkinesis. (From Otto CM: *Textbook of clinical echocardiography*, ed 4, Philadelphia, Elsevier Saunders, 2009:191, Fig. 8.9.)

- Pharmacologic agents (e.g., dobutamine) can be used to induce stress to evaluate cardiac function in selected patients who cannot exercise on a treadmill or bicycle because of orthopedic or other problems.
- When stress echocardiography is used for preoperative assessment, the presence of one or more regional wall motion abnormalities with stress is associated with an increased risk of cardiac complications.
- Contraindications to stress testing are unstable angina with recent rest pain, acute myocarditis or pericarditis, uncompensated congestive heart failure (CHF), uncontrolled hypertension, critical aortic stenosis, untreated life-threatening cardiac arrhythmias, advanced AV block, and severe hypertrophic obstructive cardiomyopathy.
- Cost: $$$

2. Cardiovascular Radionuclide Imaging (Thallium, Sestamibi, Dipyridamole [Persantine] Scan), Positron Emission Tomography (PET), Single-Photon Emission Computed Tomography (SPECT)

Indications
- Suspected myocardial ischemia based on ECG changes, history
- Post-MI, post-CABG, postangioplasty risk stratification
- Evaluation of chest pain in patients with Wolff-Parkinson-White syndrome
- Evaluation of young women with chest pain (high rates of false-positive results with conventional stress test)
- Evaluation of adequacy of therapy while patient on medication
- Evaluation of patients with significant abnormalities on resting ECG (e.g., LBBB or paced rhythm, LVH and baseline ST-segment or T-wave abnormalities, sloping ST segment secondary to digitalis administration)
- Preoperative risk assessment

Strengths
- Useful in patients with underlying bundle branch block or paced rhythm
- Useful in patients with LVH and baseline ST-segment or T-wave abnormalities

- Significantly higher sensitivity for diagnosing CAD than conventional treadmill exercise test
- Advantages of stress perfusion imaging over stress echocardiography are higher sensitivity, especially for one-vessel CAD, and better accuracy in evaluating possible ischemia when multiple left ventricular wall motion abnormalities are present

Weaknesses

- Expensive
- Lower sensitivity in women than in men; artifacts caused by breast attenuation may affect interpretation of scans in women
- Major disadvantage of Tc-99m sestamibi is the need to administer separate stress and rest injections to identify regions of reversible ischemia because of its negligible delayed redistribution over time after single IV injection
- Symmetric three-vessel disease may result in false negative

Comments

- Viable myocardial cells extract the labeled radionuclide from the blood. An absent uptake (cold spot on scan) is an indicator of an absence of blood flow to an area of the myocardium. A fixed defect indicates MI at that site, whereas a defect that reperfuses suggests ischemia.
- **Positron emission tomography (PET)** has been evaluated extensively as a noninvasive imaging modality for viability assessment. PET viability studies involve the determination and comparison of both myocardial blood flow and the metabolic status of the myocardium. Perfusion is usually assessed with N-13–labeled NH_3, and glucose utilization is typically assessed with F-18 fluorodeoxyglucose (FDG), which when taken up by the myocardium reflects glucose transport across the myocyte membrane. FDG accumulates in myocytes in proportion to glucose uptake, but it undergoes phosphorylation by hexokinase to FDG-6-phosphate in the first step of glycolysis. FDG-6-phosphate is a form of deoxyglucose that becomes "trapped" in the myocyte and is not metabolized further. As such, FDG uptake reflects exogenous glucose utilization. Normal myocardium preferentially utilizes fatty acids for energy but switches to increased glucose utilization during periods of ischemia. In myocardial regions with ischemic dysfunction, myocardial glucose uptake may be increased, and thus FDG uptake will be enhanced, reflecting viability. Conversely, FDG will not accumulate in areas of fibrosis or scar. FDG uptake is then compared with resting perfusion imaging. Perfusion-metabolism is considered "matched" if areas of preserved flow show normal metabolic activity and areas of reduced flow have diminished FDG uptake (scar); however, perfusion and metabolism may be discordant, or "mismatched." In this scenario, FDG uptake will be present in areas of hypoperfusion, indicating that despite decreased blood flow, the myocardium is still metabolically active and hence viable. This mismatched defect is the most predictive of functional recovery after revascularization (Fig. 1.40).[1]
- Dipyridamole can be used in conjunction with thallium imaging in patients who are unable to exercise adequately on a treadmill or bicycle because of orthopedic or other problems. Dipyridamole injection is followed by thallium injection and subsequent imaging. IV dipyridamole increases coronary flow significantly over the resting level without major change in the heart rate blood pressure product, with less angina, and with less ST depression than with exercise. Dipyridamole may cause bronchospasm and is contraindicated in patients with bronchospastic disease.
- If vasodilating agents are contraindicated, inotropic agents (e.g., dobutamine) can be used instead. They increase myocardial oxygen demand by increasing heart rate, systolic blood pressure, and contractility, and secondarily increase blood flow.
- Newer agents such as sestamibi (Cardiolite, Myoview) are chemically bound to technetium. Advantages are better imaging characteristics, decreased attenuation, and faster imaging (Fig. 1.41). Disadvantages are higher cost and lower sensitivity compared with thallium in detecting viable myocardium.

- Contraindications to stress testing are unstable angina with recent rest pain, acute myocarditis or pericarditis, uncompensated CHF, uncontrolled hypertension, critical aortic stenosis, untreated life-threatening cardiac arrhythmias, advanced AV block, severe reaction to stress agent, severe pulmonary hypertension, cocaine use within 24 hours, pregnancy, combination of low ejection fraction (EF >20%) and documented recent ventricular fibrillation or ventricular tachycardia, and obstructive hypertrophic cardiomyopathy.
- Cost: $$$

3. Cardiac MRI (CMR)
Indications
- Evaluation of pericardial effusion
- Constrictive pericarditis

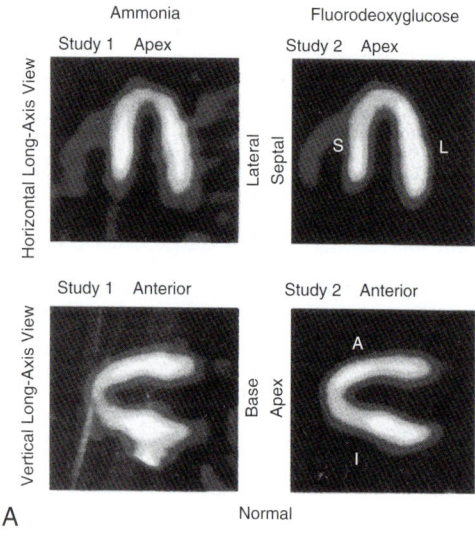

Fig. 1.40 A positron emission tomography viability study. For each panel of four views, perfusion imaging with N-13–labeled ammonia uptake is on the left, and metabolic imaging with F-18 fluorodeoxyglucose (FDG) uptake is on the right. The upper row of each panel is the horizontal long-axis slice with the apex at the top, and the lower row is a vertical long-axis slice with the apex pointing to the left. The yellow and white areas are sites of maximal activity, and the blue areas are sites of minimal activity. The red areas indicate intermediate activity. A, Normal images. B, An example of ischemia and viable tissue. C, An example of myocardial scarring. Note the homogeneous uptake of both ammonia and FDG in the normal study. In the presence of ischemia (B), there is a mismatch (*arrows*), with decreased perfusion and augmented metabolic activity in the apex and the anterior wall. With scarring (C), there is equally decreased uptake of FDG and ammonia in the inferior and lateral walls. A, Anterior wall; I, inferior wall; L, lateral wall; S, septum. (From Sellke FW, del Nido PJ, Swanson SJ: *Sabiston & Spencer surgery of the chest*, ed 9, Philadelphia, Elsevier, 2016.)

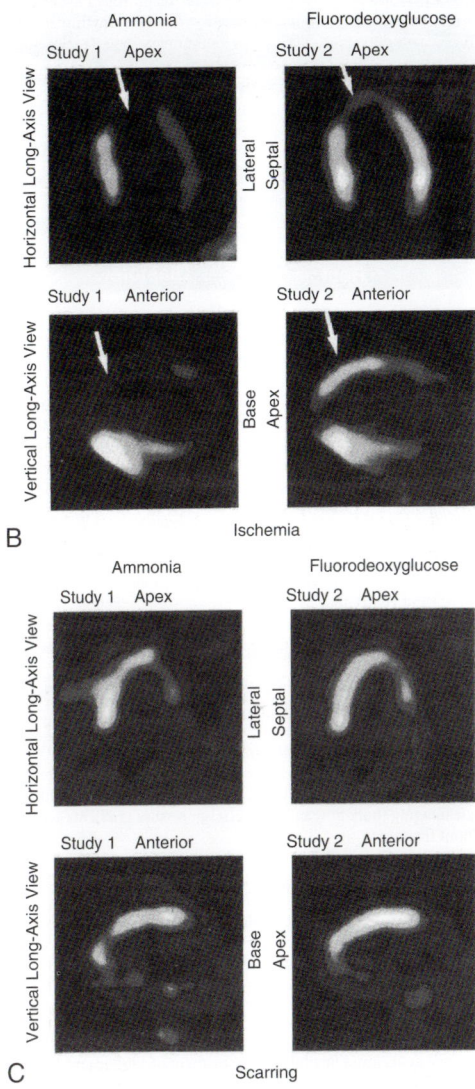

Fig. 1.40, cont'd

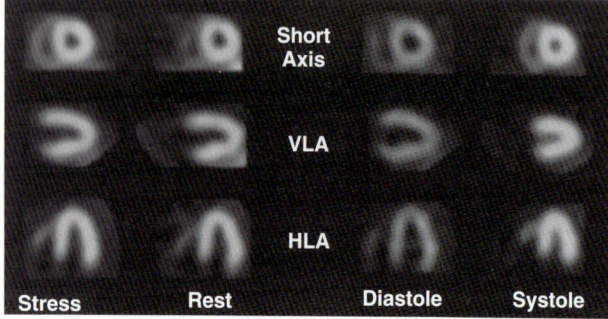

Fig. 1.41 Stress and rest single-photon emission computed tomography (SPECT) studies (*left two columns*) in a normal patient, showing representative short-axis, vertical long-axis (VLA), and horizontal long-axis (HLA) images. Note the uniform uptake of Tc-99m sestamibi on both the stress and the rest tomograms, consistent with homogeneous regional myocardial blood flow. The right two columns show the end-diastolic and end-systolic images acquired during stress and demonstrate uniform systolic thickening in all myocardial segments. The left ventricular cavity size is greater on images acquired during diastole compared with systole, consistent with a normal left ventricular ejection fraction. The "brightness" of the images at end-systole correlates directly with the degree of systolic thickening. (From Goldman L, Bennet JC: *Cecil textbook of medicine*, ed 21, Philadelphia, Saunders, 2000.)

- Evaluation of distribution of hypertrophy in hypertrophic cardiomyopathy
- Evaluation of right ventricular dysplasia
- Thoracic aorta abnormalities (dissection, coarctation, aneurysm, hematoma)
- Assessment of anatomy in congenital heart disease (intracardiac shunt, anomalous coronary arteries)
- Imaging of intracardiac masses such as thrombi and tumors (atrial myxoma, rhabdomyosarcoma, metastatic disease)
- Suspected cardiac involvement from sarcoidosis
- Suspected cardiac hemochromatosis, amyloidosis
- CAD (MI, myocardial ischemia): first-pass perfusion measurements for assessment of myocardial ischemia, dobutamine stress cine-MRI to detect stress-induced ischemia
- Physiologic imaging (bulk flow in large vessels, pressure gradients across stenotic lesions, shunt fraction)
- Quantified cavity volumes, EF, ventricular mass
- Assessment of morphologic findings and wall motion abnormalities in cases in which echocardiography is not diagnostic (20% of echocardiography is not diagnostic because of insufficient image quality [e.g., patients with emphysema, obese patients])
- Assessment of bypass grafts (includes magnetic resonance angiography [MRA])

Strengths
- Noninvasive
- No ionizing radiation
- Superior image quality and flexibility in assessment of cardiac anatomy, coronary blood flow, and myocardial perfusion
- Images can be generated in any planar orientation
- Less operator dependent than echocardiogram
- Unlike echocardiography, images are not limited by an acoustic window

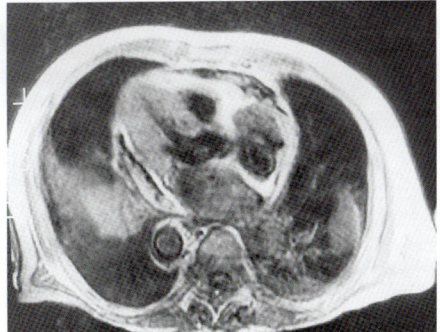

Fɪɢ. 1.42 Magnetic resonance imaging scan of constrictive pericarditis in rheumatoid arthritis. The dense white infiltrate between the pericardium and gray myocardium is pericardial fluid. (From Hochberg MC et al [eds]: *Rheumatology*, ed 3, St. Louis, Mosby, 2003.)

Weaknesses

- Expensive
- Needs cooperative patient
- Time-consuming
- Not readily available
- Cannot be performed in patients with aneurysm clips that are not compatible with magnetic resonance or in patients with a pacemaker, cochlear implants, or metallic foreign body in the eyes; safe in women with intrauterine devices (IUDs), including copper ones, and in patients with surgical clips and staples
- Suboptimal images in patients with irregular rhythm (e.g., atrial fibrillation, frequent ectopy)
- Image distortion in the region immediately surrounding the prosthesis in patients with bioprosthetic and mechanical heart valves
- Image distortion in patients with sternotomy wires and thoracic vascular clips

Comments

- CMR is an excellent imaging technique for evaluation of the pericardium (Fig. 1.42) and pericardial effusions. It can also be used for imaging of thoracic aorta and great vessels (Fig. 1.43), cardiac tumors and masses, cardiomyopathies, and for quantitative assessment of ventricular volumes and mass. Its major limiting factor is its cost disadvantage compared with ultrasound.
- Myocardial perfusion can be evaluated with MRI by giving an IV contrast agent (e.g., Gd-DPTA), which is taken up by viable myocardial cells concomitantly with dipyridamole or other pharmacologic stress agent.
- Anxious patients (especially those with claustrophobia) should be premedicated with an anxiolytic agent, and their imaging should be done with "open MRI" whenever possible.
- Cost: $$$$

4. Multidetector CT Scan

Indications

- This test can be used to identify and measure coronary artery calcifications. Calcification levels can be related to the extent and severity of underlying atherosclerosis and can potentially improve cardiovascular risk prediction. In clinically selected

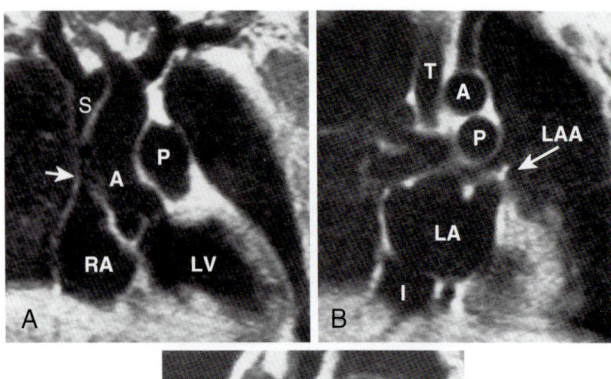

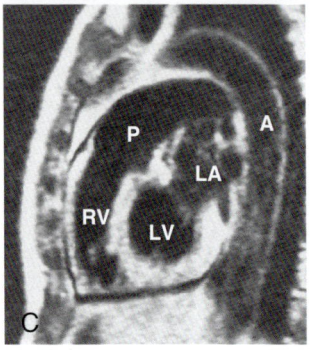

Fig. **1.43** Normal radiographic anatomy, magnetic resonance images. A, Coronal section at the level of the aortic valve. The right border of the cardiac silhouette is formed by the superior vena cava (*S*) and the right atrium (*RA*). The *arrow* indicates the caval-atrial junction. The lower portion of the left cardiac border is formed by the left ventricle (*LV*). *A*, Ascending aorta; *P*, main pulmonary artery. B, Coronal section at the level of the left atrium. The upper portion of the left cardiac border is formed by the aorta (*A*), main pulmonary artery (*P*), and left atrial appendage (*LAA*); see *arrow*. *I*, Inferior vena cava; *LA*, left atrium; *T*, trachea. C, Sagittal section near the midline. The right ventricle (*RV*) forms the anterior surface of the heart, which abuts the sternum. The pulmonary artery (*P*) extends upward and posteriorly from the ventricle. The posterior border of the heart is formed by the left atrium (*LA*) and left ventricle (*LV*). The aorta (*A*) is behind the heart. (From Goldman L, Schafer AI: Radiology of the heart. *Goldman's Cecil medicine*, ed 24, Philadelphia, Saunders, 2012.)

intermediate-risk patients, it may be reasonable to measure the atherosclerosis burden using multidetector CT (Fig. 1.44) to refine clinical risk prediction and to select patients for more aggressive target values for lipid-lowering therapies

- Coronary calcium measurements are not indicated in patients at low or high risk of cardiovascular disease
- Multidetector CT is useful in excluding coronary disease in selected patients in whom a false-positive or inconclusive stress test is suspected
- Coronary calcium assessment may be considered in symptomatic patients to determine the cause of cardiomyopathy

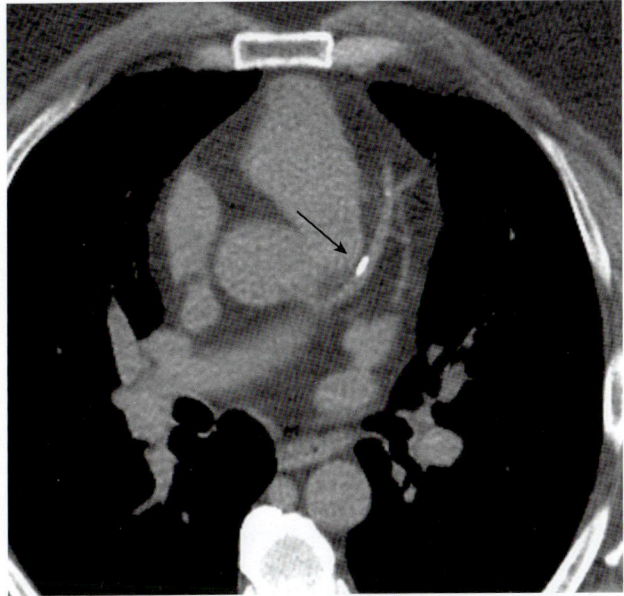

Fɪɢ. 1.44 Calcium (*arrow*) in the left anterior descending artery territory. Calcium score is determined by assigning a score to each focus of calcium in the coronary arteries with Hounsfield density of more than 130 (1 for lesions with a maximum density of 130-199 HU, 2 for lesions 200-299 HU, 3 for lesions 300-399 HU, 4 for HU >400) and multiplying the area of calcification with the score. The summated score over the entire coronary artery system yields the total calcium score. The calcium score is indexed to race, age, and gender to yield a percentile value, which is predictive of future coronary events for that patient. (From Boxt LM, Abbara S: *The requisites, cardiac imaging*, ed 4, Philadelphia, Elsevier, 2016.)

Strengths
- Speed
- Safety (less invasive than angiography)
- Lower cost than angiography
- High sensitivity and negative predictive value

Weaknesses
- Limited to patients with regular rhythm and slow rates (the image quality is inversely related to the heart rate)
- Poor images in morbidly obese patients
- Inaccurate visualization of the coronary artery within a stent
- Coronary calcification interferes with images obtained by CT; decreased diagnostic accuracy in older patients because of the prevalence and severity of coronary calcifications with increasing age
- Insufficient breath holding or slightly different positions of the heart at subsequent heartbeats are artifacts that can be more difficult to control
- High radiation exposure

Comments

- If calcification is detected in the coronary arteries, a "calcium score" is computed for each of the coronary arteries based on the size and density of the regions identified to contain calcium. Although the calcium score does not correspond directly to narrowing in the artery caused by atherosclerosis, it correlates with the severity of coronary atherosclerosis present. For example, a calcium score of 1 to 10 indicates minimal plaque burden and low likelihood of CAD, whereas a score of 101 to 400 indicates moderate plaque burden and high likelihood of moderate nonobstructive CAD. Calcium scores of more than 300 are associated with an increased risk of MI and cardiac death at every risk level. The calcium score can also be used to compare the patient's results with those of others of the same age and gender to determine a percentile ranking.
- A multidetector CT scan is a reasonable test to assess patients who have equivocal treadmill or functional test results and to assess patients with chest pain who have equivocal or normal echocardiography findings and negative cardiac enzyme results.
- Research data is currently insufficient on the use of serial cardiac CT in assessing subclinical atherosclerosis over time and in detecting noncalcified plaque.
- Cost: $$$$

5. Transesophageal Echocardiogram (TEE)

Indications

- Suspected subacute bacterial endocarditis (SBE) (Fig. 1.45)
- Evaluation of prosthetic valves

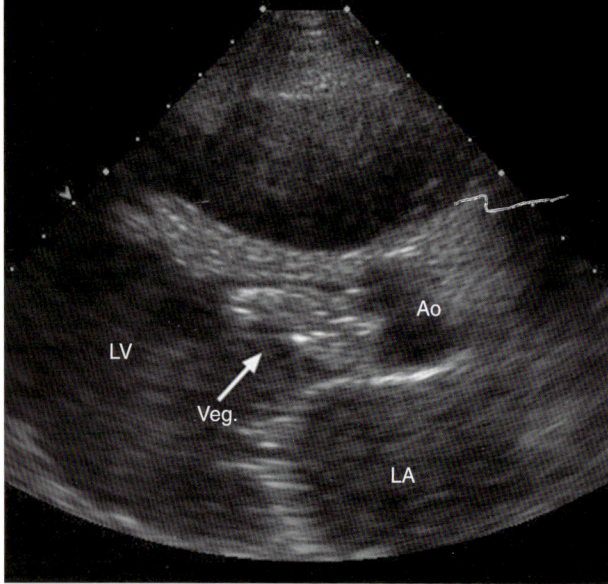

Fig. 1.45 Native aortic valve endocarditis in a patient with severe aortic regurgitation and cardiogenic shock. There is a large, mobile vegetation that is adherent to the aortic valve and that prolapses into the LV outflow tract (*arrow*). *Ao,* aorta; *LA,* left atrium; *LV,* left ventricle; *Veg.,* vegetation. (From Sellke FW, del Nido PJ, Swanson SJ: *Sabiston & Spencer surgery of the chest,* ed 9, Philadelphia, Elsevier, 2016.)

- Evaluation of embolic source
- Suspected aortic dissection
- Identification of intracardiac shunts
- Visualization of atrial thrombi
- Diseases of aorta
- Intracardiac mass
- Intraoperative use in valve repair, congenital heart surgery, repair of hypertrophic obstructive cardiomyopathy

Strengths
- Image quality superior to that of transthoracic echocardiogram (TTE)
- No ionizing radiation

Weaknesses
- Invasive
- Requires patient preparation, monitoring
- Complication rate of 0.2% to 0.5% (e.g., esophageal trauma, aspiration, cardiac dysrhythmias, respiratory depression secondary to sedation)

Comments
- TEE is performed by mounting an ultrasound transducer at the end of a flexible tube to image the heart from the esophagus (Fig. 1.46).
- Useful modality for assessing valvular pathologic findings and diseases of the aorta.
- Cost: $$$

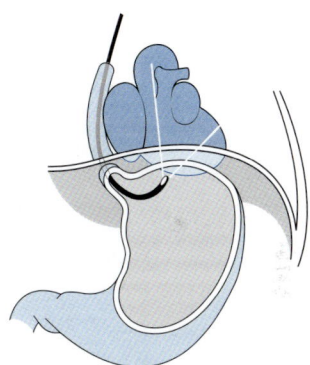

Fig. 1.46 Transesophageal echocardiography. Diagram illustrates location of the transesophageal endoscope within the esophagus. (From Pagana KD, Pagana, TJ: *Mosby's diagnostic and laboratory test reference,* ed 8, St. Louis, Mosby, 2007.)

6. Transthoracic Echocardiography (TTE)
Indications
- Evaluation of heart murmur
- Chest pain
- Evaluation of EF
- Systemic embolus
- Syncope of suspected cardiac etiologic factors
- Suspected endocarditis

- Pericardial effusion
- Abnormal heart size on chest film
- Atrial septal defect (ASD)
- Ventricular septal defect (VSD)
- Valvulopathy
- Cardiomyopathy
- Guidance of needle placement for pericardiocentesis

Strengths
- Noninvasive
- Fast
- Can be performed at bedside
- No need for patient preparation, premedication, or monitoring
- No ionizing radiation

Weaknesses
- Less sensitive than TEE for SBE, prosthetic valves
- Limited use in obese patients, patients with chronic obstructive pulmonary disease (COPD), those with chest deformities
- Resting echocardiogram not sensitive for detecting CAD

Comments
- Fig. 1.47 demonstrates standard acoustic windows for transthoracic echocardiography. Figs. 1.48 and 1.49 illustrate the four basic image planes used in TTE. Echocardiography zones are shown in Fig. 1.50. Table 1.3 summarizes key echocardiographic findings and limitations of echocardiography in various disorders.
- TEE is preferred over TTE in evaluation of prosthetic valve, embolic source, and SBE.
- Doppler echocardiogram is useful for evaluation of shunts and stenotic or regurgitant valves and for measurement of cardiac output (Tables 1.4, 1.5, 1.6).
- Contrast echocardiography uses commercially produced microbubbles or agitated saline and air to obtain a better definition when evaluating for intracardiac shunts.
- Fig. 1.51 shows the echocardiographic appearance of hypertrophic cardiomyopathy.
- Cost: $$

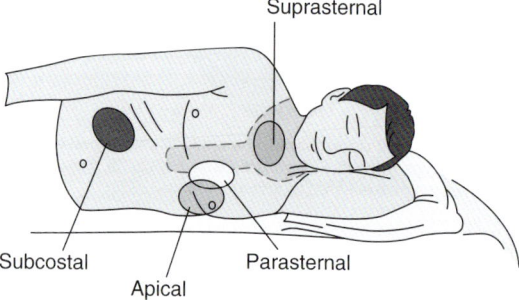

Fig. 1.47 Standard acoustic windows for transthoracic echocardiography. Parasternal and apical windows are obtained with the patient in a left lateral position. The subcostal window is obtained with the patient supine and legs flexed, to relax the abdominal wall. The suprasternal notch window is obtained with the patient's head tilted to the side to aid transducer placement. (From Otto CM, Schwaegler RG, Freeman RV: *Echocardiography review guide*, ed 2, Philadelphia, Saunders, 2011, Fig. 2.6.)

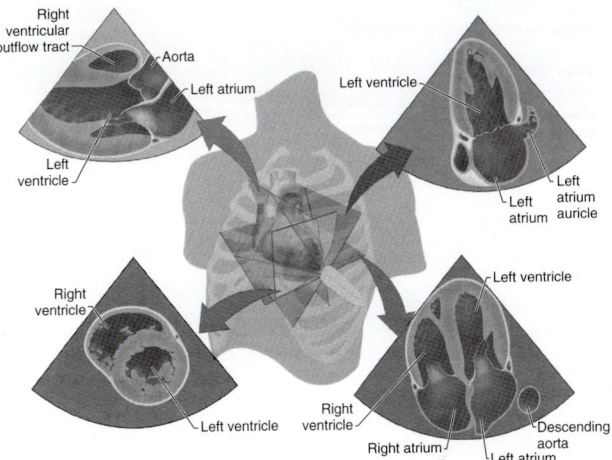

Fig. 1.48 The four basic image planes used in transthoracic echocardiography. A parasternal transducer position or "window" is used to obtain long- and short-axis views. The long-axis view extends from the left ventricular apex through the aortic valve plane. The short-axis view is perpendicular to the long-axis view, resulting in a circular view of the left ventricle. The transducer is placed at the ventricular apex to obtain the two-chamber and four-chamber views, each of which is approximately a 60-degree rotation from the long-axis view and perpendicular to the short-axis view. The four-chamber view includes both ventricles and both atria. The two-chamber view includes the left ventricle and left atrium; sometimes the atrial appendage is visualized. (From Otto CM: *Textbook of clinical echocardiography,* ed 4, Philadelphia, Elsevier Saunders, 2009:32, Fig. 2.1.)

7. Intravascular Ultrasound (IVUS)

Indications
- Left main CAD evaluation when doubt persists about the severity of stenosis on angiography
- Discordance between symptoms or noninvasive test results and coronary angiogram
- Ambiguous lesions (ostial lesion, bifurcation lesion, aneurysm, hazy lesion)
- Transplant vasculopathy
- In-stent restenosis or thrombosis
- Stent implantation in medium-sized (2.5–3.25 mm) arteries
- Unsatisfactory angiographic or symptomatic results of percutaneous coronary interventions

Strengths
- IVUS provides cross-sectional images of both the arterial wall and lumen with excellent resolution, reveals the diffuse nature of atherosclerosis and the involvement of reference segments, and takes into account vessel wall remodeling

Weaknesses
- Invasive
- Requires full anticoagulation
- Coronary spasm occurs in 2% of patients (usually responds to intracoronary nitroglycerin)

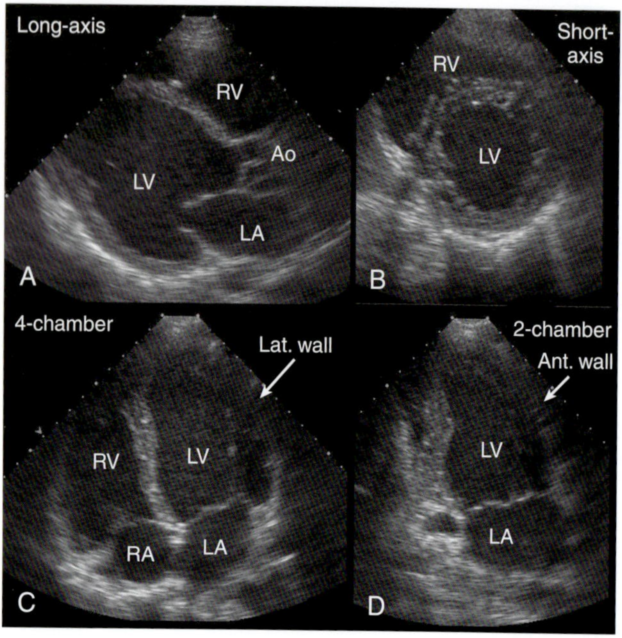

Long-axis

RV

Ao

LV

LA

A

Short-axis

RV

LV

B

4-chamber

Lat. wall

RV   LV

RA   LA

C

2-chamber

Ant. wall

LV

LA

D

Fig. 1.49 A, Parasternal long-axis view in a patient with dilated cardiomyopathy shows the anteroseptum and posterior wall of the left ventricle. B, Parasternal short-axis view in a patient shows a cross-section through the midportion of ventricles. C, Apical four-chamber view shows the lateral (*Lat. wall*) and inferoseptal walls of the left ventricle (*LV*) and the free wall of the right ventricle. D, Apical two-chamber view shows anterior (*Ant. wall*) and inferior walls of the LV. *Ao,* aorta; *LA,* left atrium; *LV,* left ventricle; *RA,* right atrium; *RV,* right ventricle. (From Sellke FW, del Nido PJ, Swanson SJ: *Sabiston & Spencer surgery of the chest,* ed 9, Philadelphia, Elsevier, 2016.)

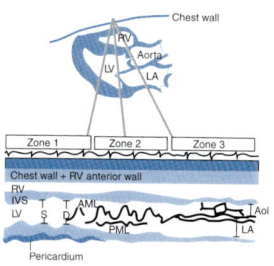

Fig. 1.50 Echocardiogram. (From Weissleder R, Wittenberg J, Harisinghani MG, Chen JW: *Primer of diagnostic imaging,* ed 4, St. Louis, Mosby, 2007.)

TABLE 1.3   Indications for Echocardiography

DIAGNOSIS	KEY ECHOCARDIOGRAPHIC FINDINGS	LIMITATIONS OF ECHOCARDIOGRAPHY
Cardiomyopathy		
	LV size, wall thickness, and systolic function RV size and function PA pressure estimate Valvular anatomy and dysfunction	
Valvular Heart Disease		
Valve stenosis	Etiology and severity of stenosis LV and RV size and systolic function PA pressure estimate	May underestimate stenosis severity
Valve regurgitation	Mechanism and etiology of regurgitation Severity of regurgitation PA pressure estimate	TEE and 3-D imaging may be needed to evaluate MR and valve anatomy
Prosthetic valves	Evidence for stenosis or regurgitation PA pressure estimate	Imaging prosthetic valves limited by artifact TEE needed for suspected prosthetic MR
Endocarditis	Detection of vegetations, abscess Presence, degree of valve dysfunction	TEE increases sensitivity for detecting vegetations and abscess
Coronary Artery Disease		
Acute MI	Segmental wall motion abnormality reflects "myocardium at risk" Global LV function (EF) Evaluate for MI complications	Coronary artery anatomy not directly visualized
Angina	Global and segmental LV function Exclude other causes of angina (e.g., AS, HOCM)	Resting wall motion may be normal despite significant CAD Stress echocardiography needed to induce ischemia
End-stage ischemic disease	Global and segmental LV function RV function, PA pressures	
Pericardial Disease		
	Pericardial thickening Detection, size, location of pericardial fluid Signs of tamponade physiology	Tamponade is a hemodynamic and clinical diagnosis (blood pressure, heart rate)
Aortic Disease		
Aortic root dilation	Etiology of aortic dilation Accurate aortic diameter measurements Associated aortic regurgitation	
Aortic dissection	Accurate aortic diameter measurements Imaging of dissection "flap" (dissection) Associated aortic regurgitation	TEE more sensitive and specific Cannot assess distal vascular beds

Continued

TABLE 1.3	Indications for Echocardiography—cont'd	
DIAGNOSIS	**KEY ECHOCARDIOGRAPHIC FINDINGS**	**LIMITATIONS OF ECHOCARDIOGRAPHY**
Cardiac Masses		
LV/LA thrombus	TTE has high sensitivity and specificity for LV thrombus	TEE is needed to detect LA thrombus reliably
Cardiac tumors	Size, location, and physiologic consequences of tumor mass	Extracardiac involvement not well seen Cannot distinguish benign from malignant, or tumor from thrombus
Pulmonary HTN		
	RV size and systolic function PA pressure estimate	
Congenital Heart Disease		
	Detection of anatomic abnormalities Quantitation of physiologic abnormalities LV and RV size and systolic function	No direct intracardiac pressure measurements Saline contrast may be needed to assess intracardiac shunts

3-D, Three-dimensional; *AS,* aortic stenosis; *CAD,* coronary artery disease; *EF,* ejection fraction; *HOCM,* hypertrophic obstructive cardiomyopathy; *HTN,* hypertension; *LA,* left atrial; *LV,* left ventricular; *MI,* myocardial infarction; *MR,* mitral regurgitation; *MV,* mitral valve; *PA,* pulmonary artery; *RV,* right ventricular; *TEE,* transesophageal echocardiography; *TTE,* transthoracic echocardiography.
Adapted from Otto CM, Schwaegler RG, Freeman RV: *Echocardiography review guide,* ed 2, Philadelphia, 2011, Saunders, pp 89–91.

TABLE 1.4	Echocardiographic Assessment of Aortic Regurgitation Severity		
FINDING	**MILD**	**MODERATE**	**SEVERE**
Central jets: color jet width/LVOT diameter	<25%	25%–64%	≥65%
Vena contracta	<0.3 cm	0.3–0.6 cm	>0.6 cm
Pressure half-time	>500 ms	200–500 ms	<200 ms
LV end-diastolic dimension	Normal	5–6 cm	>6 cm
Regurgitant volume	<30 mL	30–59 mL	≥60 mL
Regurgitant fraction	<30%	30%–49%	≥50%
Regurgitant orifice area	<0.1 cm²	0.1–0.29 cm²	≥0.3 cm²

LV, left ventricle; *LVOT,* left ventricular outflow tract.
From Parrillo JE, Dellinger RP: *Critical care medicine: principles of diagnosis and management in the adult,* ed 4, Philadelphia, 2014, Saunders.

Comments

- IVUS imaging is conducted by experienced interventional physicians in the cardiac catheterization laboratory. A 0.0014-inch guidewire is first advanced to the distal segment of the artery of interest. IVUS imaging is then performed with 2.5- to 3.5-French catheters with miniaturized (<1 mm diameter) ultrasound transducers at their tips (Fig. 1.52).
- Intracoronary nitroglycerin (150–300 mcg) should be administered (unless contraindicated) before IVUS examination to minimize dynamic fluctuations in vasomotor tone.
- Cost: $$$$

TABLE 1.5   Echocardiographic Assessment of Mitral Regurgitation Severity

FINDING	MILD	MODERATE	SEVERE
Central jets: jet area by color Doppler/LA area	<20%	20%–39%	>40%
Vena contracta	<0.3 cm	0.3–0.69 cm	≥0.7 cm
Pulmonary veins	Predominant systolic	Blunted systolic	Systolic reversal
LA size	Normal	Normal or dilated	Dilated
LV size	Normal	Normal or dilated	Dilated
Regurgitant volume	<30 mL	30–59 mL	≥60 mL
Regurgitant fraction	<30%	30%–49%	≥50%
Regurgitant orifice area	<0.2 cm²	0.2–0.39 cm²	≥0.4 cm²

LA, left atrium; *LV,* left ventricle.
From Parrillo JE, Dellinger RP: *Critical care medicine: principles of diagnosis and management in the adult,* ed 4, Philadelphia, 2014, Saunders.

TABLE 1.6   Echocardiographic Assessment of Tricuspid Regurgitation Severity

FINDING	MILD	MODERATE	SEVERE
Central jets: jet area by color Doppler	<5 cm²	5–10 cm²	>10 cm²
Vena contracta	Not defined	Not defined, but <0.7 cm	>0.7 cm
PISA radius	≤0.5 cm	0.6–0.9 cm	>0.9 cm
RA size	Normal	Normal or dilated	Dilated
RV size	Normal	Normal or dilated	Dilated
Jet density	Soft	Dense	Dense
Jet contour	Parabolic	Variable	Triangular with early peak
Hepatic veins	Predominant systolic	Blunted systolic	Systolic reversal

PISA, proximal isovelocity surface area; *RA,* right atrium; *RV,* right ventricle.
From Parrillo JE, Dellinger RP: *Critical care medicine: principles of diagnosis and management in the adult,* ed 4, Philadelphia, 2014, Saunders.

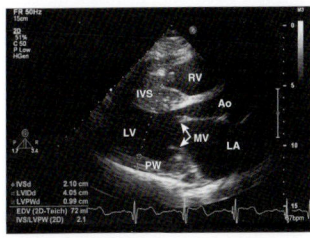

FIG. 1.51 Echocardiographic appearance of hypertrophic cardiomyopathy. Parasternal long-axis view from a patient with hypertrophic cardiomyopathy demonstrating asymmetric septal hypertrophy. The interventricular septum (*arrow*) measures 2.1 cm, the posterior wall measures 0.99 cm. *Ao,* aorta; *IVS,* interventricular septum; *LA,* left atrium; *LV,* left ventricle; *MV,* mitral valve; *PW,* posterior wall; *RV,* right ventricle. (From Issa Z, Miller JM, Zipes DP: *Clinical arrhythmology and electrophysiology,* ed 2, Philadelphia, Saunders, 2012.)

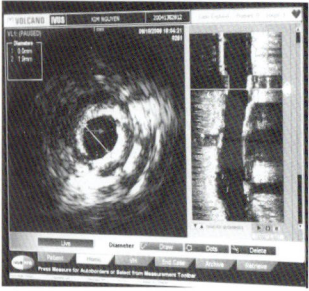

Fig. 1.52 Angiographic monitor display of intravascular ultrasound image of coronary artery during percutaneous coronary intervention. Precise dimensions of the artery and lumen are provided. (From Goldman L, Schafer AI: *Goldman's Cecil medicine*, ed 24, Philadelphia, Saunders, 2012.)

D. Chest Imaging

1. Chest Radiograph
Figs. 1.53 and 1.54 describe normal anatomy on chest radiograph.

Indications
- Dyspnea
- Chest trauma
- Chest pain
- Chronic cough
- Hemoptysis
- Suspected lung neoplasm (primary or metastatic)
- Suspected infectious process (e.g., tuberculosis [TB], pneumonia, abscess [Fig. 1.55])
- Inhalation injury
- Pulmonary nodule
- Suspected pleural effusion
- Pneumothorax
- Pulmonary plaques
- Pneumonia follow-up
- Assessment before cardiopulmonary surgery
- Confirmation of feeding tube placement, Swan-Ganz catheter, central venous catheter, endotracheal tube, transvenous ventricular pacemaker
- Suspected acute respiratory distress syndrome (ARDS), CHF
- Mesothelioma
- Interstitial lung disease
- Sarcoidosis (Fig. 1.56)
- Lymphoma

Strengths
- Low cost
- Readily available
- Low radiation
- Can be performed at bedside

Weaknesses
- Low diagnostic yield
- For portable chest x-radiographs, poor results in obese patients, heart overmagnification caused by film taken anteroposteriorly, poor respiratory effort, and poor positioning

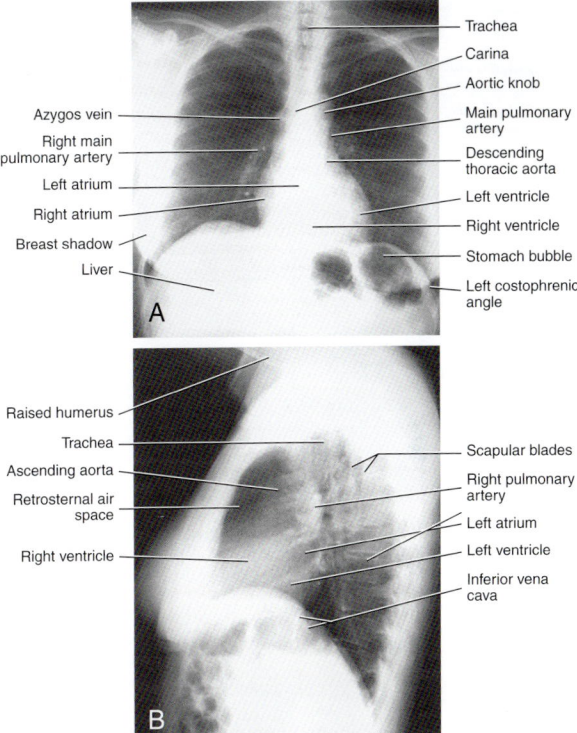

Trachea
Carina
Aortic knob
Main pulmonary artery
Descending thoracic aorta
Left ventricle
Right ventricle
Stomach bubble
Left costophrenic angle

Azygos vein
Right main pulmonary artery
Left atrium
Right atrium
Breast shadow
Liver

Raised humerus
Trachea
Ascending aorta
Retrosternal air space
Right ventricle

Scapular blades
Right pulmonary artery
Left atrium
Left ventricle
Inferior vena cava

Fig. 1.53 Normal anatomy on the female chest radiograph in (A) the upright postero-anterior projection and (B) the lateral projection. (From Mettler FA: *Primary care radiology*, Philadelphia, Saunders, 2000.)

of patients; the degree of inspiration is likely to be substantially less in a portable chest x-ray than with an erect film, making the heart appear relatively larger and providing less optimal visualization of the lungs because they are not optimally expanded

Comments
- Proper exposure for evaluating the cardiac structures is present when the spine is just visible behind the heart.
- Radiographs are the most cost-effective method of staging chest pathologic conditions.
- Cost: $

2. CT Scan of Chest
Indications
- Nondiagnostic, abnormal plain chest radiograph (e.g., characterization of chest masses [Fig. 1.57], pleural masses, abscess [Fig. 1.58], cavitary or cystic changes, or nonspecific infiltrates noted on plain chest films)
- Follow-up of pulmonary nodule

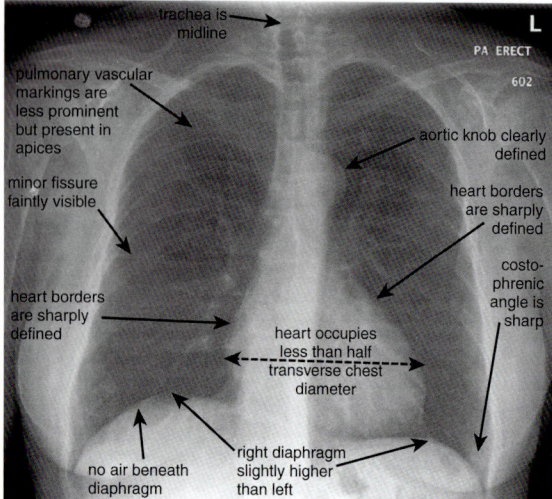

Fig. 1.54 The normal frontal chest x-ray. Normal features of the posterior-anterior upright chest x-ray are demonstrated. (From Broder JS: *Diagnostic imaging for the emergency physician*, Philadelphia, Saunders, 2011.)

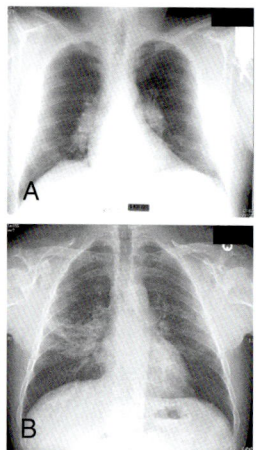

Fig. 1.55 A, Stage I sarcoidosis: prominent hilar lymphadenopathy and normal lungs. B, Stage III sarcoidosis: interstitial infiltrate without hilar lymphadenopathy. (From Mason RJ, Broaddus VC, Murray JF, et al: *Murray and Nadel's textbook of respiratory medicine*, ed 4, Philadelphia, Saunders, 2004.)

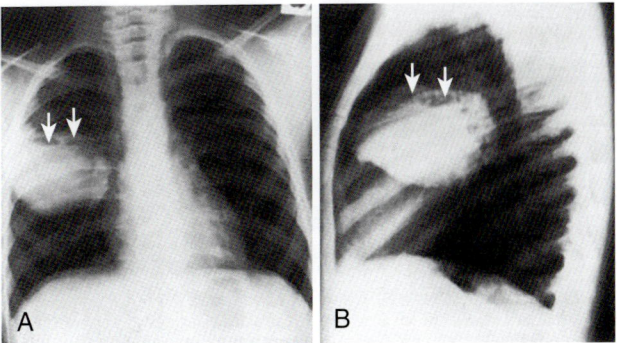

Fig. 1.56 A and B, Multiloculated lung abscess (*arrows*). (From Brook I: Lung abscess and pulmonary infections due to anaerobic bacteria. In Chernick V, Boat TF, Wilmott RW, et al, [eds]: *Kendig's disorders of the respiratory tract in children*, ed 7, Philadelphia, Saunders, 2006:482.)

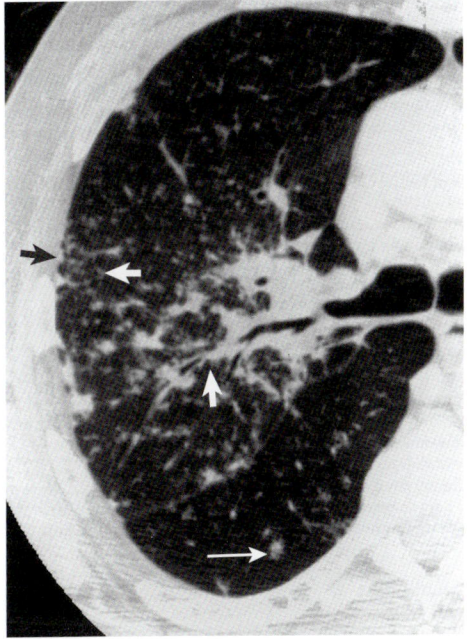

Fig. 1.57 Pulmonary sarcoidosis on high-resolution computed tomography scan. There are nodules of varying sizes: small interstitial nodules 1 to 2 mm in diameter (*intermediate white arrow*), irregular and larger nodules (*thin white arrow*), and subpleural nodules (*black arrow*). Many of the nodules have a subpleural distribution (particularly along the oblique fissure) as well as a bronchocentric distribution, giving an irregular bronchovascular-lung interface (*broad white arrow*). (Courtesy Dr. F. Gleeson, Oxford, England.)

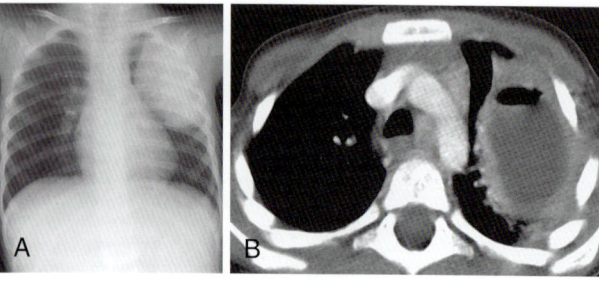

Fig. 1.58 Pulmonary abscess in a 2-year-old boy with persistent cough. A, Chest radiograph shows large oval mass in the left upper lobe. B, Computed tomography scan demonstrates an abscess with a thick enhancing wall that contains both air and fluid. (From Slovis TL [ed]: *Caffey's pediatric diagnostic imaging,* ed 11, Philadelphia, Mosby, 2008:1297.)

- Staging of lung carcinoma
- Mediastinal widening
- Evaluation of bronchiectasis and interstitial lung disease
- Chest trauma
- Characterization of suspected thoracic aortic pathologic conditions
- Abnormal hilum
- Differentiation of pleural from parenchymal abnormalities
- Evaluation of thymus in patients with myasthenia gravis
- Suspicion of aortic dissection

Strengths
- Fast. Spiral or helical CT allows the entire chest to be easily imaged during a single breath hold.
- Preferred method for examining the hila and the mediastinum
- Excellent method for evaluation of pleura and chest wall

Weaknesses
- Potential for significant contrast reaction
- Expensive
- Radiation exposure (Table 1.7)

Comments
- Table 1.8 summarizes general chest CT principles
- Contrast enhancement may be necessary for the evaluation of known or suspected vascular abnormalities (e.g., aortic aneurysm or dissection), abnormal hilum, and certain abnormalities of the pleura.
- High-resolution chest CT uses a specific algorithm of very thin slices to evaluate for interstitial lung diseases, bronchiectasis, or lymphangitic spread of carcinoma.
- Cost: CT of chest without contrast $$$; CT of chest with and without contrast $$$$

3. MRI Scan of Chest
Indications
- Evaluation of chest wall disease when CT is inconclusive
- Assessment for aortic dissection (Fig. 1.59)
- Assessment of hilar and mediastinal pathologic conditions when CT is inconclusive
- Superior sulcus carcinoma
- Posterior mediastinal masses

TABLE 1.7	Radiation Dose for Chest CT Protocols	
SITUATION		**DOSE (MSV)**
Normal yearly background radiation		2.5–3.2
Chest radiograph (single view)		0.05
Routine chest CT (300 mA)		5–7
Routine chest CT (modulated, approximately 100–150 mA)		1.5–2
High-resolution CT with volumetric imaging (supine, expiratory) (modulated, approximately 100–150 mA)		1.5–2
High-resolution CT with spaced axial images (supine, prone, expiratory)		1
Low-dose volumetric CT (40 mA)		< 0.5–1

mSv, millisieverts.
From Webb WR, Brant WE, Major NM: *Fundamentals of body CT,* ed 4, Philadelphia, 2015, Saunders.

TABLE 1.8	Chest CT: General Principles
Scan levels	Lung apices to the posterior costophrenic angles
Patient position	Supine; prone scans sometimes used for diagnosis of lung disease or pleural effusions
Lung volume	Full inspiration, single breath hold; expiratory scans sometimes used to diagnose air trapping
Gantry rotation time	Approximately 0.5 s in most instances
Scan duration	Approximately 2.5 s for the thorax using multidetector CT and fast scanning
Detector width	The thinnest detectors (e.g., 0.625 mm) are typically used for image acquisition
Pitch (table excursion)	Depends on tolerable image noise; increased if noise is OK, decreased if there is a desire for high resolution
Reconstruction algorithm	High-resolution algorithm used for most studies; standard or soft-tissue algorithm usually used for vascular studies
Two- or three-dimensional reconstructions	Not routine; occasionally useful for lung, airway, or vascular studies
Contrast agents	Intravenous contrast injection in some cases; oral contrast agents only for gastrointestinal abnormalities

From Webb WR, Brant WE, Major NM: *Fundamentals of body CT,* ed 4, Philadelphia, 2015, Saunders.

- Follow-up lymphoma
- Brachial plexus lesions
- Contraindications to contrast medium in patients with mediastinal or vascular abnormality

Strengths
- Noninvasive
- Safe contrast agent
- No ionizing radiation
- Soft-tissue resolution
- Multiplanar
- Excellent imaging of mediastinum and chest wall

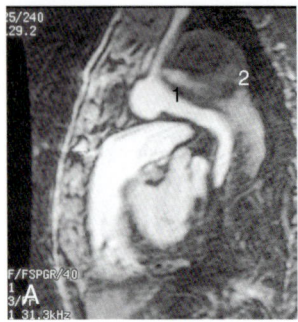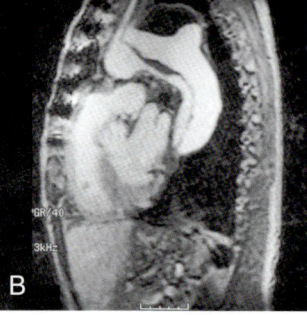

Fɪɢ. **1.59** Longitudinal magnetic resonance imaging scan in a chronic type B dissection extending into the arch. A, Early phase showing (1) the entry point and (2) a faint visualization of the outline of the aneurysm. B, Late phase demonstrating partial opacification of the aneurysm and the extension of the dissection along the subclavian artery. (Courtesy Dr. Loren Ketai, University of New Mexico.) (From Crawford MH, DiMarco JP, Paulus WJ, [eds]: *Cardiology*, ed 2, St. Louis, Mosby, 2004.)

Weaknesses
- Expensive
- Needs cooperative patient
- Time-consuming
- Motion artifacts secondary to cardiac and respiratory movements
- Inadequate imaging of lung (normal lung does not produce a magnetic resonance signal because of magnetic susceptibility effects)
- Cannot be performed in patients with aneurysm clips that are not compatible with magnetic resonance or in patients with a pacemaker, cochlear implants, or metallic foreign body in the eyes; safe in women with intrauterine devices (IUDs), including copper ones, and in patients with surgical clips and staples

Comments
- Used predominantly as a problem-solving tool if CT is inconclusive.
- Anxious patients (especially those with claustrophobia) should be premedicated with an anxiolytic agent, and their imaging should be done with "open MRI" whenever possible.
- Cost: MRI with contrast $$$$, MRI without contrast $$$, MRI with and without contrast $$$$$

E. Endocrine Imaging

1. Adrenal Medullary Scintigraphy (Metaiodobenzylguanidine [MIBG] Scan)

Indications
- Evaluation of suspected intraadrenal paraganglioma or pheochromocytoma (Fig. 1.60)
- Survey of the entire body for the presence of extraadrenal and metastatic lesions from paragangliomas or pheochromocytomas

Strengths
- Sensitivity for detection of pheochromocytoma is greater than 90%, specificity greater than 95%

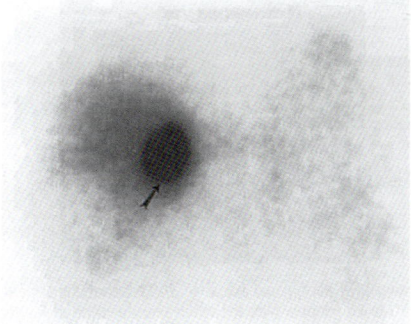

Fig. 1.60 Anterior abdominal view on ^{123}I-MIBG scan (72 hours postinjection) of a right adrenal pheochromocytoma (*arrow*). (From Besser CM, Thorner MO: *Comprehensive clinical endocrinology*, ed 3, St. Louis, Mosby, 2002.)

Weaknesses
- Interference from several drugs (e.g., tricyclic antidepressants, labetalol, cocaine, reserpine) and barium

Comments
- Adrenal medullary scintigraphy uses the tracer MIBG, an analog of guanethidine. Uptake occurs in the adrenal medulla, neuroblastic tumor tissues, and other organs with rich adrenergic innervation (e.g., heart, spleen). Commonly used radiolabels are I-131 and I-123. When using I-131 MIGB, initial images are usually obtained at 24 hours, delayed images at 48 and 72 hours after injection. When using I-123 MIGB, initial images are obtained at 2 to 3 hours, delayed ones at 24 and 48 hours.
- Paragangliomas or pheochromocytomas demonstrate unilateral focal uptake.
- Scintigraphy with MIBG should not be used as a screening procedure for pheochromocytoma. MIGB scintigraphy is indicated only after biochemical tests suggest the diagnosis.
- Additional imaging modalities for pheochromocytoma of adrenal glands are CT and MRI. Both modalities can detect up to 90% of functional tumors.
- Cost: $$

2. Parathyroid (PTH) Scan
Indications
- Hypercalcemia with elevated PTH level
- Presurgical localization of source of PTH production

Strengths
- Noninvasive
- Best test to rule out PTH adenoma (Fig. 1.61)

Weaknesses
- PTH hyperplasia may result in nondiagnostic scan
- Recent ingestion of iodine (food, meds) or recent tests with iodine content may interfere with interpretation of results
- Pregnancy is a relative contraindication

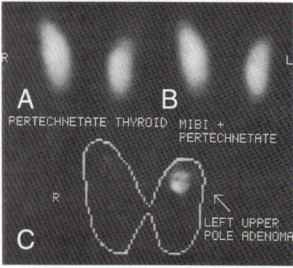

Fig. 1.61 Imaging of a parathyroid adenoma. A, Pertechnetate thyroid scan obtained conventionally 10 to 20 minutes after the intravenous injection of Tc-99m pertechnetate through an indwelling needle. B, Without the patient moving, Tc-99m methoxyisobutyl-isonitrile (MIBI) is injected, and a further series of images is taken. Tc-99m MIBI is taken up both by the parathyroid adenoma and by normal thyroid so that a combined composite image is seen. Using a change detection algorithm, the change between the two images is determined, and the statistical degree of that difference is plotted as a probability. C, The higher intensity in the upper pole of the left lobe of the thyroid indicated a change between the two images with a significance of more than 1 in 1000. This is the site of the upper-pole parathyroid adenoma. The outline of the thyroid is also shown. A small area of increased probability of change also seen in the upper-pole thyroid adenoma was removed, and a right upper-pole hyperplastic gland (100 mg) was also removed. Before imaging, it is important biochemically to confirm that hypercalcemia is due to hyperparathyroidism. Imaging of the parathyroid is intended to localize the site of adenomas or hyperplastic glands. Visualization of a gland depends on its size. A normal parathyroid gland of less than 20 mg will not be visualized by this technique. Earlier attempts to image parathyroid glands using thallium in a similar way have proved less successful than the use of MIBI. (From Besser CM, Thorner MO: *Comprehensive clinical endocrinology,* ed 3, St. Louis, Mosby, 2002.)

Comments
- It is necessary to look for an ectopic location in the chest or other location in the neck when evaluating a PTH scan.
- Cost: $$

3. Thyroid Scan (Radioiodine Uptake Study)

Indications
- Thyroiditis
- Hyperthyroidism
- Thyroid nodule (Fig. 1.62)
- Detection of lingual thyroid
- Thyroglossal cyst

Strengths
- Noninvasive

Weaknesses
- Contraindicated in pregnancy because radioiodine crosses the placenta; significant exposure of the fetal thyroid can occur and may result in cretinism
- Radioiodine is excreted in human breast milk; nursing should be stopped after diagnostic studies with radioiodine
- Significant interference from iodine contained in foods and medications can interfere with imaging

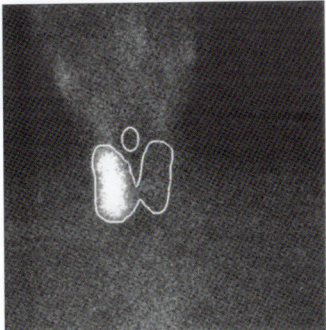

Fɪɢ. 1.62 An I-131 scan demonstrates an area of increased uptake in the right lobe of a 32-year-old woman with increased thyroid function tests and a palpable nodule. This scan is consistent with a toxic or hyperfunctioning nodule. (From Townsend CM, Beauchamp RD, Evers BM, Mattox KL [eds]: *Sabiston textbook of surgery,* ed 17, Philadelphia, Saunders, 2004.)

Comments
- In the normal euthyroid subject, distribution of radiotracer is homogeneous and uniform throughout the gland.
- In Graves disease, concentration of activity is uniformly increased.
- In Hashimoto disease, radioiodine uptake values are variable depending on the stage of disease.
- In nontoxic nodular goiter, several areas experience relatively increased activity, whereas activity is decreased in other areas.
- In toxic hot nodule, a rounded area of markedly increased concentration of activity is seen.
- Cost: $$

4. Thyroid Ultrasound
Indications
- Thyroid nodules (Fig. 1.63)
- Thyromegaly
- Multinodular goiter
- PTH abnormalities
- To direct image-guided biopsy

Strengths
- Noninvasive
- Fast
- No ionizing radiation

Weaknesses
- Fine-needle aspiration biopsy necessary for definitive diagnosis
- May miss nodules less than 1 cm in diameter
- Interpretation of large cysts (>4 cm) often difficult because of presence of areas of cystic or hemorrhagic degeneration

Comments
- Ultrasound is an excellent modality to demonstrate thyroid gland anatomy and to guide biopsy or cyst aspiration.
- Thyroid ultrasound is also useful to detect PTH abnormalities.

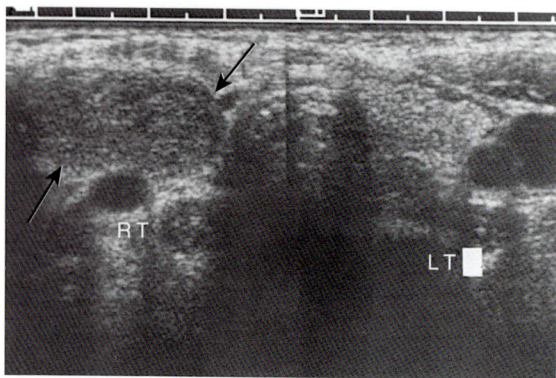

Fig. 1.63 Preoperative ultrasound of a patient with a 4 × 2 cm homogeneous right thyroid (RT) mass (*arrows*). Resection demonstrated a follicular adenoma. *LT,* left thyroid. (From Townsend CM, Beauchamp RD, Evers BM, Mattox KL [eds]: *Sabiston textbook of surgery,* ed 17, Philadelphia, Saunders, 2004.)

- Approximately 70% of PTH lesions are evident on ultrasound.
- Cost: $

F. Genitourinary Imaging

1. Obstetric Ultrasound

Indications

- Determination of gestational age (size–date discrepancy ≥2 weeks, before elective pregnancy termination)
- Multiple gestation (determination of fetal number)
- Suspected fetal demise
- Suspected abortion
- Incomplete abortion
- Spontaneous abortion
- Determination of fetal presentation
- Fetal anatomy survey
- Placental evaluation
- Diagnosis of fetal abnormalities
- Umbilical cord evaluation
- Vaginal bleeding
- Suspected congenital abnormality
- Assistance in obtaining amniotic fluid
- Maternal disease (e.g., hypertension, diabetes mellitus [DM], rubella, cytomegalovirus [CMV], human immunodeficiency virus [HIV])
- Preterm labor or rupture of membranes before 36 weeks
- Suspected placental abruption
- First-degree relative with congenital anomaly
- Evaluate for fetal growth and for intrauterine growth retardation (IUGR)

Strengths

- Fast
- Noninvasive

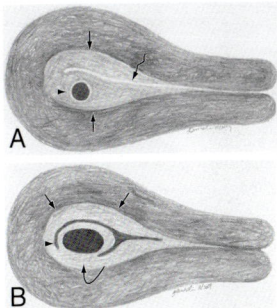

Fɪɢ. **1.64** Diagram of early gestational sac. A, Intradecidual sac sign. The eccentric sac (*arrowhead*) is located within the endometrium (*arrows*). Note the decidual reaction around the sac. The *curly arrow* points to the endometrial cavity. B, Double decidual sac sign. Echogenic decidua parietalis (*arrows*) and capsularis (*arrowhead*) form the double decidual sign. Decidua basalis develops at the implantation of the gestational sac into the endometrium (*curved arrow*). (From Fielding JR et al: *Gynecologic imaging,* Philadelphia, 2011, Saunders.)

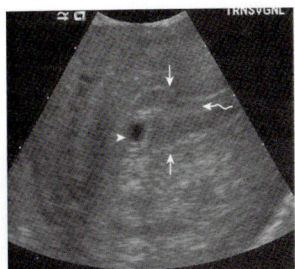

Fɪɢ. **1.65** Intradecidual sac sign. Transvaginal ultrasound shows the eccentric sac (*arrowhead*) located within the endometrium (*arrows*). Note the echogenic wall to the sac. The *curly arrow* points to the endometrial cavity. (From Fielding JR et al: *Gynecologic imaging,* Philadelphia, 2011, Saunders.)

- Readily available
- No ionizing radiation
- Can be repeated serially

Weaknesses
- 20% false-negative rate in ectopic pregnancy
- May miss placental abruption if there is no retroplacental hemorrhage at time of scan or inadequate visualization of hemorrhage; blood may be isoechoic to placenta and difficult to evaluate

Comments
- Figs. 1.64 and 1.65 show gestational sac on obstetric sonogram.
- Cost: $$

2. Pelvic Ultrasound

Indications
- Pelvic mass
- Pelvic pain
- Infertility
- Uterine or ovarian mass
- Uterine (Müllerian) anomalies
- Pregnancy, including ectopic
- Abnormal vaginal bleeding

Strengths
- Fast
- Can be performed at bedside
- Low cost
- Noninvasive
- No radiation
- No need for contrast agent
- Examination not affected by renal function
- Readily available

Weaknesses
- Shadowing by gas, bone, or calculi often obscures views behind acoustic shadowing
- Ultrasonographic studies are operator dependent
- Affected by body habitus
- Requires patient cooperation and full bladder

Comments
- Typically consists of transabdominal and transvaginal scanning (Fig. 1.66).
- Cost: $$

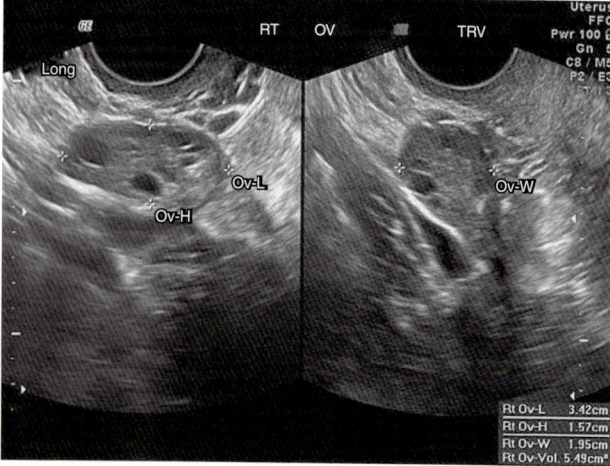

FIG. 1.66 Normal adult ovary in longitudinal (Long) and transverse (TRV) plane on transvaginal ultrasound. The ovary is located medial to the internal iliac vessels. Calipers indicate ovarian measurements, and the ovarian volume is normal. (From Fielding JR et al: *Gynecologic imaging*, Philadelphia, 2011, Saunders.)

3. Prostate Ultrasound

Indications

- Guidance for biopsy (Fig. 1.67)
- Abnormal digital rectal examination
- Voiding difficulty
- Elevated prostate-specific antigen (PSA)
- Infertility workup

Strengths

- No ionizing radiation
- Imaging modality of choice as a guide to needle biopsy in suspected prostate carcinoma
- Useful to assess bladder volume and estimate size of prostate

Weaknesses

- Prostate ultrasound is invasive and uncomfortable for patients
- Fecal material in rectum will interfere with sonogram results

Comments

- Cost: $$

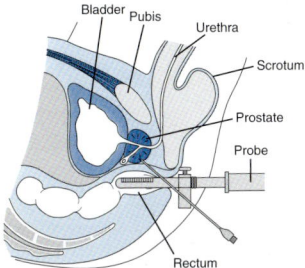

Fig. 1.67 Rectal ultrasonography. Diagram demonstrating transrectal biopsy of the prostate. (From Pagana KD, Pagana, TJ: *Mosby's diagnostic and laboratory test reference*, ed 8, St. Louis, Mosby, 2007.)

4. Renal Ultrasound

Indications

- Renal insufficiency
- Nephrolithiasis
- Renal mass
- Polycystic kidney disease
- Acquired cystic renal disease
- Hydronephrosis
- Pyelonephritis (Fig. 1.68)
- Renal infarction
- Acute or chronic renal failure

Strengths

- Noninvasive
- Can be performed at bedside
- Fast
- No ionizing radiation
- Readily available

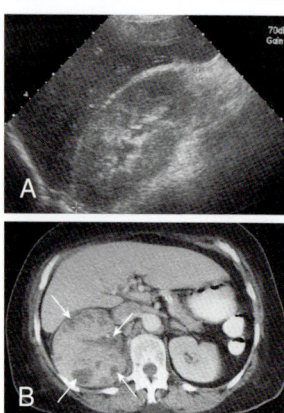

Fig. 1.68 Acute pyelonephritis. A, Ultrasound image demonstrates an enlarged echogenic kidney. Bipolar length of kidney is 12.9 cm. B, Computed tomography scan with contrast enhancement obtained 24 hours later demonstrates multiple nonenhancing abscesses (*arrows*). (From Floege J, Johnson RJ, Feehally J: *Comprehensive clinical nephrology*, ed 4, Philadelphia, Saunders, 2010.)

Weaknesses
- Affected by body habitus
- Retained barium from radiographs may interfere with results

Comments
- Cost: $$

5. Scrotal Ultrasound
Indications
- Testicular pain or swelling
- Testicular trauma
- Testicular or other scrotal masses, hydrocele (Fig. 1.69)
- Search for undescended testicle
- Infertility evaluation
- Testicular torsion
- Infection (Fig. 1.70)

Strengths
- Noninvasive
- Fast
- Requires no patient preparation
- No ionizing radiation

Weaknesses
- Not useful for staging of testicular neoplasm (MRI is preferred)

Comments
- Color Doppler ultrasound is the imaging modality of choice when suspecting testicular torsion; however, nuclear medicine flow study can also be used for diagnosis.
- Cost: $

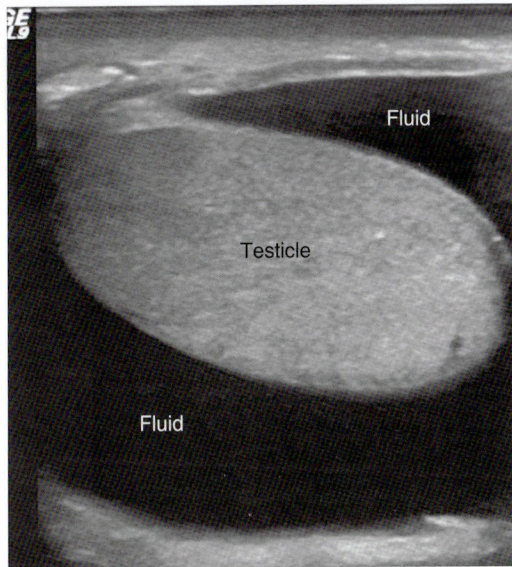

Fig. 1.69 Hydrocele. A longitudinal image shows the normal testicle completely surrounded by fluid. (From Mettler FA: *Essentials of radiology,* ed 3, Philadelphia, Elsevier, 2014.)

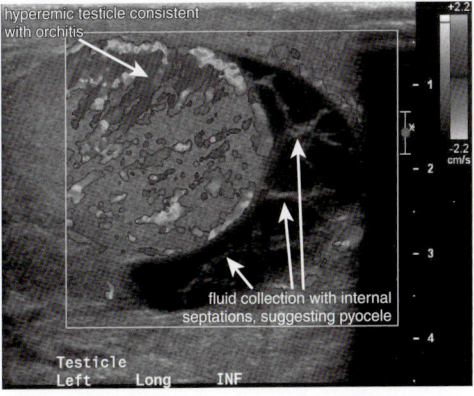

Fig. 1.70 Orchitis with pyocele: testicular ultrasound. The left testicle is markedly hyperemic on color Doppler ultrasound compared with the right testicle (not shown), consistent with orchitis. There is complex fluid collection with internal septations and internal echoes that is predominantly located inferior to the testicle and is concerning for a pyocele. Simple hydroceles are homogeneously black without internal echoes. Fluid collections with internal structure are more likely to represent infection or blood products, whereas simple fluid is more likely serous. (From Broder JS: *Diagnostic imaging for the emergency physician,* Philadelphia, Saunders, 2011.)

6. Transvaginal (Endovaginal) Ultrasound

Indications

- Evaluation of adnexa and ovaries
- Evaluation of pregnancy (Fig. 1.71)
- Infertility workup
- Suspected ectopic pregnancy
- Suspected endometrial abnormalities
- Guidance for aspiration or biopsy of pelvic fluid collection or mass
- Lost IUD

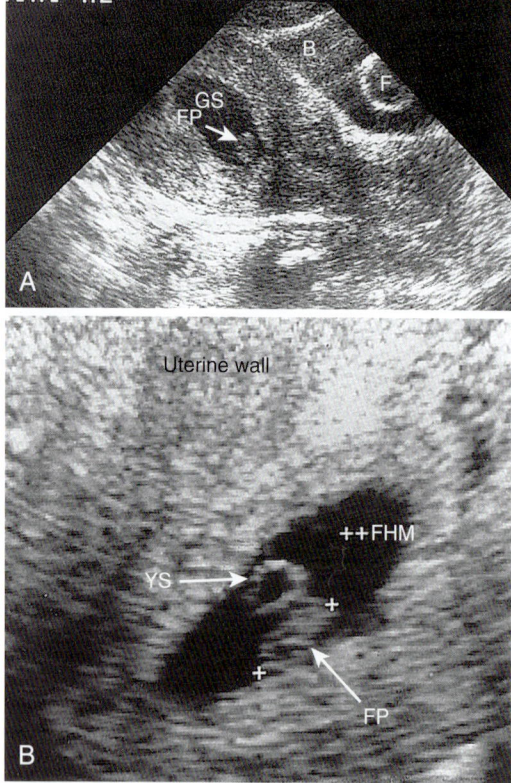

Fɪɢ. 1.71 Early normal obstetric ultrasound. A longitudinal transabdominal ultrasound image (A) demonstrates the bladder (*B*) and a Foley catheter (*F*) within it. Superior to and behind the bladder is the uterus with a gestational sac (*GS*) centrally and a fetal pole (*FP*) within it. More detail can be obtained by using transvaginal ultrasound (B). In this case the fetal pole can be measured; a yolk sac (*YS*) also is seen, and the technician has indicated that fetal heart motion (*FHM*) was seen. (From Mettler FA: *Essentials of radiology,* ed 3, Philadelphia, Elsevier, 2014.)

Strengths
- Useful for evaluation of obese or gaseous patients
- Very good resolution images of pelvis
- No ionizing radiation
- Full bladder not required
- Earlier detection of pregnancy compared with transabdominal ultrasound

Weaknesses
- Unable to evaluate false pelvis
- Limited field of view secondary to high-frequency transducer
- Will miss any structures or abnormalities more than 10 cm away from transducer

Comments
- Contraindications include imperforate hymen, patient refusal, premature membrane rupture (increased risk of infection)
- Cost: $$

7. Urinary Bladder Ultrasound
Indications
- Hematuria
- Recurrent cystitis
- Bladder neoplasm (primary or metastatic) (Fig. 1.72)
- Urinary incontinence
- Bladder diverticula
- Bladder calculi
- Evaluation of masses posterior to the bladder

Strengths
- Noninvasive
- Fast
- No ionizing radiation
- Can be performed at bedside

Weaknesses
- Affected by overlying bowel gas and body habitus (e.g., in morbidly obese patient, fat overlying bladder impedes visualization)
- Operator dependent; results may be affected by skill of technician
- Requires cooperative patient and full bladder distention

Comments
- Cost: $$

8. Hysterosalpingography (HSG)
Indications
- Primary and secondary infertility (confirmation of tubal patency in infertile patients)
- Diagnosis of tubal anomalies (including diverticula and accessory ostia) (Fig. 1.73)
- Evaluate tube patency after tubal ligation (confirmation or assessment for reversal of sterilization procedures)
- Evaluation of filling defects within the endometrial canal

Strengths
- Less expensive than laparoscopy
- Rapid (takes 10 minutes to perform) and relatively safe (complications occur in less than 3% of patients)
- Allows for visualization of the internal morphologic characteristics of the endocervical canal, uterine cavity, and fallopian tubes

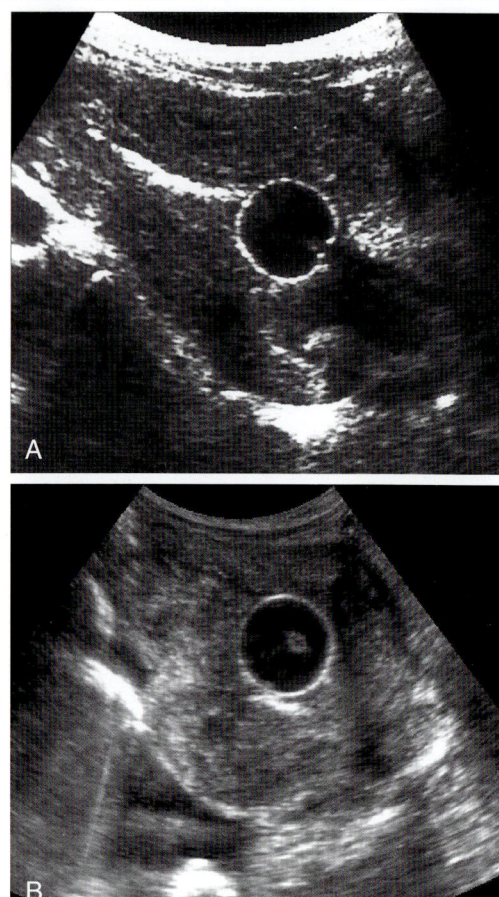

Fig. 1.72 Diffuse bladder wall thickening. Two patients presenting with urinary reten-tion. A, Interstitial cystitis. B, Diffuse transitional cell carcinoma in second patient. Both sonograms show marked circumferential bladder wall thickening after Foley catheterization. Cystoscopy and biopsy are required for differentiation. (From Rumack CM, Wilson SR, Charboneau JW, Levine D: *Diagnostic ultrasound*, ed 4, Philadelphia, Mosby, 2011.)

Weaknesses
- Limited diagnostic use (discovers only 50% of peritubal disease diagnosed by direct visualization via laparoscope)
- Radiation exposure

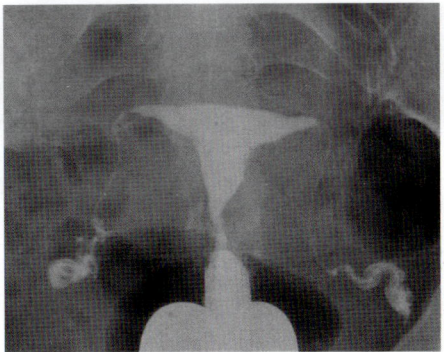

Fig. 1.73 A normal hysterosalpingogram. The triangular outline of the uterine cavity is seen, with passage of dye along the fallopian tubes and spill into the peritoneal cavity. (From Greer IA, Cameron IT, Kitchener HC, Prentice A: *Mosby's color atlas and text of obstetrics and gynecology*, London, Harcourt, 2001.)

Comments

- HSG is an imaging modality in which the uterine cavity and the lumina of the fallopian tubes are visualized by injecting contrast material through the cervical canal.
- Contraindications to HSG are acute pelvic infection, pregnancy, active uterine bleeding, recent uterine surgery, and contrast media allergy.
- HSG should be performed only on days 6 to 12 after last menstrual period (LMP).
- Complications of HSG include pain, lymphatic or venous intravasation, pelvic infection, allergic reaction to contrast media, vasovagal reaction, and uterine perforation.
- Cost: $$

9. Intravenous Pyelography (IVP) and Intravenous Retrograde Pyelography

Indications

- Hematuria
- Suspected urolithiasis (Fig. 1.74)
- Renal cell carcinoma
- Renal and ureteral anomalies (Fig. 1.75), strictures
- Bladder tumors, diverticula, cystocele, calculi
- Enterovesical fistulas (e.g., Crohn's disease, diverticulitis, trauma, surgery)
- Retrograde urethrogram mainly used to evaluate strictures or anterior urethral disease in men and to confirm equivocal findings on IVP

Strengths

- Inexpensive
- Provides both functional and anatomic information (may identify anatomic abnormalities that predispose to stone formation)
- Able to image entire urinary tract
- Shows precise site of obstruction in urolithiasis

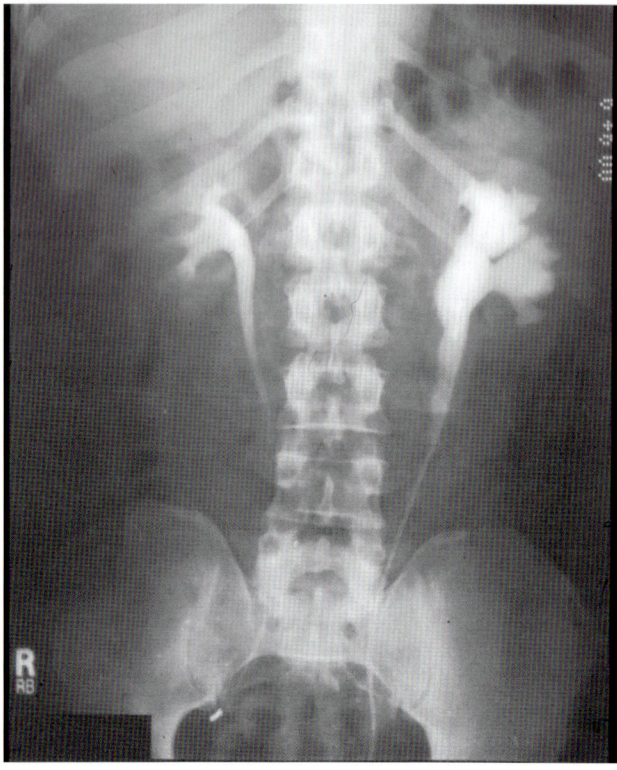

Fɪɢ. **1.74** Intravenous pyelography. Normal right kidney and dilated collecting system on the left. The obstruction was relieved with a stent. (From from Skorecki K, Chertow GM, Marsden PA, et al. *Brenner and Rector's The Kidney*, 10 ed, Philadelphia, 2016, Elsevier.)

Weaknesses

- Potential for significant IV contrast reaction
- Gas in the rectum can mimic filling defect in bladder
- Requires patient preparation to minimize intestinal gas and feces, which may mask abnormal findings
- Contraindicated in pregnancy
- Examination affected by renal function
- Compression of proximal bulbar urethra by prominent bulbocavernous muscles may be mistaken for urethral stricture
- Retained barium from previous barium examination can interfere with interpretation
- Radiation exposure

Comments

- Ultrasonography and CT have largely replaced IVP as the initial studies in urologic imaging during the past two decades.
- Risk of contrast-induced nephrotoxicity is 3% to 7%. Increased risk in patients with dehydration, DM, and creatinine 1.4 mg/dL or higher.

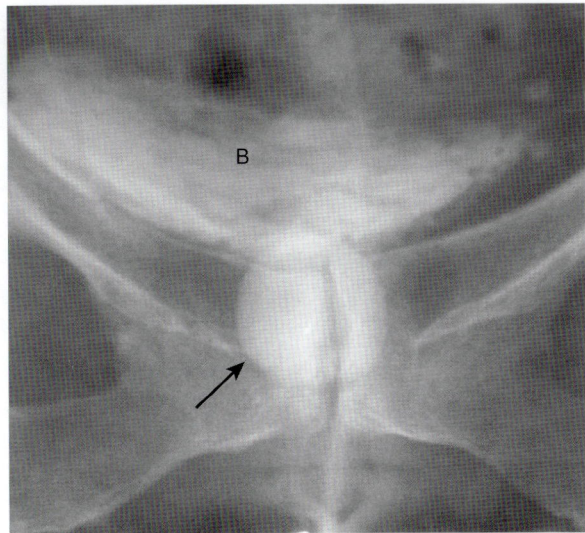

Fig. 1.75 Urethral diverticulum in a 55-year-old woman. Postvoid view of intravenous urogram shows a round contrast collection (*arrow*) below the bladder base (B). (From Fielding JR et al: *Gynecologic imaging,* Philadelphia, Saunders, 2011.)

- Increased risk of contrast reaction in patients with prior reaction, history of asthma, or severe allergies. Risk of IV contrast reaction is much lower when using nonionic contrast; however, nonionic contrast is much more expensive.
- Cost: $$

G. Musculoskeletal and Spinal Cord Imaging

1. Plain X-Ray Films of Skeletal System
Indications
- Trauma
- Infections (osteomyelitis, TB)
- Scoliosis and other developmental abnormalities
- Rheumatoid arthritis (RA), psoriatic arthritis, ankylosing spondylitis, Reiter's syndrome
- Paget's disease of bone
- Compression fractures
- Osteoarthritis
- Tumorlike processes (fibrous dysplasia, bony cysts)
- Bone neoplasms (primary or metastatic)
- Bone pain
- Multiple myeloma
- Legg-Calvé-Perthes syndrome
- Osgood-Schlatter's disease
- Gout
- Hyperthyroidism
- Hemochromatosis

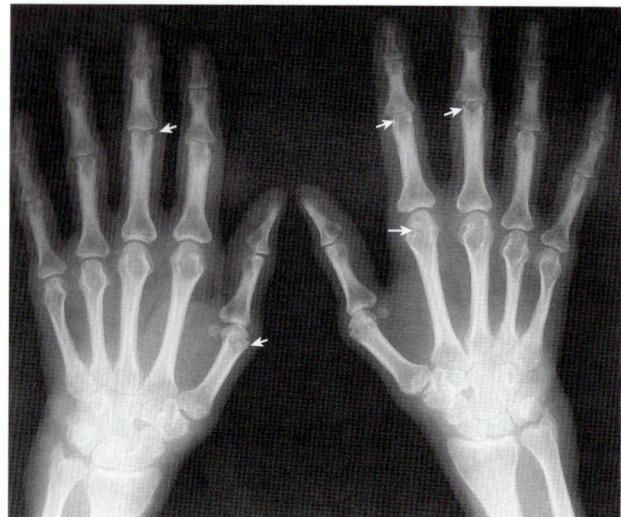

Fig. 1.76 Rheumatoid arthritis. Bilateral posteroanterior hand radiographs demonstrate multiple erosions involving predominantly the proximal interphalangeal and metacarpophalangeal joints. Assessment of bilateral involvement and symmetry is helpful in narrowing the differential diagnosis. (From Hochberg MC et al: *Rheumatology,* ed 5, St. Louis, Elsevier, 2011.)

- Evaluation of bone alignment
- Evaluate prosthesis

Strengths
- Inexpensive
- Readily available

Weaknesses
- May miss stress fractures
- May miss aseptic vascular necrosis
- May miss early osteomyelitis, septic arthritis

Comments
- Fig. 1.76 describes a plain x-ray of the hands in a patient with RA.
- Cost: $

2. Bone Densitometry (Dual-Energy X-Ray Absorptiometry [DEXA] Scan)
Indications
- Postmenopausal women 65 years or older, regardless of additional risk factors
- Postmenopausal women younger than 65 years and with additional risk factors for osteoporotic fractures (parental history of hip fracture, current cigarette smoking, a body weight less than 58 kg, use of [or plans to use] corticosteroids longer than 3 months, or serious long-term conditions thought to increase fracture risk, such as hyperthyroidism or malabsorption)
- Follow-up hormone therapy

TABLE 1.9	Causes of Erroneous Bone Mineral Density Measures by Dual-Energy X-ray Absorptiometry in the Lumbar Spine
Overestimation of bone mineral density	
Extraneous calcification (lymph nodes, aorta)	
Degenerative disk and spine disease (osteophytes)	
Ankylosing spondylitis	
Vertebral fracture	
Sclerotic metastases	
Vertebral hemangioma	
Overlying metal artifacts (navel rings)	
Surgical interventions (metallic rods, spinal fusion)	
Vertebroplasty	
Paget's disease	
Treatment with strontium ranelate	
Underestimation of bone mineral density	
Laminectomy	

From Pope TL, Bloem HL, Beltran J, et al: *Musculoskeletal imaging*, ed 2, Philadelphia, Elsevier Saunders, 2015.

Strengths
- Readily available
- Noninvasive
- Faster and less radiation than quantitative computed tomography (QCT)
- Can be performed serially to assess disease progression

Weaknesses
- Less sensitive than QCT for detecting early trabecular bone loss
- Table 1.9 summarizes causes of erroneous bone density measures by DEXA

Comments
- The decision to test for bone mineral density (BMD) should be based on an individual's risk profile, and testing is never indicated unless the results are likely to influence a treatment decision.
- Bone density measurement at a specific skeletal site predicts fractures at that site better than bone density measurements made at a different skeletal site (Fig. 1.77).
- Cost: $$

3. MRI Scan of Spine
Indications
- Suspected neoplasm (primary or metastatic)
- Radiculopathy
- Acute myelopathy
- New or progressive neurologic deficit
- High-impact trauma
- Suspected spinal infection (Fig. 1.78), osteomyelitis, discitis
- Neurogenic claudication (onset with prolonged standing, relief with back flexion)
- Spinal hematoma
- Syringohydromyelia

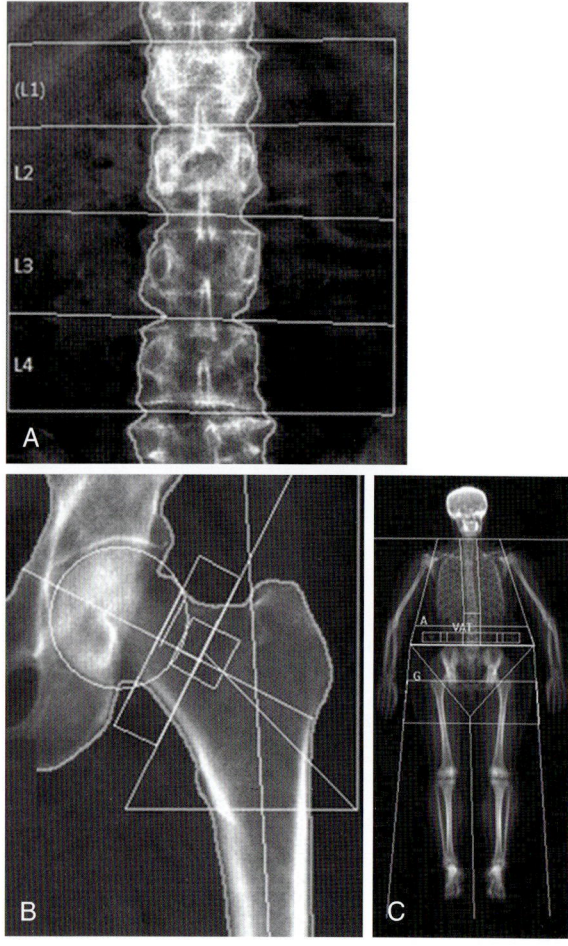

Fig. 1.77 Dual-energy x-ray absorptiometry (DXA) provides "areal" bone mineral density (BMD) (g/cm²) and is currently the gold standard for diagnosis of osteoporosis by bone densitometry (World Health Organization definition T score −2.5 or below) in (A) postero-anterior lumbar spine (L1-4) or (B) hip (femoral neck or total). C, DXA of the whole body can provide information on total and regional BMD and body composition (fat and muscle mass). Recent additional parameters measured are android A/gynoid G ratio and visceral adipose tissue (VAT). (From Pope TL, Bloem HL, Beltran J, et al: *Musculoskeletal imaging*, ed 2, Philadelphia, Elsevier Saunders, 2015.)

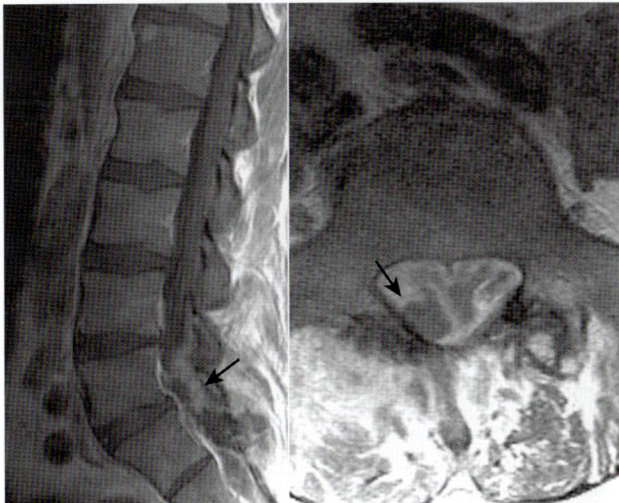

Fig. 1.78 An epidural abscess. A sagittal postgadolinium T1-weighted image (*left*) shows abnormal enhancing soft tissue (*arrow*) in the epidural space. The axial postgadolinium T1-weighted image (*right*) more clearly shows a nonenhancing fluid collection (*arrow*) in this patient with recent intravenous drug use and sciatica. (From Soto JA, Lucey BC: Emergency radiology: the requisites, ed 2, Philadelphia, Elsevier, 2017.)

- Failure of conservative therapy for back pain
- Spinal stenosis
- Back pain in patient with cancer

Strengths
- Noninvasive
- Safe contrast agent (MRI uses gadolinium, an IV agent that is not nephrotoxic)
- No ionizing radiation
- Soft-tissue resolution
- Multiplanar
- Best for identifying disk changes and evaluating extent of injury
- Excellent modality for evaluation of intradural metastases and intramedullary tumors

Weaknesses
- Expensive
- Needs cooperative patient
- Time-consuming
- Cannot be performed in patients with non–magnetic resonance–compatible aneurysm clips, pacemaker, cochlear implants, or metallic foreign body in eyes; safe in women with IUDs (including copper ones), and those with surgical clips and staples

Comments
- Imaging for back pain generally should be considered only after conservative management fails. Exceptions are back pain with neurologic symptoms (e.g., sphincter disturbances, reflex changes), HIV infection, IV drug use, and history of cancer.
- The sensitivity and specificity of MRI for disk herniations are similar to those of myelography; however, MRI is the best imaging test for suspected lateral disk herniation because of its multiplanar capabilities.

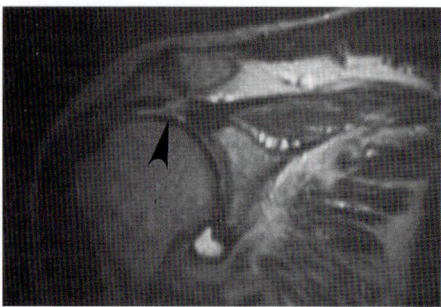

Fɪɢ. 1.79 Magnetic resonance imaging (MRI) scan of complete rotator cuff tear. T2-weighted MRI scan of the shoulder shows discontinuity of the supraspinatus tendon indicative of complete tear (*arrow*). The proximal tendon margin is frayed and retracted 1.5 cm. Focal swelling and increased signal in opposing articular cartilages of the glenohumeral joint are evidence of early degenerative disease. (From Hochberg MC et al [eds]: *Rheumatology*, ed 3, St. Louis, Mosby, 2003.)

- For evaluation of intraaxial and extraaxial spinal lesions, MRI is the procedure of choice because of its high soft-tissue resolution and multilinear capabilities.
- MRI is the procedure of choice in patients with suspected spinal stenosis.
- Enhanced images are indicated when infection, inflammation, neoplasia, intrinsic spinal cord lesions, or extradural spinal cord lesions from primary neoplastic or metastatic lesions are suspected or after spinal surgery to separate scar from recurrent disk.
- Unenhanced images are indicated in suspected degenerative disease of spine and in spinal cord trauma.
- May be of limited use after back surgery because of metallic artifact from hardware.
- Anxious patients (especially those with claustrophobia) should be premedicated with an anxiolytic agent, and their imaging should be done with "open MRI" whenever possible.
- Cost: MRI without contrast $$$$; MRI with and without contrast $$$$$

4. MRI Scan of Shoulder
Indications
- Rotator cuff tear (Fig. 1.79)
- Glenohumeral dislocation
- Glenoid labral tear
- Persistent shoulder pain despite conservative treatment when shoulder surgery is contemplated

Strengths
- Sensitivity and specificity for suspected rotator cuff tears are 85% for partial tears, 95% for full tear
- Sensitivity for suspected glenoid labral tear is greater than 90% (same as arthrography)

Weaknesses
- Expensive
- Needs cooperative patient
- Time-consuming
- Cannot be performed in patients with non–magnetic resonance–compatible aneurysm clips, pacemaker, cochlear implants, or metallic foreign body in eyes; safe in women with IUDs (including copper ones), and those with surgical clips and staples

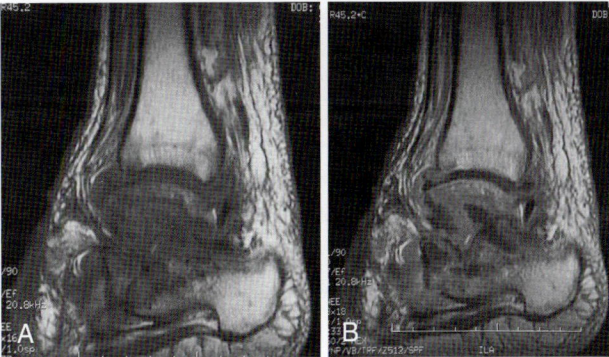

Fig. 1.80 Septic arthritis of the ankle complicated with osteomyelitis. A, Sagittal view on T1-weighted spin-echo magnetic resonance imaging (MRI) scan shows severe destruction of the subtalar joint with bone involvement. B, Contrast-enhanced T1-weighted spin-echo MRI scan shows a generalized enhancement of both synovial and bone tissues and abscess formation into talus and calcaneus. (From Hochberg MC et al: *Rheumatology,* ed 5, St. Louis, Elsevier, 2011.)

Comments
- Excellent imaging modality for evaluation of cartilage, tendons, ligaments, and soft-tissue abnormalities.
- Anxious patients (especially those with claustrophobia) should be premedicated with an anxiolytic agent, and their imaging should be done with "open MRI" whenever possible.
- Cost: MRI without contrast $$$$

5. MRI Scan of Hip and Extremities
Indications
- Aseptic necrosis of hip
- Nondisplaced hip fracture
- Legg-Calvé-Perthes disease
- Hip pain with negative plain films
- Transient osteoporosis of the hip
- Suspected neoplasm
- Suspected osteomyelitis (Fig. 1.80)

Strengths
- Most sensitive imaging modality for early aseptic necrosis

Weaknesses
- Expensive
- Needs cooperative patient
- Time-consuming
- Cannot be performed in patients with non–magnetic resonance–compatible aneurysm clips, pacemaker, cochlear implants, or metallic foreign body in eyes; safe in women with IUDs (including copper ones), and those with surgical clips and staples

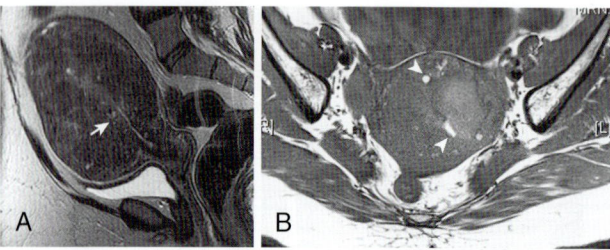

Fig. 1.81 Adenomyosis on magnetic resonance imaging. Sagittal T2-weighted image (A) shows an enlarged uterus with thickened junctional zone and cystic areas in the myometrium (*arrow*) consistent with adenomyosis. Axial T1-weighted image (B) shows hyperintense areas (*arrowheads*) in the myometrium, which are caused by blood collecting within the cystic areas in the myometrium. (From Fielding JR et al: *Gynecologic imaging*, Philadelphia, Saunders, 2011.)

Comments
- Excellent imaging modality for evaluation of cartilage, tendons, ligaments, and soft-tissue abnormalities.
- Anxious patients (especially those with claustrophobia) should be premedicated with an anxiolytic agent, and their imaging should be done with "open MRI" whenever possible.
- Cost: $$$$

6. MRI Scan of Pelvis
Indications
- Leiomyoma location
- Endometriosis
- Adenomyosis (Fig. 1.81)
- Congenital abnormalities
- Presurgical planning
- Suspected neoplasm
- Suspected osteomyelitis

Strengths
- Noninvasive
- Safe contrast agent (MRI uses gadolinium, an IV agent that is less nephrotoxic)
- No ionizing radiation

Weaknesses
- Expensive
- Needs cooperative patient
- Time-consuming
- Cannot be performed in patients with aneurysm clips that are not compatible with magnetic resonance or in patients with a pacemaker, cochlear implants, or metallic foreign body in the eyes; safe in women with intrauterine devices (IUDs), including copper ones, and in patients with surgical clips and staples

Comments
- Anxious patients (especially those with claustrophobia) should be premedicated with an anxiolytic agent, and their imaging should be done with "open MRI" whenever possible.
- Cost: $$$$

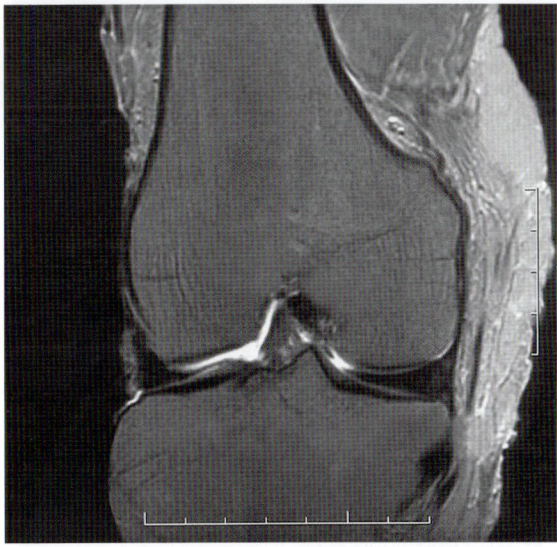

Fɪɢ. **1.82** Magnetic resonance image of a knee joint demonstrating the diversity of connective tissue structures. The image was obtained with a 1.5-Tesla magnet using a fast spin-echo T2-weighted imaging protocol. Tissues with a higher water content, such as adipose and articular cartilage, appear lighter, as does the synovial fluid. The subchondral bone is dark because of a high mineral and low water content and can be seen just below the very thin layer of articular cartilage. (Courtesy Dr. Carol Muehleman, Rush Medical College, Chicago.)

7. MRI Scan of Knee

Figs. 1.82 and 1.83 illustrate structures evaluated by MRI of knee.

Indications
- Cruciate ligament tear
- Medial collateral ligament tear
- Posterior collateral ligament tear (Fig. 1.84)
- Meniscal tear
- Patellar dislocation or fracture
- Loose body
- Occult knee fracture
- Septic arthritis

Strengths
- Excellent imaging modality for evaluation of cartilage, tendons, ligaments, and soft-tissue abnormalities

Weaknesses
- Expensive
- Needs cooperative patient
- Time-consuming

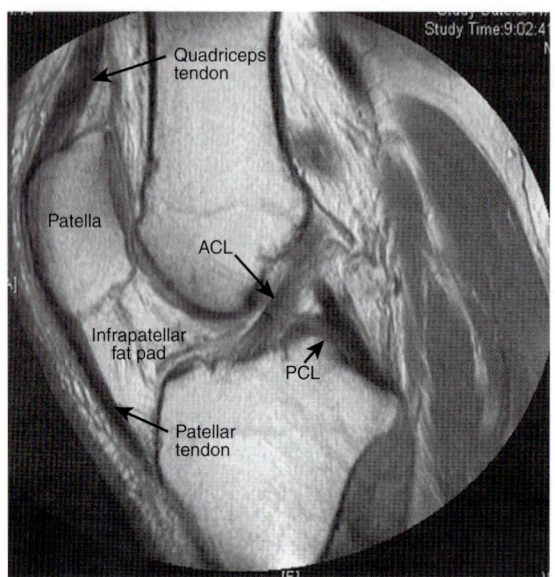

Fig. 1.83 Sagittal section of the knee joint. (From Hochberg MC et al: *Rheumatology*, ed 5, St. Louis, Mosby, Elsevier, 2011.)

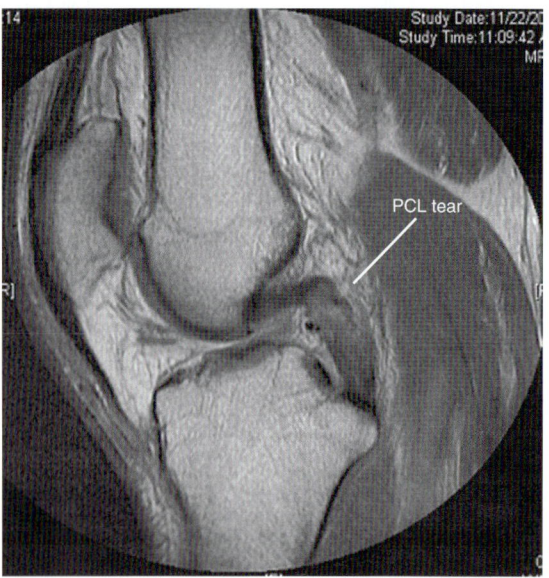

Fig. 1.84 Magnetic resonance image showing posterior cruciate ligament (PCL) rupture. (From Hochberg MC et al: *Rheumatology*, ed 5, St. Louis, Mosby, Elsevier, 2011.)

- Cannot be performed in patients with aneurysm clips that are not compatible with magnetic resonance or in patients with a pacemaker, cochlear implants, or metallic foreign body in the eyes; safe in women with intrauterine devices (IUDs), including copper ones, and in patients with surgical clips and staples.

Comments
- MRI is not indicated when physical examination is unequivocal. It should be performed only when physical examination is inconclusive or equivocal and the physician strongly suspects a tear or other significant abnormalities.
- Anxious patients (especially those with claustrophobia) should be premedicated with an anxiolytic agent, and their imaging should be done with "open MRI" whenever possible.
- Cost: MRI without contrast $$$

8. CT Scan of Spinal Cord
Indications
- Radiculopathy
- Intervertebral disk disease
- Spondylolisthesis
- Primary or metastatic spinal cord neoplasms
- Spinal nerve tumors
- Syringohydromyelia
- High-impact trauma
- Infection

Strengths
- Fast
- Easy to monitor patients
- Useful to identify difficult anatomic regions not well visualized by plain films (e.g., C1–C2, C7–T1) (Fig. 1.85)

Weaknesses
- Potential for significant contrast reaction
- Less sensitive than MRI in identifying intrinsic damage to the spinal cord, extrinsic compression, and ligamentous injury

Comments
- Imaging for back pain generally should be considered only after conservative management fails. Exceptions are back pain with neurologic symptoms (e.g., sphincter disturbances, reflex changes), HIV infection, IV drug use, and history of cancer.
- Enhanced images are indicated when infection, inflammation, neoplasia, intrinsic spinal cord lesions, or extradural spinal cord lesions from primary neoplastic or metastatic lesions are suspected and after spinal surgery to separate scar from recurrent disk.
- Unenhanced images are indicated in suspected degenerative disease of spine and in spinal cord trauma.
- Cost: $$$

9. Arthrography
Indications
- Cartilage injury
- Implant loosening
- Ligament and tendon tears (Fig. 1.86)
- Suspected intraarticular loose body

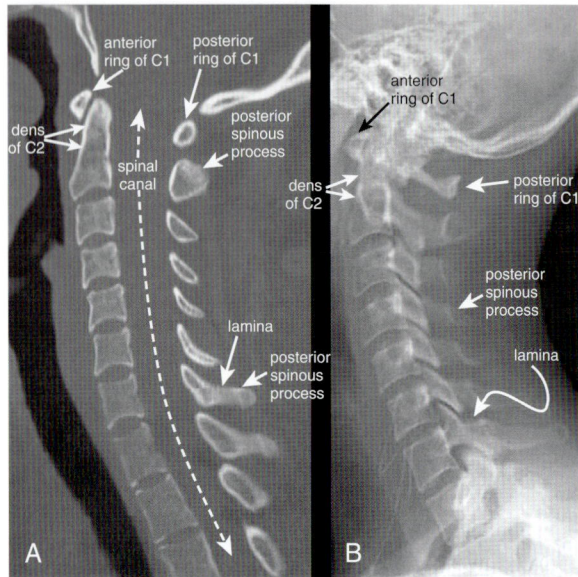

FIG. 1.85 Normal midsagittal computed tomography (CT) image, compared with normal lateral cervical spine x-ray. This 24-year-old woman hydroplaned on a wet road at 45 mph and complained of midline cervical spine tenderness. Her CT scan is reviewed to demonstrate normal findings. A, When reviewing CT sagittal images, follow the paradigm of the lateral x-ray. First, select bone windows. Next, select the midsagittal CT image and note the alignment of vertebral bodies, using the same four lines used for evaluating alignment on the lateral x-ray. Then inspect each vertebra for fractures (shown in subsequent figures). Notice how the facet joints are not visible in the midsagittal plane. Also notice how large the dens is and how small the ring of C1 is. Prevertebral soft tissues can be evaluated using the same criteria as for x-ray. B, Lateral x-ray for comparison. (From Broder JS: *Diagnostic imaging for the emergency physician,* Philadelphia, Saunders, 2011.)

Strengths
- Excellent visualization of ligament, tendon, cartilage injury in shoulder, knee, elbow, wrist, hip, and ankle

Weaknesses
- Invasive
- Potential reaction to contrast media
- Expensive

Comments
- Arthrography is contraindicated in patients with a history of reaction to contrast media or with skin infection at the site of injection.
- Cost: CT arthrogram $$$, MRI arthrogram $$$$

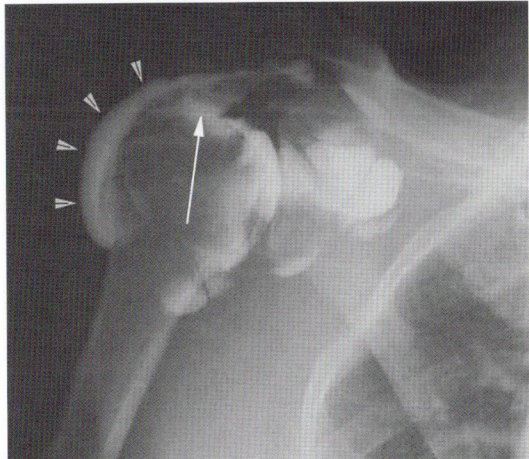

Fig. 1.86 A single image from a conventional single-contrast shoulder arthrogram demonstrates contrast medium extension from the glenohumeral joint into the subacromial/subdeltoid bursa (*arrowheads*) through a large full-thickness rotator cuff tear (*arrow*). (From Pope TL, Bloem HL, Beltran J, et al: *Musculoskeletal imaging*, ed 2, Philadelphia, Elsevier Saunders, 2015.)

10. CT Myelography

Indications
- Evaluation of spinal vascular malformations
- Evaluation of suspected spine lesions (e.g., small osteophytes impinging on nerve roots)

Strengths
- Excellent for evaluation of small osteophytic lesions and nerve roots
- Can visualize bony stenosis

Weaknesses
- Invasive
- Poor visualization of intramedullary lesions
- Poor soft-tissue resolution
- Side effects (e.g., hypersensitivity reactions, headaches, seizures, aseptic meningitis, nausea; headaches in approximately 10%–25% of patients)

Comments
- This imaging modality is now less commonly used with general availability of MRI. If the patient cannot have an MRI or the MRI is limited (e.g., after spinal fusion), myelogram is the next study of choice.
- Imaging for back pain generally should be considered only after conservative management fails. Exceptions are back pain with neurologic symptoms (e.g., sphincter disturbances, reflex changes), HIV infection, IV drug use, and history of cancer.
- CT myelography is the radiographic examination of the spinal canal and spinal cord with nonionic contrast injected in the subarachnoid space via lumbar puncture or occasionally lateral cervical puncture at C1–C2 level.
- Cost: $$$$

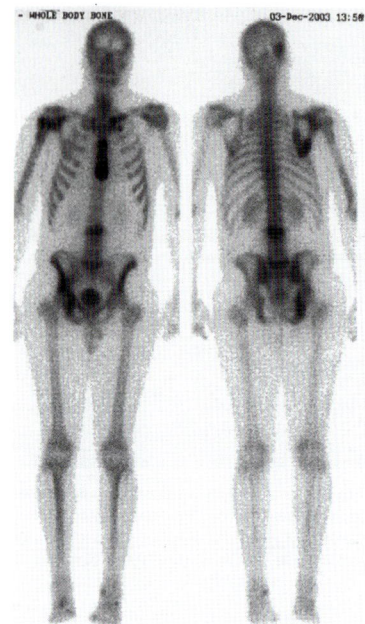

RT ANTERIOR LT   LT POSTERIOR RT
23.1 mCi 99mTc-MDP TOTAL BODY BONE IMAGING

FIG. 1.87 Tc-99m methyl diphosphonate (MDP) bone scan shows polyostotic Paget's disease involving the sternum, right scapula, two lumbar vertebral bodies, and right hemipelvis. There may be early pagetic disease in the right skull as well. (From Hochberg MC et al: *Rheumatology,* ed 5, St. Louis, Mosby, Elsevier, 2011.)

11. Nuclear Imaging (Bone Scan, Gallium Scan, White Blood Cell [WBC] Scan)

Indications

- Infection
- Metastatic disease
- Unexplained bone pain
- Avascular necrosis
- Evaluation of bone lesions seen on other imaging studies
- Seronegative spondyloarthropathies
- Paget's disease of bone (Fig. 1.87)
- Metabolic bone diseases (e.g., osteomalacia, hypervitaminosis A or D)
- Assessment of bone graft viability
- Stress fractures or shin splints
- Temporomandibular joint (TMJ) derangement
- Prosthetic loosening
- Reflex sympathetic dystrophy

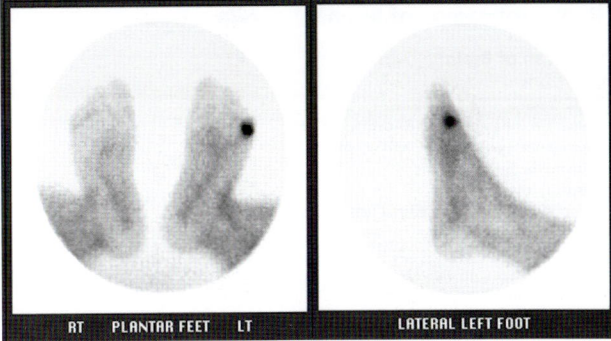

FIG. 1.88 Osteomyelitis. Intense accumulation of Tc-99m white blood cell in the proximal phalanx of the fifth digit of the left foot at 4 hours postinjection. (From Specht N: *Practical guide to diagnostic imaging*, St. Louis, Mosby, 1998.)

Strengths

- Highly sensitive for bone lesions
- Three-phase bone imaging is very useful in suspected infection, osteonecrosis, and stress fractures; gallium and WBC scans can also be performed to look for infection (Fig. 1.88)

Weaknesses

- Nonspecific
- Requires availability of current plain bone radiographs for side-by-side comparison
- "Flare phenomenon"—apparent worsening of serial bone scans yet clinical improvement after chemotherapy often seen with metastatic breast and prostate carcinoma caused by chemotherapy-related tumor suppression
- Expensive
- Less specific than MRI for osteomyelitis
- False-negative scans in multiple myeloma (better detectability with plain radiograph than with radionuclide scan)

Comments

- In bone scanning, a diphosphonate compound (methyl diphosphonate [MDP]) is labeled (e.g., with Tc-99m), becomes incorporated into the mineral phase of bone, and is used to demonstrate pathologic conditions of the bone. The major factors affecting the uptake of the tracer are osteoblastic activity and blood flow.
- *Total body bone imaging* should be performed when suspected disease may involve more than one site.
- *Three-phase bone imaging* is recommended for evaluation of localized area when detection of regional hyperemia is crucial to diagnosis. Common indications are infection, assessment of bone graft viability, and reflex sympathetic dystrophy.
- *Single-photon emission computed tomography (SPECT) bone imaging* should be requested when a lesion in lumbar spine, knees, skull, or facial bones is suspected. It can also be performed on children to evaluate for spondylolysis.
- Bone scintigraphy should precede CT with contrast when both studies are requested because contrast for CT can affect bone images.
- Cost: limited to one area $$, whole-body bone scan $$$

H. Neuroimaging of Brain

1. CT Scan of Brain

Indications

- Head trauma
- Suspected subarachnoid hemorrhage (SAH)
- Central nervous system (CNS) neoplasm
- Cerebral hemorrhage
- Cerebral infarct
- Suspected subdural or epidural hematoma
- Hypoxic encephalopathy
- Cranial nerve tumors
- Cerebral imaging in patients with contraindications to MRI (e.g., pacemaker, metallic foreign body)

Strengths

- Fast, widely available
- Can easily detect acute parenchymal and subarachnoid hemorrhage (SAH) and calcifications
- Easy to monitor patients
- Noncontrast CT is fastest, most sensitive, and most specific modality to demonstrate SAH
- More sensitive than MRI for detection of calcifications (lesions that have a strong tendency to calcify are caused by toxoplasmosis, craniopharyngioma, chondrosarcoma, retinoblastoma, tuberous sclerosis, and Sturge-Weber syndrome)
- Preferred to MRI in facial or head trauma (excellent for bones), postoperative craniotomy, and sinusitis patients
- Inflammatory or congenital lesions of the temporal bone also better visualized with CT than MRI

Weaknesses

- Less sensitive to parenchymal lesions and leptomeningeal processes, particularly white matter lesions, than MRI
- Potential for significant contrast reaction
- Not useful for evaluation of tissue perfusion, metabolism, or vessel blood flow

Comments

- Noncontrast CT of the brain is often used as the initial imaging modality for patients suspected of having had a stroke and in patients with suspected SAH.
- Enhanced images are indicated when infection, inflammation or neoplasia, and seizures are suspected.
- Unenhanced images are indicated in hemorrhagic or ischemic events, head trauma, congenital anomalies, and degenerative diseases.
- On CT of the brain, hemorrhage will be denser than surrounding brain tissue but not as dense as calcium or bone.
- Patients with acute infarcts often have a "normal" CT scan initially; therefore cerebral infarcts cannot be excluded on the basis of a negative CT. Early findings are due to edema in the affected portion of the brain. Subacute findings include an increase in the mass effect. A wedge-shaped, hypodense area extending to the cortex or involving the basal ganglia, thalamus, or brainstem will develop.
- In *subdural hematoma* (Fig. 1.89 *A*), CT reveals a concave blood collection between skull and brain. It will cross suture lines but will not cross the midline. Contrast-enhanced CT will accentuate the nonenhancing subdural blood.
- In *epidural hematoma* (Fig. 1.89 *B*), CT reveals a biconvex hyperdense blood collection with significant mass effect and brain edema. It will not cross suture lines.
- In cases in which the detection of calcium is important, noncontrast CT is preferred.

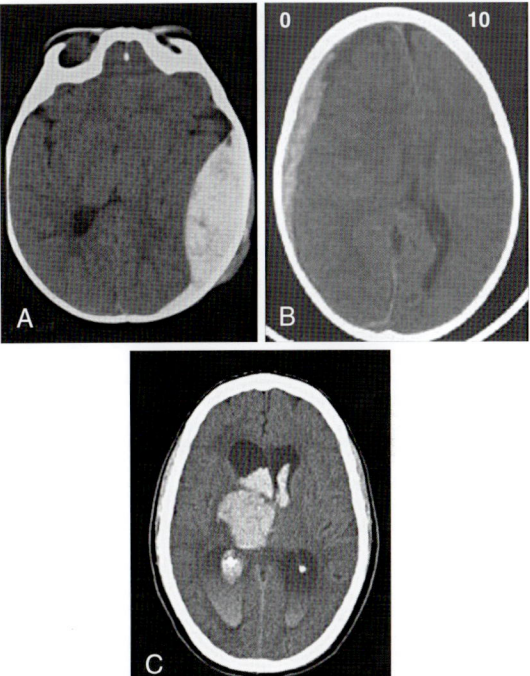

Fig. 1.89 Computed tomographic appearance of blood in patients with hematoma. A, Epidural hematoma. B, Subdural hematoma. C, Intraparenchymal and intraventricular hematomas. (From Adams JG et al: *Emergency medicine, clinical essentials*, ed 2, Philadelphia, Elsevier, 2013.)

- New Orleans Criteria recommend CT after minor head injury if the patient meets one or more of the following criteria: headache, vomiting, age older than 60 years, drug or alcohol intoxication, persistent anterograde amnesia (deficits in short-term memory), visible trauma above clavicle, or seizure.
- Cost: CT of head without contrast $$; CT of head with contrast $$$; CT of head with and without contrast $$$

2. MRI Scan of Brain
Indications
- Suspected brain neoplasm (primary or metastatic)
- Suspected demyelinating diseases of brain (e.g., multiple sclerosis [MS])
- Suspected sellar and parasellar abnormalities
- Suspected brain abscess and cerebritis
- Suspected granulomatous, fungal, and parasitic encephalopathies
- Suspected encephalitis (much more sensitive than CT)

- Suspected congenital malformations (e.g., Chiari malformations, corpus callosum abnormalities, cephaloceles)
- Stroke evaluation (more sensitive than CT, especially in the brainstem and deep white matter)
- After trauma to evaluate for diffuse axonal injury

Strengths

- Noninvasive
- Generally safe contrast agent; gadolinium-based, noniodinated IV contrast agents are used to assess vascular integrity
- No ionizing radiation
- Soft-tissue resolution
- Multiplanar

Weaknesses

- Expensive
- Needs cooperative patient; critically ill patients are often unable to tolerate MRI until they are more hemodynamically stable
- Time-consuming: most routine MRI studies take approximately 45 minutes to perform, and studies of the entire neuroaxis with and without contrast require more than 90 minutes
- The need for preprocedural preparation and screening limits its use in critical care patients
- Not as sensitive as CT for detection of SAH
- Insensitive to the presence of calcification and bone
- CT preferred to MRI in facial or head trauma, postoperative craniotomy, and sinusitis patients
- Inflammatory or congenital lesions of temporal bone also better visualized with CT than MRI
- Cannot be performed in patients with aneurysm clips that are not compatible with magnetic resonance or in patients with a pacemaker, cochlear implants, or metallic foreign body in the eyes; safe in women with intrauterine devices (IUDs), including copper ones, and in patients with surgical clips and staples. Respirators and physiologic monitors must also be MRI compatible. Only oxygen and nitrogen tanks composed of aluminum can enter the magnet suite. The website www.MRIsafety.com is a useful resource for checking the MRI compatibility of particular medical devices

Comments

- MRI is the imaging procedure of choice for evaluation of suspected brain tumor, intracranial mass, suspected pituitary and juxtasellar lesions, cerebellar and brainstem symptoms, hydrocephalus, lesions of visual system, congenital CNS abnormalities, and suspected structural abnormalities related to epilepsy.
- MRI is superior to CT in the initial diagnosis of acute stroke. Most of the superiority of MRI is attributed to its ability to detect acute ischemic stroke. It is now rapidly available in many centers and is preferred by most physicians as the initial imaging modality in patients with suspected acute stroke.
- Enhanced images are indicated when infection, inflammation, neoplasia, or seizures are suspected. Contrast may also be required in the evaluation of demyelinating disorders to identify small plaques in the region of optic nerves.
- Unenhanced images are indicated in hemorrhagic or ischemic events, head trauma, congenital anomalies, and degenerative diseases.
- Figs. 1.90 and 1.91 describe imaging parameters and normal anatomy of the brain on MRI. When white matter disease (e.g., MS) is suspected, the best imaging test is MRI with fluid attenuated inversion recovery (FLAIR) images.
- When SAH is suspected, MRI is not preferred as the initial study. Noncontrast CT is the fastest, most sensitive, and most specific modality to demonstrate SAH.
- Dedicated protocols can evaluate the orbits, sella, and cerebellopontine angles with thin cuts and superior detail to CT.

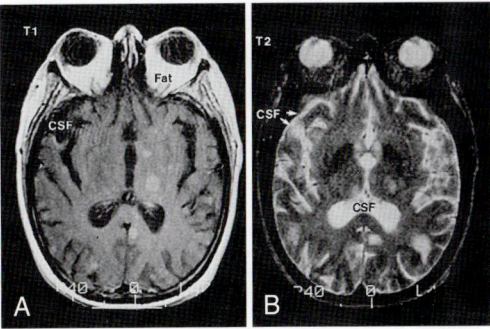

Fig. 1.90 Magnetic resonance imaging (MRI) scan of the brain. A wide variety of imaging parameters can make tissues appear different. The two most common presentations are T1-weighted images (A), in which fat appears white, water and cerebrospinal fluid appear black, and brain and muscle appear gray. In almost all MRI images, bone gives off no signal and appears black. B, With T2-weighted imaging, fat is dark and water and cerebrospinal fluid have a high signal and appear bright or white. The brain and soft tissues still appear gray. (From Mettler FA: *Primary care radiology,* Philadelphia, WB Saunders, 2000.)

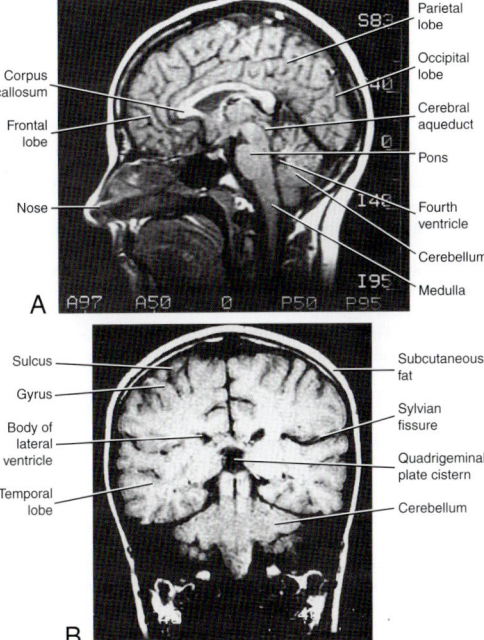

Fig. 1.91 A and B, Normal magnetic resonance anatomy of the brain in coronal and sagittal projections. (From Mettler FA: *Primary care radiology,* Philadelphia, Saunders, 2000.)

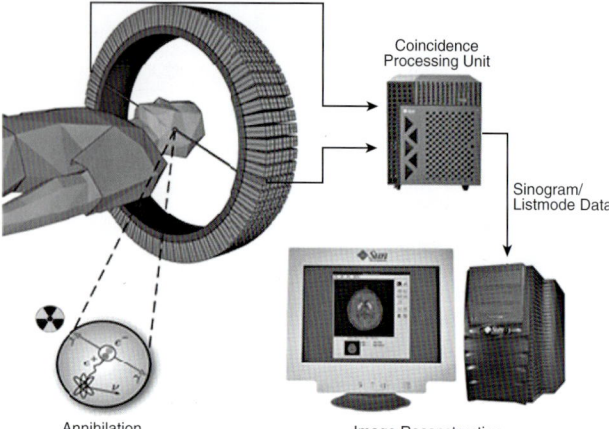

Fɪɢ. 1.92 Formation of the positron emission tomography (PET) image. The F-18 atom of the intravenously injected fluorodeoxyglucose molecule undergoes an annihilation reaction, with the release of two photons at 180 degrees. These are detected by the PET scanner gantry detectors, processed and localized by a computer, and lead to the formation of a PET image. (Courtesy GE Healthcare.)

- Anxious patients (especially those with claustrophobia) should be premedicated with an anxiolytic agent, and their imaging should be done with "open MRI" whenever possible.
- Functional MRI (fMRI) can detect changes in blood flow and map performance of cognitive tasks to areas of increased or decreased brain activity, but its use outside of research is limited to the presurgical workup of patients with tumors or focal epilepsy.
- Diffusion-weighted imaging (DWI) is useful for the early identification of acute ischemic stroke, for the differentiation of necrotic tumor from pyogenic abscess, and for the evaluation of cellularity and grade of tumors.
- Cost: MRI of brain without contrast $$$$, MRI of brain with contrast $$$$, MRI of brain with and without contrast $$$$$

I. Positron Emission Tomography (PET)

Fig. 1.92 illustrates the formation of the PET image.

Indications

- Diagnosis, staging, and restaging of lung cancer (non–small cell), esophageal cancer, colorectal cancer, head and neck cancer (excluding CNS and thyroid), lymphoma (Fig. 1.93), melanoma
- Breast cancer: as an adjunct to standard imaging modalities in staging (patients with distant metastasis) and restaging (patients with local or regional recurrence or metastasis)
- Breast cancer: as an adjunct to standard imaging modalities for monitoring tumor response to treatment for women with locally advanced and metastatic breast cancer when a change in therapy is anticipated
- Presurgical evaluation for refractory seizures

FDG-PET at diagnosis

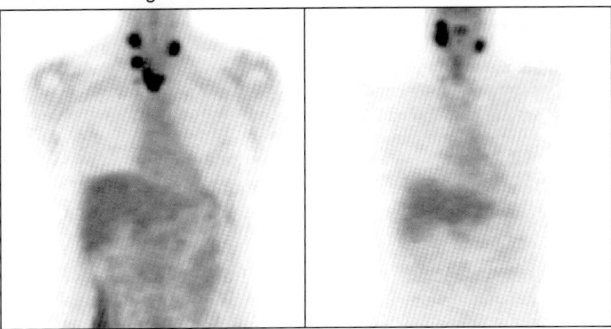

FDG-PET 2 months later after 3 cycles of R-CHOP

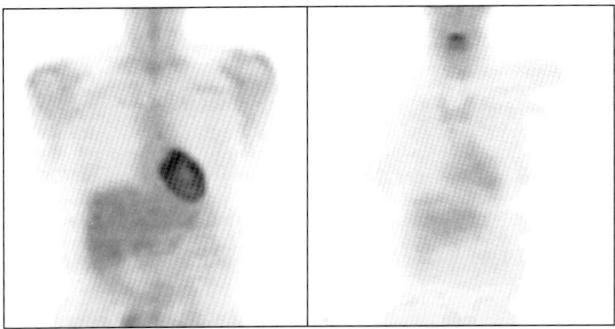

Fɪɢ. 1.93 Fluorodeoxyglucose–positron emission tomography (FDG-PET) scan before and after treatment with rituximab, cyclophosphamide, doxorubicin, Oncovin, and prednisolone (R-CHOP). Resolution of cervical and upper mediastinal disease is shown. (From Young NS, Gerson SL, High KA [eds]: *Clinical hematology*, St. Louis, Mosby, 2006.)

- Evaluation for myocardial viability before revascularization
- Evaluation for the presence of malignancy in a pulmonary nodule
- Evaluation of Alzheimer's disease and frontal lobe dementias

Strengths
- Can detect malignant involvement of small nodes (unlike CT, which looks for lymph node enlargement to detect malignancy)
- Can detect malignancy in tumor sites that have same appearance as adjacent normal structures on CT
- Useful in detecting unknown sites of metastatic disease and reducing futile thoracotomies in patients with non–small cell lung cancer
- Useful in evaluation of pulmonary nodules (a nodule that demonstrates F-18 fluorodeoxyglucose [FDG] uptake on PET typically warrants biopsy)
- Useful to differentiate residual masses caused by scar tissue versus active lymphoma in post–lymphoma therapy patients
- Useful to identify Alzheimer's disease and frontal lobe dementias (e.g., Pick's disease)

Weaknesses

- Poor visualization of neoplasms that are not very metabolically active or are unable to retain FDG (e.g., hepatocellular carcinoma, prostate cancer, bronchoalveolar lung cancer)
- Decreased sensitivity in patients with diabetes
- Poor imaging of brain metastases caused by an overall significant amount of increased FDG uptake in the brain (high background activity)
- Poor visualization of very small lung metastases (<5 mm)
- Unable to distinguish malignancy from inflammatory disease (e.g., sarcoidosis, TB) because of increased FDG uptake by active granulomatous disease
- Not useful in defining regional draining lymph nodes in melanoma (sentinel node biopsy is superior)
- Expensive
- Limited availability

Comments

- PET measures the distribution of radioisotope-containing compounds (e.g., 18F-FDG) that are given intravenously and can study cerebral perfusion as well as cerebral energy metabolism. PET captures chemical and physiologic changes related to metabolism, as opposed to gross anatomy and structure. Many types of neoplasms demonstrate an increased uptake of glucose. A PET camera measures radiolabeled phosphorylated FDG to determine increased cellular activity typical of many types of neoplasms.
- Rubidium-82, a tracer used for PET myocardial perfusion imaging, localizes in the myocardium in proportion to regional blood flow. It can be used to evaluate myocardial viability to accurately predict which patients will have significant improvement in left ventricular function after revascularization.
- PET imaging in early Alzheimer's disease reveals decreased metabolic activity in the mesial temporal, posterior temporal, and parietal lobes. This is useful to distinguish Alzheimer's dementia from Pick's disease and other frontal lobe dementias (decreased metabolic activity in frontal lobes).
- Costs associated with PET scanning and its availability are major limiting factors to its use.
- Cost: $$$$ for PET of body, extremity, abdomen, brain metabolic, brain perfusion; $$$$ for cardiac PET

J. Single-Photon Emission Computed Tomography (SPECT)

Indications

- Evaluation of CAD
- Evaluation of CNS diseases, including epilepsy, cerebrovascular disease, and psychiatric disorders
- Can help provide 3-D evaluation of lesions in bone scanning, hepatobiliary imaging, and other areas of nuclear medicine

Strengths

- Contrast resolution higher than with planar images
- Useful for epilepsy and CAD
- Easier to implement than PET
- Improved sensitivity of diagnosis of coronary disease and delineation of the size of ischemic or infarcted myocardium
- Regional cerebral blood flow (CBF) brain SPECT useful to identify patients at risk for stroke who may benefit from neurosurgical revascularization procedures
- Tracer iodine-123 IBZM can map the dopamine D2 receptor and is useful in evaluation of movement disorders and schizophrenia
- Cerebral necrosis after brain tumor therapy cannot be distinguished from residual or recurrent tumor by either CT or MRI; SPECT useful to distinguish between new or residual glioblastoma (Fig. 1.94) after surgery and brain necrosis

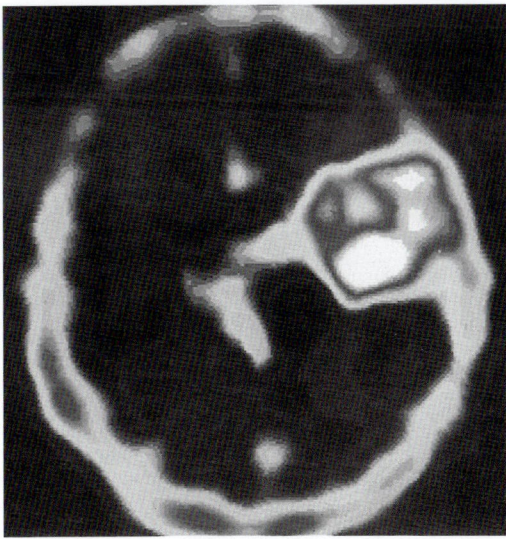

Fɪɢ. **1.94** Tl-201 SPECT in a 40-year-old man with a left frontotemporal mass on magnetic resonance imaging scan. High uptake typical of high-grade glioma, which was confirmed on biopsy to be a glioblastoma. (Image courtesy Professor Donald M. Hadley, Glasgow, Scotland.) (From Grainger RG, Allison D: *Grainger and Allison's diagnostic radiology: a textbook of medical imaging,* ed 4, Sydney, Churchill Livingstone, 2001.)

Weaknesses
- Resolution not as good as MRI and CT
- Expensive
- Not readily available

Comments
- SPECT is a technique that uses one, two, or three gamma cameras to record activity emitted from multiple projections around the patient. It can study the distribution of isotopes incorporated into other biologically active compounds, allowing measurements of other aspects of tissue metabolism.
- Brain SPECT applications use the radionuclide Tc-99m HMPAO, a brain flow tracer that is extracted by the brain in proportion to regional CBF.
- Cost: $$$

K. Vascular Imaging

1. Angiography
Indications
- Cardiac: evaluation of coronary anatomy (Fig. 1.95); severity of valvular disease; diseases of the myocardium, pericardium, or endocardium; congenital defects; pulmonary hypertension; therapeutic procedures (angioplasty, embolization, selective drug therapy)
- Cerebral angiography: evaluation for vascular lesions of the brain (Fig. 1.96) and great vessels of the neck (aneurysm, AVM, vasculitis)
- Thoracic aorta (Fig. 1.97): chest trauma, suspected dissection, aneurysm, evaluation of vascular anatomy and anomaly, aortic mass (neoplasm, nonneoplastic), vasculitis

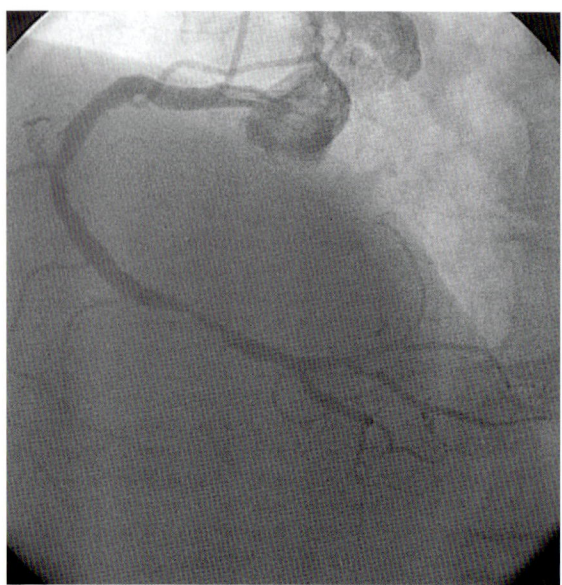

Fig. 1.95 Example of a significant stenosis in the right coronary artery. (From Goldman L, Schafer AI: *Goldman's Cecil medicine*, ed 24, Philadelphia, Saunders, 2012.)

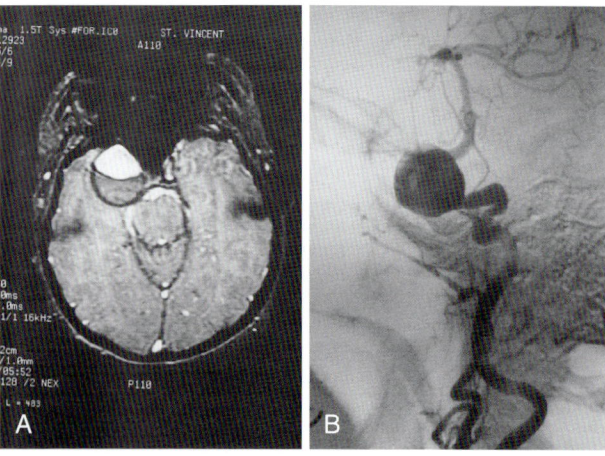

Fig. 1.96 Cerebral aneurysm. A, Axial, noncontrast gradient-echo image demonstrates a giant aneurysm in the medial portion of the right middle cranial fossa. The area of increased signal intensity represents the lumen of the aneurysm. B, Right carotid arteriogram (oblique view of the same patient) demonstrates a giant aneurysm of the distal right internal carotid artery. (From Specht N: *Practical guide to diagnostic imaging*, St. Louis, Mosby, 1998.)

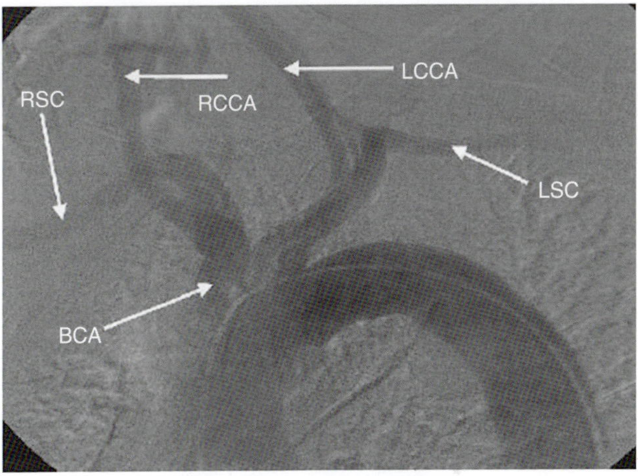

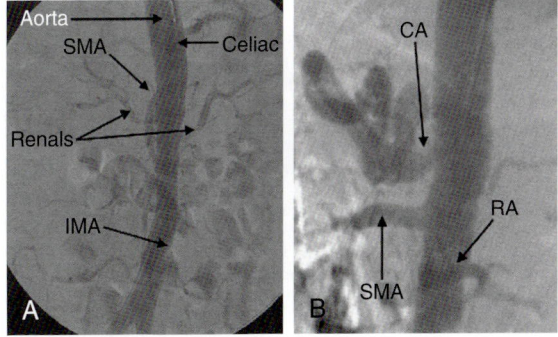

Fig. 1.97 Digital subtraction angiography of the thoracic arch and great vessels. *BCA,* brachiocephalic artery; *LCCA,* left common carotid artery; *LSC,* left subclavian artery; *RCCA,* right common carotid artery; *RSC,* right subclavian artery. (From Sellke FW, del Nido PJ, Swanson SJ: *Sabiston & Spencer surgery of the chest,* ed 9, Philadelphia, Elsevier, 2016.)

Fig. 1.98 A, Anteroposterior view of digital subtraction angiography of the abdominal aorta and major branches. B, Lateral digital subtraction angiography of the abdominal aorta and major branches. *CA,* celiac artery; *IMA,* inferior mesenteric artery; *RA,* renal artery; *SMA,* superior mesenteric artery. (From Sellke FW, del Nido PJ, Swanson SJ: *Sabiston & Spencer surgery of the chest,* ed 9, Philadelphia, Elsevier, 2016.)

- Abdominal aorta (Fig. 1.98): abdominal trauma, aneurysm, vasculitis, preoperative evaluation, peripheral vascular disease
- Renal: renal artery stenosis, renovascular hypertension, trauma, renal vein thrombosis, vasculitis, neoplasm, transplanar, AVM
- Mesenteric: GI hemorrhage (Fig. 1.99), AVM, angiodysplasia, intestinal ischemia, splenic and splanchnic aneurysm, neoplasm

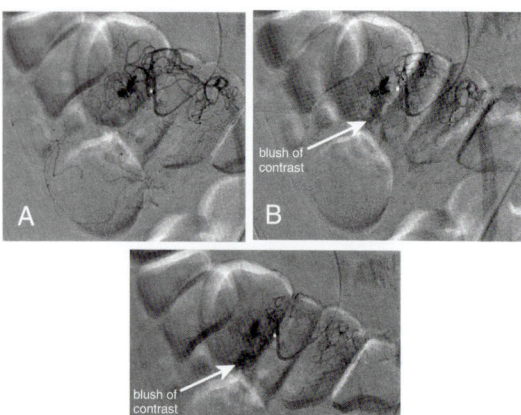

Fig. 1.99 Gastrointestinal hemorrhage imaged with mesenteric angiography. A, B, C, Close-up fluoroscopic images of the cecum during angiography, separated in time by a few seconds. This patient presented with rectal bleeding and underwent a tagged red blood cell study showing radiotracer accumulation in the right lower quadrant. Mesenteric angiography was performed and shows active extravasation of contrast in the region of the cecum arising from a branch of the ileocolic artery. This was selectively embolized with a combination of particles and coils. Selective arteriography after embolization showed no further active extravasation of contrast. These images show sequential frames from a digital subtraction angiogram, cropped to show just the region of the cecum. A faint blush of contrast is seen developing over a few seconds, representing active gastrointestinal hemorrhage. (From Broder JS: *Diagnostic imaging for the emergency physician*, Philadelphia, Saunders, 2011.)

- Hepatic: neoplasm, pretransplant assessment, AVM, focal nodular hyperplasia
- Subclavian: aneurysm, AVM, arterial insufficiency, trauma
- Splenic: trauma, neoplasm, aneurysm
- Pancreatic: evaluation and staging of neoplasm, localization of islet cell tumor
- Bronchial: refractory hemoptysis, AVM, pulmonary sequestration
- Pulmonary: pulmonary embolism (PE), vasculitis, AVM, congenital vascular lesions
- Detection of superior sagittal sinus thrombosis
- Pretranssphenoidal hypophysectomy assessment

Strengths

- Most accurate method for evaluation of vasculature
- Best modality for preoperative assessment and delineation of branch vessel involvement
- Allows intervention (e.g., percutaneous transluminal angioplasty [PTA], renal artery angioplasty, intravascular stents, delivery of therapeutic agents, thrombolysis, embolization of aneurysms or intraabdominal or pelvic bleeding)

Weaknesses

- Invasive
- Low but significant risk of mortality (<0.05%) and morbidity (e.g., local complications at catheter insertion area, contrast reaction, renal insufficiency, thrombosis, embolism)
- Expensive
- Poor visualization of mural thrombus or extravascular hematoma
- Does not provide information about the disease process that takes place in the vessel wall

Comments
- Angiography is the gold standard for evaluation of vascular lesions and aortic dissection.
- Relative contraindications are coagulopathy, renal insufficiency, allergy to contrast agents, uncontrolled CHF, and metformin use within 48 hours.
- Cost: $$$$

2. Aorta Ultrasound

Fig. 1.100 illustrates a normal ultrasound of the aorta.

Indications
- Suspected aneurysm
- Arterial dissection

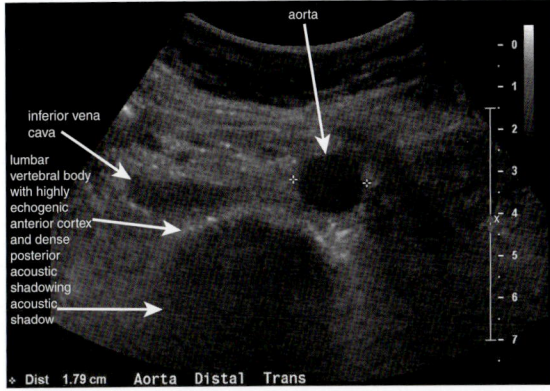

Fig. **1.100** Normal aorta: ultrasound. Bedside ultrasound in the hands of emergency physicians has been shown to be sensitive for the presence of abdominal aortic aneurysm. Ultrasound is considered poorly sensitive for rupture. In the presence of symptoms suggestive of aortic rupture, identification of an aneurysm by bedside ultrasound should be acted on as if diagnostic of rupture. This ultrasound demonstrates a normal aorta, viewed in short-axis (transverse) cross-section. The aorta should be examined in cross-section from the xiphoid process to the umbilicus, the usual location of the aortic bifurcation to form the iliac arteries. The majority of aortic aneurysms are infrarenal, and this area should be examined in detail. If measurements are made only at the subxiphoid and umbilical levels, an aneurysm might be missed, because the aortic diameter may be quite normal proximal and distal to a saccular aneurysm. Several features of the normal aorta should be recognized. First, the normal abdominal aorta lies just anterior and to the left side of the lumbar vertebral bodies. These are identified by their highly echogenic anterior surface (the bony cortex), which reflects sound waves and casts a dense posterior acoustic shadow. This should not be confused with an aortic aneurysm and is a reliable landmark for orientation. The normal aorta should be less than 3 cm in diameter and should have a circular cross-sectional contour. The inferior vena cava (IVC) should lie just anterior to the lumbar vertebral bodies, on the right side. The IVC should be the same size or smaller than the aorta. Depending on the patient's intravascular volume status, the IVC may be round or may have a flattened, elliptical shape. The IVC often varies in size and shape with respiration, but this should not be mistaken for the pulsatile aorta. (From Broder JS: *Diagnostic imaging for the emergency physician*, Philadelphia, Saunders, 2011.)

Strengths
- Fast
- Noninvasive
- No radiation
- Can be performed at bedside
- Can be repeated serially

Weaknesses
- Operator dependent; results may be affected by skill of technician
- Retained barium from x-ray procedures will interfere with interpretation

Comments
- Ultrasound is ideal for following aneurysm progression over time.
- Cost: $$

3. Arterial Ultrasound

Indications
- Suspected aneurysm
- Arterial dissection
- Arterial stenosis
- Arterial occlusion
- Suspected AV fistula or pseudoaneurysm
- Image guidance for thrombin injection of pseudoaneurysm

Strengths
- Fast
- Noninvasive
- No radiation
- Can be performed at bedside
- Can be repeated serially

Weaknesses
- Operator dependent; results may be affected by skill of technician
- Retained barium from x-ray procedures will interfere with interpretation

Comments
- Ultrasound is ideal for following aneurysm progression over time.
- Useful as a screening method for suspected AV fistula.
- Doppler ultrasound can be used to study the patency of grafts because it can provide a measurement of flow volume/unit time.
- Cost: $$

4. Captopril Renal Scan (CRS)

Indications
- Detection of renal artery stenosis in the setting of clinically suspected renovascular hypertension
- Evaluation for revascularization in patient with known renal artery stenosis

Strengths
- Noninvasive

Weaknesses
- Limited use as screening method for renal artery stenosis because of low specificity and sensitivity
- Drop in systemic blood pressure after administration of angiotensin-converting enzyme (ACE) inhibitors may interfere with test interpretation
- Poor renal function from any cause makes interpretation difficult

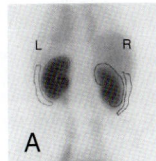

Captopril-enhanced renography

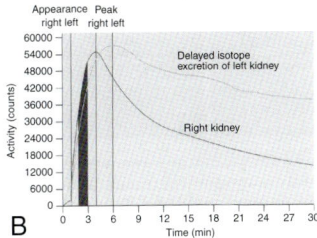

FIG. 1.101 Captopril-enhanced renography. A, Scan in a patient with newly developing hypertension. B, Renogram demonstrates delayed arrival and excretion of isotope (MAG3) in the affected left kidney. (From Johnson RJ, Feehally J: *Comprehensive clinical nephrology*, ed 2, St. Louis, Mosby, 2000.)

Comments

- CRS involves two radionuclide studies, one with and one without captopril. In patients with hemodynamically significant renal artery stenosis, delayed uptake and cortical retention in the affected kidney will occur (Fig. 1.101).
- Because a change in the glomerular filtration rate (GFR) on renal scan after administration of captopril suggests that renal artery stenosis is more likely to be hemodynamically significant, the appropriate use of CRS should be in patients with known renal artery stenosis (by renal arteriography or MRA) who are considered for revascularization.
- Cost: $$

5. Carotid Ultrasonography

Fig. 1.102 illustrates a normal carotid artery ultrasound.

Indication

- Screening for extracranial vascular disease primarily at the carotid bifurcation

Strengths

- Noninvasive
- Safe
- No ionizing radiation

Weaknesses

- Operator dependent; can be significantly influenced by skills and bias of operator
- Of limited utility for disease above the region of the carotid bifurcation
- In 10% of patients, carotid bifurcation lies above angle of jaw, making ultrasound difficult or impossible
- Subject to errors of interpretation in cases of high-grade stenoses or complete occlusion

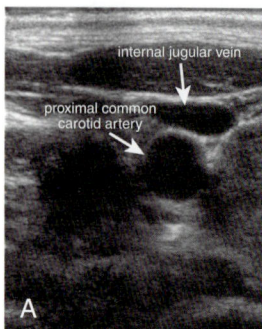 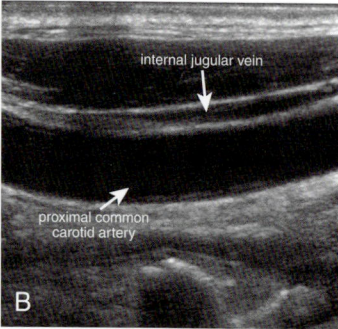

Fɪɢ. 1.102 Normal carotid artery ultrasound. Ultrasound is useful in assessment of the carotid arteries and internal jugular veins. It can detect carotid stenosis with high sensitivity and can be used to assess for vascular dissection. Thrombosis of the internal jugular veins or carotid arteries can also be detected. Ultrasound has become a standard tool in the placement of internal jugular central venous catheters; these figures illustrate why ultrasound can be invaluable in preventing accidental puncture of the carotid artery. Anatomy textbooks show the carotid artery lying medial to the internal jugular vein. In this patient, however, the internal jugular vein lies superficial to the carotid artery, not medial to it. The internal jugular vein is compressible and often appears elliptic or even polygonal in short-axis cross-section. The carotid artery is far less compressible and has a circular short-axis cross-section. A, Short-axis (transverse) view. B, Long-axis (longitudinal) view. (From Broder JS: *Diagnostic imaging for the emergency physician,* Philadelphia, Saunders, 2011.)

- Calcified plaques interfere with visualization of vascular lumen
- Difficulty visualizing tandem lesions
- Echolucency of acute thrombi (indistinguishable from flowing blood)

Comments
- Carotid arteriography is necessary when surgical intervention is contemplated.
- Duplex color flow Doppler is the best initial screening test to evaluate the carotid arteries.
- Cost: $$

6. Computed Tomographic Angiography (CTA)
Indications
- Screening of asymptomatic patients at risk for cerebral aneurysms
- Rapid evaluation of symptomatic patients for aneurysms and dissection (Fig. 1.103)
- PE (Fig. 1.104)

Strengths
- Artifact is less than with conventional angiography
- Fast
- High sensitivity
- Noninvasive
- Not affected by flow-related effects seen with MRA
- Does not involve MRI compatibility problems (e.g., pacemakers, metallic clips)
- Can detect intraluminal thrombus, calcification in the neck of an aneurysm, and extravascular hematoma
- Can reveal complications of aneurysms (e.g., compression of other structures, bone erosion)

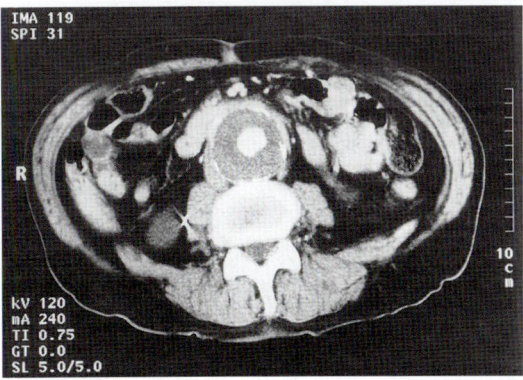

FIG. **1.103** Computed tomography scan of an abdominal aortic aneurysm. Contrast in the lumen of the vessel, a thick layer of mural thrombus, and calcification in the aneurysm wall are shown. (From Crawford MH, DiMarco JP, Paulus WJ [eds]: *Cardiology,* ed 2, St. Louis, Mosby, 2004.)

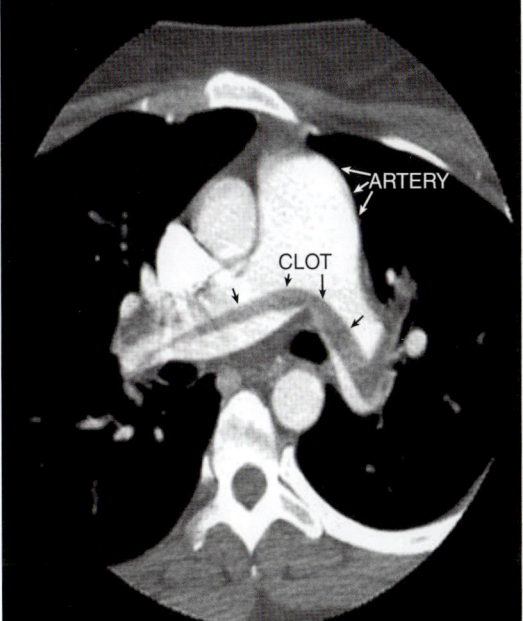

FIG. **1.104** Computed tomographic pulmonary arteriogram demonstrating a pulmonary embolus. (From Goldman L, Schafer AI: *Goldman's Cecil medicine,* ed 24, Philadelphia, Saunders, 2012.)

Weaknesses
- Use of ionizing radiation
- Injection of IV iodine (potential for contrast reaction)
- Excessive length of time for processing of data after test
- Poor visualization of vessels at base of skull
- May produce artifacts caused by calcifications in walls of vessels and aneurysm clips
- Cannot be repeated (unlike MRA)
- Does not reliably delineate involvement of branch vessels
- Poor visualization of entry and reentry sites when evaluating aortic dissection

Comments
- In CTA, IV contrast and a thin-slice helical CT are used to create a 3-D view of blood vessels.
- CTA is an alternative to MRA for evaluation of extracranial and intracranial vasculature.
- Cost: $$$

7. Magnetic Resonance Angiography (MRA)

Indications
- Evaluation of extracranial vasculature for the presence of lesions of the carotid artery, extracranial vasculitis (e.g., giant cell arteritis), congenital vascular abnormalities (e.g., fibromuscular disease), dissection of vertebral and carotid arteries, and extracranial traumatic fistula
- Evaluation of cerebral vasculature (Fig. 1.105), suspected intracranial aneurysm

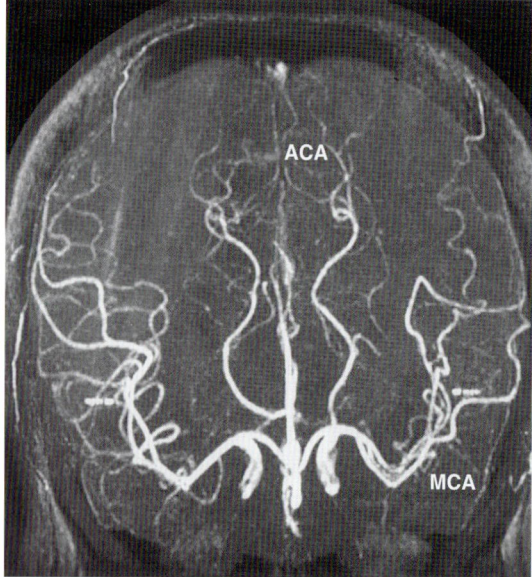

Fig. 1.105 Magnetic resonance angiogram. An anterior view of the head shows intracerebral vessels, including the anterior cerebral artery (*ACA*) and the middle cerebral artery (*MCA*). These images were obtained without injection of any contrast agent. (From Mettler FA: *Primary care radiology,* Philadelphia, Saunders, 2000.)

- Follow-up of unruptured intracranial aneurysm
- Follow-up of treated aneurysm when conventional angiography is contraindicated
- Workup of intracranial vasculitis
- Intracranial venous occlusive disease
- Intracranial vascular compression syndromes
- Definition of blood supply to vascular neoplasms
- Evaluation of AVMs
- Aortic coarctation (Fig. 1.106)

Strengths

- Noninvasive and readily repeatable
- Safe, no ionizing radiation
- Useful screening tool for both extracranial and intracranial vascular disease
- May be performed without contrast
- Images can be reconstructed in any plane
- Can be used safely in patients with renal insufficiency
- Unlike CTA or conventional angiography, able to demonstrate both anatomy and flow rate
- In severe (70%-99%) carotid stenosis, is 95% sensitive and 90% specific versus 86% sensitivity and 87% specificity for duplex ultrasonography

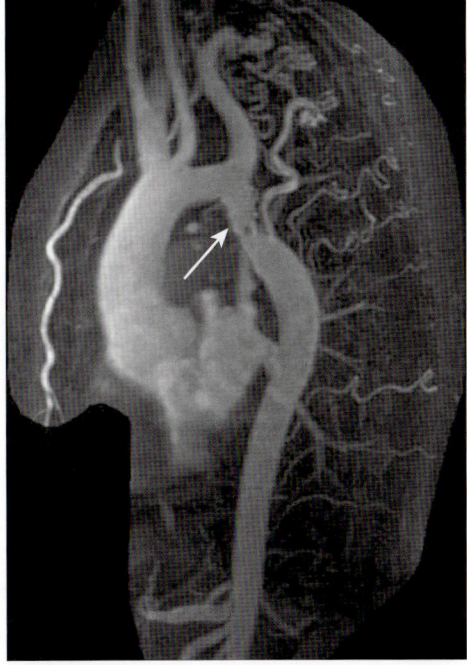

Fig. 1.106 Three-dimensional contrast-enhanced magnetic resonance angiogram in a patient with an aortic coarctation (*arrow*). (From Goldman L, Schafer AI: *Goldman's Cecil medicine,* ed 24, Philadelphia, Saunders, 2012.)

Weaknesses

- Requires cooperative patient
- Need contrast to image distal vessels adequately
- May overestimate degree of vascular stenosis
- May miss small aneurysms (<3 mm)
- Use instead of conventional angiography may preclude other diagnoses; for example, in patients with suspected carotid stenosis, additional vascular lesions such as brain AVM will be missed if only extracranial MRA is performed
- Cannot be performed in patients with aneurysm clips that are not compatible with magnetic resonance or in patients with a pacemaker, cochlear implants, or metallic foreign body in the eyes; safe in women with intrauterine devices (IUDs), including copper ones, and in patients with surgical clips and staples
- Poor visualization of ulcerations in atheromas
- Images of distal intracranial vessels generally difficult to interpret
- Slow blood flow in a vessel with high-grade stenosis may be falsely interpreted as occlusion (has surgical implications because surgery is often performed for stenosis and not for vascular occlusion)
- Sensitivity to motion (e.g., ventilation) limits use in thoracic region
- Peristaltic motion in abdomen may interfere with interpretation

Comments

- Although conventional angiography remains the definitive diagnostic modality for evaluation of intracranial aneurysm, MRA is rapidly replacing it for initial evaluation of cerebral vasculature.
- MRI contrast agents (when used) are less allergenic and nephrotoxic than conventional iodinated contrast agents.
- Anxious patients (especially those with claustrophobia) should be premedicated with an anxiolytic agent, and their imaging should be done with "open MRI" whenever possible.
- Cost: MRA without contrast $$$$

8. Magnetic Resonance Direct Thrombus Imaging (MRDTI)

Indication

- Diagnosis of deep vein thrombosis (DVT)

Strengths

- Noninvasive
- Highly reproducible interpretation
- Accurate in diagnosing isolated calf and proximal DVT
- Useful to determine thrombus age
- Does not require contrast or special patient preparation
- Can be performed in patient with full-length leg plaster cast
- Can be repeated serially to monitor thrombus progression
- Safe in pregnancy

Weaknesses

- High cost
- Not readily available
- Needs cooperative patient
- Time-consuming
- Cannot be performed in patients with aneurysm clips that are not compatible with magnetic resonance or in patients with a pacemaker, cochlear implants, or metallic foreign body in the eyes; safe in women with intrauterine devices (IUDs), including copper ones, and in patients with surgical clips and staples

Comments

- MRDTI is useful and well tolerated in pregnancy, in patients with full-length leg plaster casts, and in patients with isolated calf DVT (present in 5% of patients with suspected DVT).
- This is a reliable test for asymptomatic thrombosis because it does not depend on the filling of the lumen or distribution of blood flow and can image small-volume thrombi.
- Anxious patients (especially those with claustrophobia) should be premedicated with an anxiolytic agent, and their imaging should be done with "open MRI" whenever possible.
- Cost: $$$$

9. Pulmonary Angiography

Indications

- Diagnosis of PE in patients with inconclusive ventilation/perfusion (V/Q) scan (Fig. 1.107)
- Vasculitis
- AVM
- Congenital vascular lesion

Strengths

- Sensitivity 98%, specificity 97% for PE
- Allows simultaneous adjunctive procedures (e.g., placement of inferior vena cava [IVC] filter, thrombectomy, local catheter-directed thrombectomy)

Weaknesses

- Requires IV contrast (potential for significant contrast reaction)
- Invasive
- Expensive

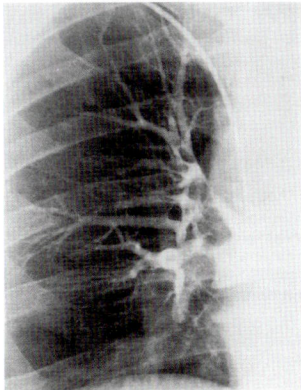

Fig. 1.107 Pulmonary angiography in an intubated patient with massive pulmonary embolism and cardiology shock. Several large filling defects indicating thrombi can be traced down the right lower and upper lobar branch of the vessel. (From Crawford MH, DiMarco JP, Paulus WJ [eds]: *Cardiology*, ed 2, St. Louis, Mosby, 2004.)

Comments
- Pulmonary angiography is the gold standard test for PE.
- Improvement in CT technology has diminished the use of pulmonary angiography for evaluation of PE.
- This test is generally used as a second-line diagnostic test in patients with an indeterminate V/Q scan.
- Relative contraindications are coagulopathy, renal insufficiency, allergy to contrast agents, uncontrolled CHF, and metformin use within 48 hours.
- LBBB and increased pulmonary artery pressure are contraindications to pulmonary angiography.
- Cost: $$$$

10. Transcranial Doppler
Indications
- Evaluation of the basal cerebral arteries
- Evaluation of subclavian steal syndrome (Fig. 1.108)
- Detection of cerebrovascular spasm (e.g., after surgery, after SAH)
- Evaluation of middle cerebral artery patency in patients with carotid stenosis

Strengths
- Noninvasive
- Can provide indirect information about extracranial arterial occlusive disease and can evaluate the intracranial carotids and the circle of Willis
- No ionizing radiation
- Fast

Weaknesses
- Operator dependent; can be significantly influenced by the skills and bias of the operator

Comments
- Transcranial Doppler uses low-frequency ultrasonography to evaluate the flow velocity spectrum of the cerebral vessels.
- Cost: $$

11. Venography
Indications
- Suspected DVT
- Performed before IVC filter placement

Strengths
- Most reliable test for diagnosis of asymptomatic thrombus and thrombus isolated within calf (Fig. 1.109) or pelvis

Weaknesses
- Potential for significant contrast reactions
- Invasive
- Inadequate imaging in the pelvis
- Insufficient delineation of the proximal extent of thrombosis in patients with above-knee DVT
- High interobserver variability (>10%)

Comments
- Venography is the gold standard for diagnosis of DVT.
- It is generally used only when other tests for DVT are inconclusive and clinical suspicion is high.
- Cost: $$

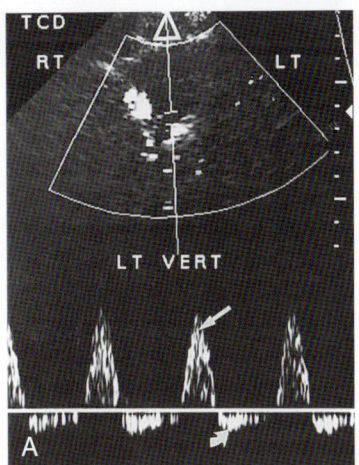

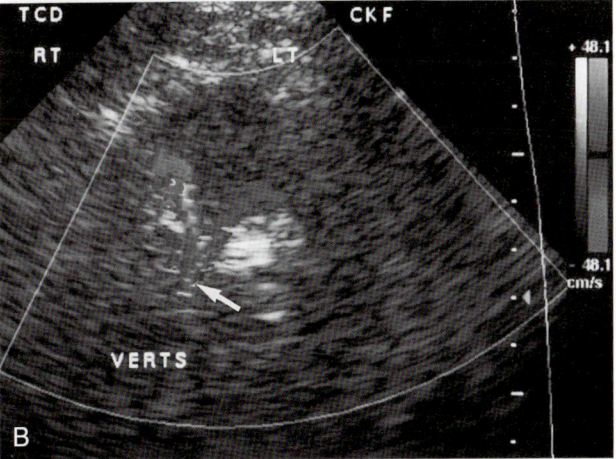

Fig. 1.108 A, Transcranial duplex scan of the posterior fossa in a patient with an incomplete left subclavian steal syndrome demonstrates retrograde systolic flow (*arrow*) and antegrade diastolic flow (*curved arrow*). The scan is obtained in a transverse projection from the region of the foramen magnum (*open arrowhead*). B, Color Doppler image obtained in the same patient demonstrates that there is retrograde flow not only within the left vertebral artery but also within the basilar artery (*arrow*). (From Rumack CM, Wilson SR, Charboneau JW, Levine D: *Diagnostic ultrasound*, ed 4, Philadelphia, Mosby, 2011.)

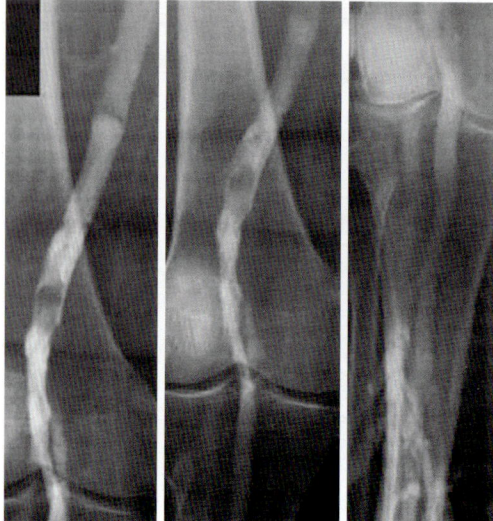

Fɪɢ. **1.109** Abnormal venogram demonstrates a persistent (two or more different views) intraluminal filling defect in the popliteal vein. (From Goldman L, Schafer AI: *Goldman's Cecil medicine*, ed 24, Philadelphia, Saunders, 2012.)

12. Compression Ultrasonography and Venous Doppler Ultrasound

Indications
- Suspected DVT (Fig. 1.110)
- Follow-up examination to evaluate propagation of venous thrombus
- Source evaluation for known or suspected PE
- Chronic venous insufficiency
- Swollen or painful leg

Strengths
- Noninvasive
- Fast
- Can be repeated serially
- Readily available
- Inexpensive
- Can be performed at bedside
- No ionizing radiation

Weaknesses
- Poor visualization of pelvic and calf veins
- Negative test does not conclusively "rule out" DVT at presentation; repeat examination after 5 to 7 days may be necessary to exclude clinically suspected thrombosis in symptomatic patients
- Operator dependent; results may be affected by skill of technician

Comments
- Venous Doppler is the initial test of choice in suspected DVT.
- Cost: $

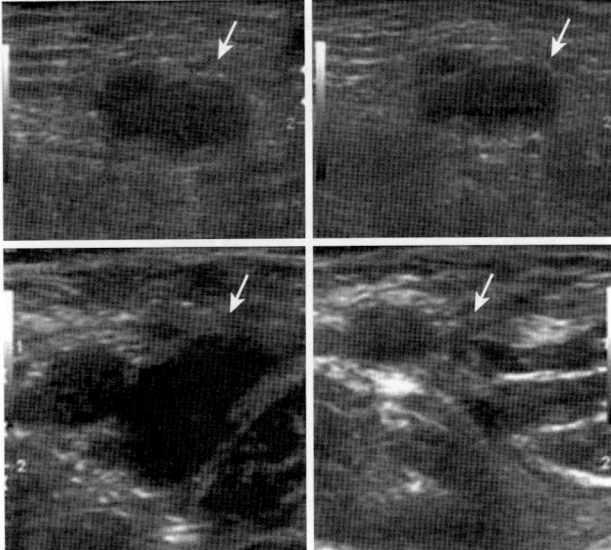

Fig. 1.110 Compression venous ultrasonography demonstrates thrombosis of the popliteal vein. The sonograms in the *top row* demonstrate examination without (*left side*) and with (*right side*) gentle probe compression of the skin overlying the popliteal vein. The lack of compressibility is diagnostic of deep vein thrombosis. The *bottom row* shows analogous views of the femoral vein, which shows partial compressibility. (From Goldman L, Schafer AI: *Goldman's Cecil medicine,* ed 24, Philadelphia, Saunders, 2012.)

13. Ventilation/Perfusion (V/Q) Lung Scan

Indications
- Suspected PE
- Suspected right-to-left shunt
- Evaluation of relative lung function before surgery (quantitative V/Q scan)

Strengths
- Noninvasive
- Widely available
- High sensitivity (>95%)

Weaknesses
- Low specificity
- Cannot be used in patients with baseline chest film abnormalities
- Cannot differentiate acute from chronic PE
- More than 70% of scans nondiagnostic
- Results can be equivocal in patients with preexisting lung disease

Comments
- A "matched defect" (area not ventilated and not perfused) can occur with COPD, tumors, lung infarction, airspace disease, asthma.
- A "V/Q mismatch" (area ventilated but not perfused) occurs with PE, vasculitis, radiation therapy, tumor compressing the pulmonary artery, and fibrosing mediastinitis compressing the pulmonary artery.

- A "reverse mismatched defect" (area perfused but not ventilated) can occur with mucus plugging, pneumonia, and alveolar pulmonary edema.
- A "normal" V/Q scan has a high predictive value for PE (<5% probability).
- A "high-probability" V/Q scan (two or more large mismatched segmental defects or any combination of mismatched defects equivalent to two large segmental defects; Fig. 1.111) has a high positive predictive value (>85% for PE).
- A nondiagnostic "intermediate" V/Q scan should be interpreted in the context of clinical suspicion and additional testing (e.g., D-dimer test).
- Patient must be cooperative enough to perform the ventilation portion of the examination. If defects are seen on the perfusion study and the patient cannot tolerate or perform the ventilation portion, the examination is then indeterminate, and further evaluation is necessary.
- CT is an excellent alternative to V/Q because CT can provide more information, is more specific, and can provide alternate reasons for patients' dyspnea or chest pain. Computed tomographic pulmonary angiography (CTPA) has in many centers replaced V/Q lung scanning because it provides a clear result (either positive or negative) and because it may detect alternative nonthrombotic causes of patients' symptoms.
- Cost: $$$

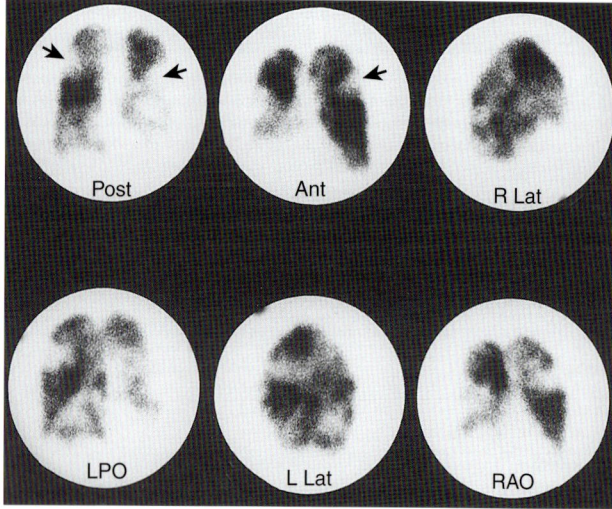

Fig. **1.111** Multiple pulmonary emboli. This young lady with shortness of breath had a normal chest radiograph and a normal ventilation lung scan. The images here are from the perfusion portion of the nuclear medicine lung scan. Note the multiple segmental and subsegmental areas without perfusion (*arrows*) throughout both lungs. This is indicative of a high probability of pulmonary emboli. *Post,* Posterior; *Ant,* anterior; *R Lat,* right lateral; *LPO,* left posterior oblique; *L Lat,* left lateral; *RAO,* right anterior oblique. (From Mettler FA: *Primary care radiology,* Philadelphia, Saunders, 2000.)

L. Oncology

1. Whole-Body Integrated (Dual-Modality) PET-CT

Indications

- Identifying and determining the extent of malignant disease and monitoring the therapy of numerous cancers (Fig. 1.112)
- Workup of solitary pulmonary nodules
- Staging of lung cancer
- Evaluation of carcinoma of unknown primary origin
- Restaging cancer; valuable in two groups of patients: those in whom other imaging or laboratory studies raise the concern of relapse, and those in whom the response to treatment needs to be evaluated
- Evaluation of efficacy of cancer chemotherapy
- Medicare-accepted indications for PET scanning include diagnosis, staging, and restaging of lymphoma, melanoma, lung cancer, head and neck cancer, and esophageal cancer; also includes staging, restaging, and evaluating treatment response to breast cancer and restaging of thyroid cancer (with negative iodine-131 scan and positive thyroglobulin)

Strengths

- Noninvasive
- In addition to its application in oncology, also useful in cardiology to detect CAD and to assess whether dysfunctional myocardial tissue is viable; in neurology and psychiatry, is useful in differentiating between tumor recurrence and radiation necrosis, differentiating Alzheimer's disease from other dementias, and locating epileptic foci

Weaknesses

- Less sensitive in detecting small neoplasms because of limited spatial resolution
- Respiratory motion can result in discrepancy of spatial information from CT and PET leading to artifacts
- High-density implants and dental fillings may lead to serious artifacts on CT images

Comments

- The effective radiation dose from a single PET scan is relatively small (10 mSv). The effective dose for PET-CT is 20 mSv because a whole-body CT is performed in conjunction with PET. However, even when more than one PET-CT scan is performed during follow-up of patients with certain types of cancer after therapy, the cumulative effective dose is similar to that of the same number of "dedicated" contrast-enhanced CT scans that are often performed during follow-up.
- Cost: $$$$

2. Whole-Body MRI

Indications

- Detection of skeletal metastases as an alternative to skeletal scintigraphy
- Evaluation of total tumor burden, particularly in patients whose tumors spread preferentially to brain, bone, and liver, such as breast and lung tumors

Strengths

- Better detection of lesions in the spine and pelvis than with scintigraphy
- Noninvasive
- Safe contrast agent (MRI uses gadolinium, an IV agent that is less nephrotoxic)
- No ionizing radiation
- Soft-tissue resolution
- Multiplanar

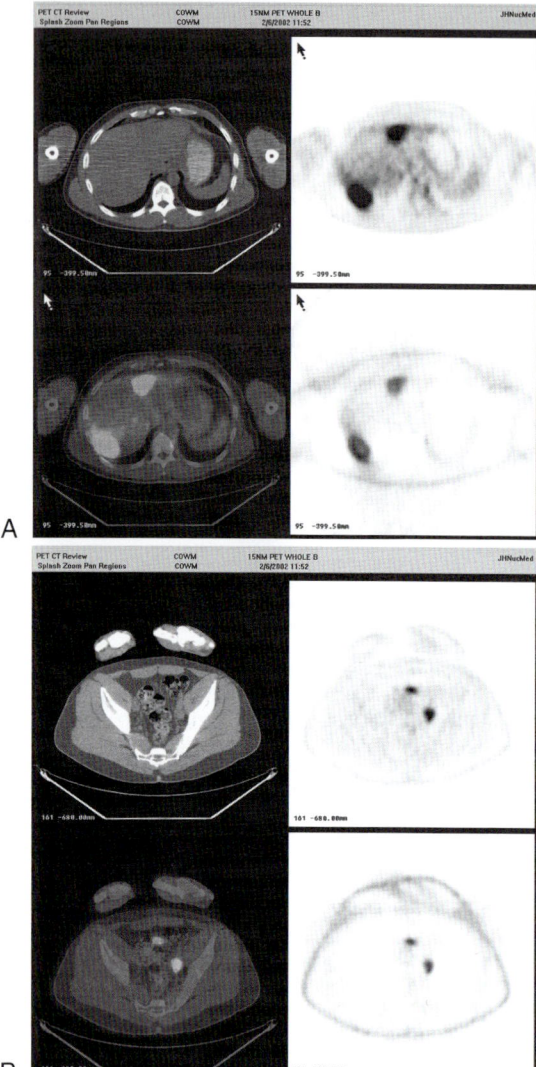

Fɪɢ. **1.112** Colon cancer. A, Positron emission tomography (PET) and computed tomography (CT) image display of a patient with two fluorodeoxyglucose (FDG)–avid lesions in the liver. These are seen on the CT (*upper left*) scan, attenuation-corrected PET (*upper right*) scan, non–attenuation-corrected PET (*lower right*) scan, and fused images (*lower left*). B, PET and CT images of the pelvis, oriented as in *A,* show increased FDG uptake in a left external iliac lymph node metastasis. (From Abeloff MD: *Clinical oncology,* ed 3, Philadelphia, Elsevier, 2004.)

Weaknesses

- Expensive
- Needs cooperative patient
- Time-consuming
- Cannot be performed in patients with aneurysm clips that are not compatible with magnetic resonance or in patients with a pacemaker, cochlear implants, or metallic foreign body in the eyes; safe in women with intrauterine devices (IUDs), including copper ones, and in patients with surgical clips and staples

Comments

- Nononcologic applications of whole-body MRI include: identifying sites suitable for percutaneous biopsy, particularly in immunocompromised hosts; evaluation of whole-body fat measurements and body composition research; and evaluation of suspected polymyositis.
- Anxious patients (especially those with claustrophobia) should be premedicated with an anxiolytic agent, and their imaging should be done with "open MRI" whenever possible.
- Cost: $$$$

References

Abeloff MD, Armitage JO, Niederhuber JE, et al. Clinical oncology. 3rd ed. Philadelphia: Elsevier; 2004.

Ali A, Santisi JM, Vargo J. Video capsule endoscopy: a voyage beyond the end of the scope. Cleve Clin J Med 2004;71(5):415–25.

Budoff MJ, Achenbach S, Blumenthal RS, et al. Assessment of coronary artery disease by cardiac computed tomography: a scientific statement from the American Heart Association Committee on Cardiovascular Imaging and Intervention, Council on Cardiovascular Radiology and Intervention, and Committee on Cardiac Imaging, Council on Clinical Cardiology. Circulation 2006;114(16):1761–91.

Anderson DR, Kahn SR, Rodger MA, et al. Computed tomographic pulmonary angiography vs ventilation-perfusion lung scanning in patients with suspected pulmonary embolism: a randomized controlled trial. JAMA 2007;298(23):2743–53.

Ballinger A. Kumar & Clark's essentials of clinical medicine. 5th ed. Edinburgh: Elsevier; 2012.

Besser GM, Thorner MO. Comprehensive clinical endocrinology. 3rd ed. St. Louis: Mosby; 2002.

Bonow RO, Mann DL, Zipes DP, Libby P. Braunwald's heart disease. 9th ed. Philadelphia: Saunders; 2012.

Broder JS. Diagnostic imaging for the emergency physician. Philadelphia: Saunders; 2011.

Cameron JL, Cameron AM. Current surgical therapy. 10th ed. Philadelphia: Saunders; 2011.

Carman TL, Deitcher SR. Advances in diagnosing and excluding pulmonary embolism: spiral CT and d-dimer measurement. Cleve Clin J Med 2002;69(9):721–9.

Chalela JA, Kidwell CS, Nentwich LM, et al. Magnetic resonance imaging and computed tomography in emergency assessment of patients with suspected acute stroke: a prospective comparison. Lancet 2007;369(9558):293–8.

Colli A, Fraquelli M, Casazza G, et al. Accuracy of ultrasonography, spiral CT, magnetic resonance, and alpha-fetoprotein in diagnosing hepatocellular carcinoma: a systematic review. Am J Gastroenterol 2006;101(3):513–23.

Crawford MH, DiMarco JP, Paulus WJ, editors. Cardiology. 2nd ed. St. Louis: Mosby; 2003.

Dumot JA. ERCP: current uses and less-invasive options. Cleve Clin J Med 2006;73(5):418, 421, 424–5.

Ebell MH. Computed tomography after minor head injury. Am Fam Physician 2006;73(12):2205–7.

Fenton JJ, Taplin SH, Carney PA, et al. Influence of computer-aided detection on performance of screening mammography. N Engl J Med 2007;356(14):1399–409.

Fielding JR, Brown DL, Thurmond AS. Gynecologic imaging. Philadelphia: Saunders; 2011.

Floege J, Johnson RJ, Feehally J. Comprehensive clinical nephrology. 4th ed. Philadelphia: Saunders; 2010.

Fraser DG, Moody AR, Morgan PS, et al. Diagnosis of lower-limb deep venous thrombosis: a prospective blinded study of magnetic resonance direct thrombus imaging. Ann Intern Med 2002;136(2):89–98.

Gay SB, Woodcock RJ. Radiology recall. Philadelphia: Lippincott Williams & Wilkins; 2007.

Goldman L, Schafer AI. Goldman's Cecil medicine. 24th ed. Philadelphia: Saunders; 2012.

Grainger RG, Allison DJ, Adam A, Dixon AK, editors. Grainger and Allison's diagnostic radiology. 4th ed. Philadelphia: Churchill Livingstone; 2001.

Greer IA, Cameron IT, Kitchener HC, Prentice A. Mosby's color atlas and text of obstetrics and gynecology. London: Harcourt; 2000.

Hassan C, Pickhardt PJ, Laghi A, et al. Computed tomographic colonography to screen for colorectal cancer, extracolonic cancer, and aortic aneurysm. Arch Intern Med 2008;168(7):696–705.

Hochberg MC, Silma AJ, Smolen JS, Weinblatt ME, Weisman MH, editors. Rheumatology. 3rd ed. St. Louis: Mosby; 2003.

Huot SJ, Hansson JH, Concaato J. Utility of captopril renal scans for detecting renal artery stenosis. Arch Intern Med 2002;162(17):1981–4.

Issa ZF, Miller JM, Zipes DP. Clinical arrhythmology and electrophysiology. 2nd ed. Philadelphia: Saunders; 2012.

Johnson RJ, Feehally J. Comprehensive clinical nephrology. 2nd ed. St. Louis: Mosby; 2000.

Juweid ME, Cheson BD. Positron-emission tomography and assessment of cancer therapy. N Engl J Med 2006;354(5):496–507.

Kim DH, Pickhardt PJ, Taylor AJ, et al. CT colonography versus colonoscopy for the detection of advanced neoplasia. N Engl J Med 2007;357(14):1403–12.

Lehman CD, Gatsonis C, Kuhl CK, et al. MRI evaluation of the contralateral breast in women with recently diagnosed breast cancer. N Engl J Med 2007;356(13):1295–303.

Mettler FA, Guiberteau MJ, Voss CM, et al. Primary care radiology. Philadelphia: WB Saunders; 2000.

Noto RB. Positron emission tomography (PET): the basics. Med Health R I 2003;86(5):129–31.

Orrison WW. Pocket medical imaging consultant. Houston: Healthhelp; 2003.

Pagana KD, Pagana TJ. Mosby's diagnostic and laboratory test reference. 7th ed. St. Louis: Mosby; 2004.

Romagnuolo J, Bardou M, Rahme E, et al. Magnetic resonance cholangiopancreatography: a meta-analysis of test performance in suspected biliary disease. Ann Intern Med 2003;139(7):547–57.

Rumack CM, Wilson SR, Charboneau JW, Levine D. Diagnostic ultrasound. 4th ed. Philadelphia: Mosby; 2011.

Sellke FW, del Nido PJ, Swanson SJ. Sabiston & Spencer surgery of the chest. 9th ed. Philadelphia: Elsevier; 2016.

Specht NT, Russo RD. Practical guide to diagnostic imaging. St. Louis: Mosby; 1998.

Stein PD, Beemath A, Kayali F, et al. Multidetector computed tomography for the diagnosis of coronary artery disease: a systematic review. Am J Med 2006;119(3):203–16.

Talley NJ, Martin CJ. Clinical gastroenterology. 2nd ed. Sidney: Churchill Livingstone; 2006.

Thrall JH, Ziessman HA. Nuclear medicine: the requisites. St. Louis: Mosby; 1995.

Townsend CM, Beauchamp RD, Evers BM, Mattox KL, editors. Sabiston textbook of surgery. 17th ed. Philadelphia: Saunders; 2004.

Weissleder R, Wittenberg J, Harisinghani MG, Chen JW. Primer of diagnostic imaging. 5th ed. St. Louis: Mosby; 2007.

Young NS, Gerson SL, High KA, editors. Clinical hematology. St. Louis: Mosby; 2006.

Laboratory Values and Interpretation of Results

This section covers more than 300 laboratory tests. Each test is approached with the following format:

- Laboratory test
- Normal range in adult patients
- Common abnormalities (e.g., positive test, increased or decreased value)
- Causes of abnormal result

The normal ranges may differ slightly, depending on the laboratory. The reader should be aware of the "normal range" of the particular laboratory performing the test. Every attempt has been made to present current laboratory test data with emphasis on practical considerations. It's important to remember that laboratory tests do not make diagnoses; doctors do. As such, any laboratory results should be integrated with the complete clinical picture and radiographic studies (if needed) to make a diagnosis.

ACE Level
Acetone (Serum or Plasma)
Acetylcholine Receptor (AChR) Antibody
Acid Phosphatase (Serum)
Acid Serum Test
Activated Clotting Time (ACT)
Activated Partial Thromboplastin Time (aPTT)
Adrenocorticotropic Hormone (ACTH)
Alanine Aminopeptidase
Alanine Aminotransferase (ALT, formerly serum glutamic-pyruvic transaminase [SGPT])
Albumin (Serum)
Alcohol Dehydrogenase
Aldolase (Serum)
Aldosterone (Plasma)
Alkaline Phosphatase (Serum)
Alpha-1-Antitrypsin (Serum)
Alpha-1-Fetoprotein (Serum)
ALT
Aluminum (Serum)
AMA
Amebiasis Serologic Test
Aminolevulinic Acid (d-ALA) (24-Hour Urine Collection)
Ammonia (Serum)
Amylase (Serum)
Amylase, Urine
Amyloid A Protein (Serum)
ANA
ANCA
Androstenedione (Serum)
Angiotensin II
Angiotensin-Converting Enzyme (ACE) Level
ANH
Anion Gap
Anticardiolipin Antibody (ACA)
Anticoagulant
Antidiuretic Hormone
Anti-DNA
Anti-ds DNA
Anti-Globulin Test, Direct
Anti-Globulin Test, Indirect
Antiglomerular Basement Antibody
Anti-HCV
Antihistone

Antimitochondrial Antibody (AMA)
Antineutrophil Cytoplasmic Antibody (ANCA)
Antinuclear Antibody (ANA)
Antiphospholipid Antibody
Anti-RNP Antibody
Anti-Scl-70
Anti-Smith (Anti-Sm) Antibody
Anti-Smooth Muscle Antibody
Antistreptolysin O Titer (Streptozyme, ASLO Titer)
Antithrombin III
Apolipoprotein A-1 (Apo A-1)
Apolipoprotein B (Apo B)
Arterial Blood Gases
Arthrocentesis Fluid
ASLO Titer
Aspartate Aminotransferase (AST, Serum Glutamic Oxaloacetic Transaminase [SGOT])
Atrial Natriuretic Hormone (ANH)
Basophil Count
Bicarbonate
Bile Acid Breath Test
Bile, Urine
Bilirubin, Direct (Conjugated Bilirubin)
Bilirubin, Indirect (Unconjugated Bilirubin)
Bilirubin, Total
Bilirubin, Urine
Bladder Tumor–Associated Antigen
Bleeding Time (Modified IVY Method)
Blood Volume, Total
Bordetella pertussis Serology
BRCA-1, BRCA-2
Breath Hydrogen Test
B-Type Natriuretic Peptide (BNP)
BUN
C3
C4
CA 15-3
CA 27-29
CA 72-4
CA 125
Calcitonin (Serum)
Calcium (Serum)
Calcium, Urine
Cancer Antigen 15-3 (CA 15-3)
Cancer Antigen 27-29 (CA 27-29)
Cancer Antigen 72-4 (CA 72-4)
Cancer Antigen 125 (CA-125)
Captopril Stimulation Test
Carbamazepine (Tegretol)
Carbohydrate Antigen 19-9
Carbon Dioxide, Partial Pressure
Carbon Monoxide
Carboxyhemoglobin (COHb)
Cardiac Markers (Serum)
Cardiac Troponins
Carcinoembryonic Antigen (CEA)
Cardio-CRP

Carotene (Serum)
Catecholamines, Urine
CBC
CCK
CCK-PZ
CD4 T-Lymphocyte Count (CD4 T-Cells)
CD40 Ligand
CEA
Cerebrospinal Fluid (CSF)
Ceruloplasmin (Serum)
Chlamydia Group Antibody Serologic Test
Chlamydia Trachomatis Polymerase Chain Reaction (PCR)
Chloride (Serum)
Chloride (Sweat)
Chloride, Urine
Cholecystokinin-Pancreozymin (CCK, CCK-PZ)
Cholesterol, Low-Density Lipoprotein
Cholesterol, High-Density Lipoprotein
Cholesterol, Total
Chorionic Gonadotropin (hCG), Human (Serum)
Chymotrypsin
Circulating Anticoagulant (Antiphospholipid Antibody, Lupus Anticoagulant)
CK
Clonidine Suppression Test
Clostridium difficile Toxin Assay (Stool)
CO
Coagulation Factors
Cold Agglutinins Titer
Complement (C3, C4)
Complete Blood Cell (CBC) Count
Conjugated Bilirubin
Coombs, Direct
Coombs, Indirect (Antiglobulin, Indirect)
Copper (Serum)
Copper, Urine
Corticotropin-Releasing Hormone (CRH) Stimulation Test
Cortisol (Plasma)
C-Peptide
CPK
C-Reactive Protein (CRP)
Creatinine Clearance
Creatine Kinase (CK), Creatine Phosphokinase (CPK)
Creatine Kinase Isoenzymes
    CK-MB
    CK-MM
    CK-BB
Creatinine (Serum)
Creatinine, Urine
Cryoglobulins (Serum)
Cryptosporidium Antigen by Enzyme Immunoassay (EIA) (Stool)
CSF
Cystatin C
Cystic Fibrosis Polymerase Chain Reaction (PCR)
Cytomegalovirus by Polymerase Chain Reaction (PCR)
DAT
d-Dimer

Dehydroepiandrosterone Sulfate
Deoxycorticosterone (11-Deoxycorticosterone, DOC), Serum
Dexamethasone Suppression Test, Overnight
Dihydrotestosterone, Serum, Urine
Direct Antiglobulin Test (DAT, Coombs Test, Direct)
Disaccharide Absorption Tests
DOC
Donath-Landsteiner (D-L) Test For Paroxysmal Cold Hemoglobinuria
Digoxin (Lanoxin)
Dilantin
Dopamine
d-Xylose Absorption Test
Electrolytes, Urine
Electrophoresis, Hemoglobin
Electrophoresis, Protein
ENA Complex
Endomysial Antibodies
Eosinophil Count
Epinephrine, Plasma
Epstein-Barr Virus (EBV) Serology
Erythrocyte Sedimentation Rate (ESR) (Westergren)
Erythropoietin (EP)
Estradiol (Serum)
Estrogens, Total
Ethanol (Blood)
Extracf Nuclear Antigen (ENA Complex, Anti-RNP Antibody, Anti-SM,
    Anti-Smith)
Factor V Leiden
FDP
FENA
Fecal FAT, Qualitative
Fecal FAT, Quantitative (72-Hour Collection)
Fecal Globin Immunochemical Test
Ferritin (Serum)
Fibrin Degradation Product (FDP)
Fibrinogen
5-Hydroxyindole-Acetic Acid, Urine
5′ Nucleotidase
Fluorescent Treponemal Antibody
Folate (Folic Acid)
Follicle-Stimulating Hormone (FSH)
Fractional Excretion of Sodium (FENA)
Free T4
Free Thyroxine Index
FSH
FTA-ABS (Serum)
Furosemide Stimulation Test
Gamma-Glutamyl Transferase (GGT)
Gastrin (Serum)
Gastrin Stimulation Test
Gliadin Antibodies, Immunoglobulin (Ig) A and IgG
Glomerular Basement Membrane Antibody
Glomerular Filtration Rate (GFR)
Glucagon
Glucose, Fasting
Glucose, Postprandial

Glucose Tolerance Test
Glucose-6-Phosphate Dehydrogenase (G6PD) Screen (Blood)
γ-Glutamyl Transferase (GGT)
Glycated (Glycosylated) Hemoglobin (HbA1c)
Glycohemoglobin
Growth Hormone
Growth Hormone–Releasing Hormone (GHRH)
Growth Hormone Suppression Test (After Glucose)
HAM Test (Acid Serum Test)
Haptoglobin (Serum)
HbA1c
HDL
Helicobacter pylori (Serology, Stool Antigen)
Hematocrit
Hemoglobin
Hemoglobin (Hb) Electrophoresis
Hemoglobin, Glycated
Hemoglobin, Glycosylated
Hemoglobin H
Hemoglobin, Urine
Hemosiderin, Urine
Heparin-Induced Thrombocytopenia Antibodies
Hepatitis A Antibody
Hepatitis B Core Antibody
Hepatitis B DNA
Hepatitis Be Antigen (HBeAg) and Antibody
Hepatitis B Surface Antibody
Hepatitis B Surface Antigen (HBsAg)
Hepatitis C Antibody (Anti-HCV)
Hepatitis C RNA
Hepatitis D Antigen and Antibody
Her-2/nue
Herpes Simplex Virus (HSV)
Heterophil Antibody
HFE Screen For Hereditary Hemochromatosis
High-Density Lipoprotein (HDL) Cholesterol
Homocysteine, Plasma
Hs-CRP
HSV
Human Herpes Virus 8 (HHV8)
Human Immunodeficiency Virus Antibody, Type 1 (HIV-1)
Human Papilloma Virus (HPV)
Huntington's Disease Polymerase Chain Reaction (PCR)
Immune Complex Assay
Immunoglobulin (Ig)
Influenza A and B Tests
INR
Insulin Autoantibodies
Insulin, Free
Insulin-Like Growth Factor I (IGF-1), Serum
Insulin-Like Growth Factor II
International Normalized Ratio (INR)
Intrinsic Factor Antibodies
Iron-Binding Capacity (Total Iron-Binding Capacity [TIBC])
Iron Saturation (% Transferrin Saturation)
Iron, Serum
Lactate (Blood)

Lactate Dehydrogenase (LDH)
Lactate Dehydrogenase (LDH) Isoenzymes
Lactose Tolerance Test (Serum)
Lanoxin
Lap Score
Lead
LDH
LDL
Legionella pneumophila Polymerase Chain Reaction (PCR)
Legionella Titer
Leukocyte Alkaline Phosphatase (LAP)
LH
Lipase
Lipoprotein (A)
Lipoprotein Cholesterol, Low Density
Lipoprotein Cholesterol, High Density
Liver Kidney Microsome Type 1 (LKM1) Antibodies
Low-Density Lipoprotein (LDL) Cholesterol
Lupus Anticoagulant (LA) Test
Luteinizing Hormone (LH), Blood
Lymphocytes
Magnesium (Serum)
Mean Corpuscular Volume (MCV)
Metanephrines, Urine
Methylmalonic Acid, Serum
Mitochondrial Antibody (Antimitochondrial antibody [AMA])
Monocyte Count
Mycoplasma pneumoniae Polymerase Chain Reaction (PCR)
Myelin Basic Protein, Cerebrospinal Fluid
Myoglobin, Urine
Natriuretic Peptide
Neisseria gonorrhoeae Polymerase Chain Reaction (PCR)
Neutrophil Count
Norepinephrine
Osmolality, Serum
Osmolality, Urine
Osmotic Fragility Test
Paracentesis Fluid
Parathyroid Hormone
Parietal Cell Antibodies
Partial Thromboplastin Time (PTT), Activated Partial Thromboplastin
    Time (aPTT)
Pepsinogen I
PFA
pH, Blood
Phenobarbital
Phenytoin (Dilantin)
Phosphatase, Acid
Phosphatase, Alkaline
Phosphate (Serum)
pH, Urine
Plasminogen
Platelet Aggregation
Platelet Antibodies
Platelet Count
Platelet Function Analysis (PFA) 100 Assay
Potassium (Serum)

Thyrotropin-Releasing Hormone (TRH) Stimulation Test
TIBC
Tissue Transglutaminase Antibody
Transferrin
Triglycerides
Triiodothyronine
Troponins, Serum
TSH
TT
Unconjugated Bilirubin
Urea Nitrogen
Uric Acid (Serum)
Urinalysis
Urine Amylase
Urine Bile
Urine Calcium
Urine cAMP
Urine Catecholamines
Urine Chloride
Urine Copper
Urine Cortisol, Free
Urine Creatinine (24-Hour)
Urine Crystals
Urine Eosinophils
Urine 5-Hydroxyindole-Acetic Acid (Urine 5-HIAA)
Urine Glucose (Qualitative)
Urine Hemoglobin, Free
Urine Hemosiderin
Urine Indican
Urine Ketones (Semiquantitative)
Urine Metanephrines
Urine Myoglobin
Urine Nitrite
Urine Occult Blood
Urine Osmolality
Urine pH
Urine Phosphate
Urine Potassium
Urine Protein (Quantitative)
Urine Sodium (Quantitative)
Urine Specific Gravity
Urine Vanillylmandelic Acid (VMA)
Varicella Zoster Virus (VZV) Serologic Testing
Vasoactive Intestinal Peptide (VIP)
Venereal Disease Research Laboratories (VDRL)
VIP
Viscosity (Serum)
Vitamin B_{12}
Vitamin D, 1,25 Dihydroxy Calciferol, Vitamin D 25(OH)D (25- Hydroxyvitamin D)
Vitamin K
Von Willebrand's Factor
WBCs
Westergren
White Blood Cell Count

ACE Level See Angiotensin-Converting Enzyme (ACE) level

Acetone (Serum or Plasma)

Normal: negative
Elevated in: diabetic ketoacidosis (DKA), starvation, isopropanol ingestion

Acetylcholine Receptor (AChR) Antibody

Normal: <0.03 nmol/L
Elevated in: myasthenia gravis. Changes in AChR concentration correlate with the clinical severity of myasthenia gravis after therapy and during therapy with prednisone and immunosuppressants. False-positive AChR antibody results may be found in patients with Eaton-Lambert syndrome.

Acid Phosphatase (Serum)

Normal range: enzymatic, prostatic 0 to 5.5 U/L; enzymatic, total 2 to 12 U/L
Elevated in: carcinoma of prostate, other neoplasms (breast, bone), Paget's disease of bone, hemolysis, multiple myeloma, osteogenesis imperfecta, malignant invasion of bone, Gaucher's disease, myeloproliferative disorders, prostatic palpation or surgery, hyperparathyroidism, liver disease, chronic renal failure, idiopathic thrombocytopenic purpura (ITP)

Acid Serum Test

See Ham Test

Activated Clotting Time (ACT)

Normal: This test is used to determine the dose of protamine sulfate to reverse the effect of heparin as an anticoagulant during angioplasty, cardiac surgery, and hemodialysis. The accepted goal during cardiopulmonary bypass surgery is usually 400 to 500 seconds.

Activated Partial Thromboplastin Time (aPTT) See Partial Thromboplastin Time (PTT), Activated Partial Thromboplastin Time (aPTT)

Adrenocorticotropic Hormone (ACTH)

Normal range: 9 to 52 pg/mL
Elevated in: Addison's disease, ectopic ACTH-producing tumors, congenital adrenal hyperplasia, Nelson's syndrome, pituitary-dependent Cushing's disease
Decreased in: secondary adrenocortical insufficiency, hypopituitarism, adrenal adenoma or adrenal carcinoma

Alanine Aminopeptidase

Normal range:
    Male: 1.11 to 1.71 mcg/mL
    Female: 0.96 to 1.52 mcg/mL
Elevated in: liver or pancreatic disease, ethanol ingestion, use of oral contraceptives, malignancy, tobacco use, pregnancy
Decreased in: abortion

Alanine Aminotransferase (ALT, formerly serum glutamic-pyruvic transaminase [SGPT])

Normal range:
Male: 10 to 40 U/L
Female: 8 to 35 U/L

Elevated in: liver disease (e.g., hepatitis, cirrhosis, Reye's syndrome), alcohol abuse, drug use (e.g., acetaminophen, statins, nonsteroidal antiinflammatory drugs [NSAIDs], antibiotics, anabolic steroids, narcotics, heparin, labetalol, amiodarone, chlorpromazine, phenytoin), hepatic congestion, infectious mononucleosis, liver metastases, myocardial infarction [MI], myocarditis, severe muscle trauma, dermatomyositis or polymyositis, muscular dystrophy, malignancy, renal and pulmonary infarction, convulsions, eclampsia, dehydration (relative increase), ingestion of Chinese herbs

Decreased in: azotemia, advanced malnutrition, chronic renal dialysis, chronic alcoholic liver disease, metronidazole use

Albumin (Serum)

Normal range: 4 to 6 g/dL

Elevated in: dehydration (relative increase), intravenous albumin infusion

Decreased in: liver disease, nephrotic syndrome, poor nutritional status, rapid intravenous hydration, protein-losing enteropathies (inflammatory bowel disease), severe burns, neoplasia, chronic inflammatory diseases, pregnancy, prolonged immobilization, lymphomas, hypervitaminosis A, chronic glomerulonephritis

Alcohol Dehydrogenase

Normal range: 0 to 7 U/L

Elevated in: drug-induced hepatocellular damage, obstructive jaundice, malignancy, inflammation, infection

Aldolase (Serum)

Normal range: 0 to 6 U/L

Elevated in: rhabdomyolysis, dermatomyositis or polymyositis, trichinosis, acute hepatitis and other liver diseases, muscular dystrophy, myocardial infarction, prostatic carcinoma, hemorrhagic pancreatitis, gangrene, delirium tremens, burns

Decreased in: loss of muscle mass, late stages of muscular dystrophy

Aldosterone (Plasma)

Normal range:
Adult supine: 3 to 16 ng/dL
Adult upright: 7 to 30 ng/dL
Adrenal vein: 200 to 800 ng/dL

Elevated in: aldosterone-secreting adenoma, bilateral adrenal hyperplasia, secondary aldosteronism (diuretics, congestive heart failure, laxatives, nephritic syndrome, cirrhosis with ascites, Bartter's syndrome, pregnancy, starvation). Table 2.1 summarizes use of plasma renin and plasma aldosterone values in the evaluation of hypokalemia or hyperkalemia.

Decreased in: Addison's disease, renin deficiency, Turner's syndrome, diabetes mellitus, isolated aldosterone deficiency, postacute alcohol intoxication (hangover phase)

TABLE 2.1	Use of Plasma Renin and Plasma Aldosterone Values to Assess the Basis of Hypokalemia or Hyperkalemia		
		RENIN	**ALDOSTERONE**
Lesions That Cause Hypokalemia			
Adrenal gland:			
Primary hyperaldosteronism		Low	High
Glucocorticoid-remediable hyperaldosteronism		Low	High
Kidney:			
Renal artery stenosis		High	High
Malignant hypertension		High	High
Renin-secreting tumor		High	High
Liddle's syndrome		Low	Low
Disorders involving 11β-hydroxysteroid dehydrogenase (HSDH)		Low	Low
Lesions That Cause Hyperkalemia			
Adrenal gland:			
Addison's disease		High	Low
Kidney:			
Pseudohypoaldosteronism type 1		High	High
Hyporeninemic hypoaldosteronism		Low	Low

From Skorecki K, Chertow GM, Marsden PA, et al: *Brenner and Rector's the kidney,* ed 10, Philadelphia, Elsevier, 2016.

Alkaline Phosphatase (Serum)

Normal range: 30 to 120 U/L
Elevated in: biliary obstruction, cirrhosis (particularly primary biliary cirrhosis), liver disease (hepatitis, infiltrative liver diseases, fatty metamorphosis), Paget's disease of bone, osteitis deformans, rickets, osteomalacia, hypervitaminosis D, hyperparathyroidism, hyperthyroidism, ulcerative colitis, bowel perforation, bone metastases, healing fractures, bone neoplasms, acromegaly, infectious mononucleosis, cytomegalovirus infections, sepsis, pulmonary infarction, hypernephroma, leukemia, myelofibrosis, multiple myeloma, drug therapy (estrogens, albumin, erythromycin and other antibiotics, cholestasis-producing drugs [phenothiazines]), pregnancy, puberty, postmenopausal females. Fig. 2.1 illustrates evaluation of isolated elevation of serum alkaline phosphatase.
Decreased in: hypothyroidism, pernicious anemia, hypophosphatemia, hypervitaminosis D, malnutrition

Alpha-1-Antitrypsin (Serum)

Normal range: 110 to 140 mg/dL
Decreased in: homozygous or heterozygous deficiency

Alpha-1-Fetoprotein (Serum)

Normal range: 0 to 20 ng/mL
Elevated in: hepatocellular carcinoma (usual values >1000 ng/mL), germinal neoplasms (testis, ovary, mediastinum, retroperitoneum), liver disease (alcoholic cirrhosis, acute hepatitis, chronic active hepatitis), fetal anencephaly, spina bifida, basal cell carcinoma, breast carcinoma, pancreatic carcinoma, gastric carcinoma, retinoblastoma, esophageal atresia

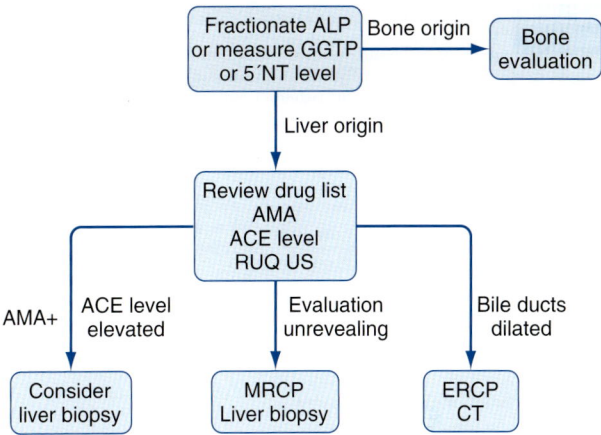

Fig. 2.1 Evaluation of an isolated elevation of the serum alkaline phosphatase level. *ACE,* angiotensin-converting enzyme; *ALP,* alkaline phosphatase; *AMA,* antimitochondrial antibodies; *5′NT,* 5′ nucleotidase; *ERCP,* endoscopic retrograde cholangiopancreatography; *MRCP,* magnetic resonance cholangiopancreatography; *RUQ,* right upper quadrant. (From Feldman M, Friedman LS, Brandt LJ: *Sleisenger and Fordtran's gastrointestinal and liver disease,* ed 10, Philadelphia, Elsevier, 2016.)

ALT See Alanine Aminotransferase

Aluminum (Serum)

Normal range: 0 to 6 ng/mL
Elevated in: chronic renal failure on dialysis, parenteral nutrition, industrial exposure

AMA See Mitochondrial Antibody

Amebiasis Serologic Test

Test description: Test is used to support diagnosis of amebiasis caused by *Entamoeba histolytica.* Serum acute and convalescent titers are drawn 1 to 3 weeks apart. A fourfold increase in titer is the most indicative result.

Aminolevulinic Acid (d-ALA) (24-Hour Urine Collection)

Normal range: 1.5 to 7.5 md/day
Elevated in: acute porphyrias, lead poisoning, diabetic ketoacidosis, pregnancy, use of anticonvulsant drugs, hereditary tyrosinemia
Decreased in: alcoholic liver disease

Ammonia (Serum)

Normal range:

Adults: 15 to 45 mcg/dL
Children: 29 to 70 mcg/dL
Elevated in: hepatic failure, hepatic encephalopathy, Reye's syndrome, portacaval shunt, drug therapy (diuretics, polymyxin B, methicillin)
Decreased in: drug therapy (neomycin, lactulose), renal failure

Amylase (Serum)

Normal range: 0 to 130 U/L
Elevated in: acute pancreatitis, macroamylasemia, salivary gland inflammation, mumps, pancreatic neoplasm, abscess, pseudocyst, ascites, perforated peptic ulcer, intestinal obstruction, intestinal infarction, acute cholecystitis, appendicitis, ruptured ectopic pregnancy, peritonitis, burns, diabetic ketoacidosis, renal insufficiency, drug use (morphine), carcinomatosis of lung, esophagus, ovary, acute ethanol ingestion, prostate tumors, post–endoscopic retrograde cholangiopancreatography, bulimia, anorexia nervosa
Decreased in: advanced chronic pancreatitis, hepatic necrosis, cystic fibrosis

Amylase, Urine See Urine Amylase

Amyloid A Protein (Serum)

Normal: <10 mcg/mL
Elevated in: inflammatory disorders (acute phase-reacting protein), infections, acute coronary syndrome, malignancies

ANA See Antinuclear Antibody

ANCA See Antineutrophil Cytoplasmic Antibody

Androstenedione (Serum)

Normal range:
Male: 75 to 205 ng/dL
Female: 85 to 275 ng/dL
Elevated in: congenital adrenal hyperplasia, polycystic ovary syndrome, ectopic adrenocorticotropic hormone–producing tumor, Cushing's syndrome, hirsutism, hyperplasia of ovarian stroma, ovarian neoplasm
Decreased in: ovarian failure, adrenal failure, sickle cell anemia

Angiotensin II

Normal range: 10 to 60 pg/mL
Elevated in: hypertension, congestive heart failure, cirrhosis, renin-secreting renal tumor, volume depletion
Decreased in: angiotensin-converting enzyme inhibitor drugs, angiotensin II receptor blocker drugs, primary aldosteronism, Cushing's syndrome

Angiotensin-Converting Enzyme (ACE) Level

Normal: <40 nmol/mL/min
Elevated in: sarcoidosis, primary biliary cirrhosis, alcoholic liver disease, hyperthyroidism, hyperparathyroidism, diabetes mellitus, amyloidosis, multiple myeloma, lung disease (asbestosis, silicosis, berylliosis, allergic alveolitis, coccidioidomycosis), Gaucher's disease, leprosy
Decreased in: ACE inhibitor therapy

ANH See Atrial Natriuretic Hormone

Anion Gap

Normal range: 9 to 14 mEq/L. The calculation of anion gap is described in Table 2.2.
Elevated in: lactic acidosis, ketoacidosis (diabetic ketoacidosis, alcoholic starvation), uremia (chronic renal failure), ingestion of toxins (paraldehyde, methanol, salicylates, ethylene glycol), hyperosmolar nonketotic coma, antibiotic therapy (carbenicillin)
Decreased in: hypoalbuminemia, severe hypermagnesemia, immunoglobulin G myeloma, lithium toxicity, laboratory error (falsely decreased sodium or overestimation of bicarbonate or chloride), hypercalcemia of parathyroid origin, antibiotic therapy (e.g., polymyxin)

TABLE 2.2 The Anion Gap	
Anion Gap = $Na^+ - (Cl^- + HCO_3^-)$ = 9 ± 3 mEq/L (Assumes Normal [Albumin])*	
Decreased Anion Gap	**Increased Anion Gap**
Increased Cations (Not Na^+)	Increased Anions (Not Cl^- or HCO_3^-)
↑ Ca^{2+}, Mg^{2+}	↑ Albumin
↑ Li^+	Alkalosis

↑ Immunoglobulin G	↑ Inorganic anions
	  Phosphate
	  Sulfate

Decreased Anions (Not Cl^- or HCO_3^-)	↑ Organic anions
	  L-Lactate
	  D-Lactate
	  Ketones
	  Uremic

Hypoalbuminemia*	
Acidosis	
Laboratory Error	

Hyperviscosity	↑ Exogenously supplied anions
	  Toxins
	  Salicylate
	  Paraldehyde
	  Ethylene glycol
	  Propylene glycol
	  Methanol
	  Toluene
	  Pyroglutamic acid (5-oxoprolene)

Bromism	

	↑ Unidentified anions
	  Other toxins
	  Uremic
	  Hyperosmolar, nonketotic states
	  Myoglobinuric acute kidney injury

	Decreased Cations (Not Na^+)
	↓ Ca^{2+}, Mg^{2+}

*For each decline in albumin by 1 g/dL from normal (4.5 g/dL), the anion gap decreases by 2.5 mEq/L.
From Skorecki K, Chertow GM, Marsden PA, et al: *Brenner and Rector's the kidney*, ed 10, Philadelphia, Elsevier, 2016.

Anticardiolipin Antibody (ACA)

Normal: negative. Test includes detection of immunoglobulin (Ig) G, IgM, and IgA antibody to phospholipid, cardiolipin.
Present in: antiphospholipid antibody syndrome, chronic hepatitis C

Anticoagulant See Circulating Anticoagulant (Antiphospholipid Antibody, Lupus Anticoagulant)

Antidiuretic Hormone

Normal range: mOsm/kg 295 to 300; 4 to 12 pg/mL
Elevated in: syndrome of inappropriate antidiuretic hormone, antipsychotic medication therapy, ectopic antidiuretic hormone from systemic neoplasm, Guillain-Barré, central nervous system infections, brain tumors, nephrogenic diabetes insipidus
Decreased in: central diabetes insipidus, nephritic syndrome, psychogenic polydipsia, demeclocycline, lithium therapy, phenytoin use, alcohol use

Anti-DNA

Normal: absent
Present in: systemic lupus erythematosus, chronic active hepatitis, infectious mononucleosis, biliary cirrhosis

Anti-ds DNA

Normal: <25 U
Elevated in: systemic lupus erythematosus

Anti-Globulin Test, Direct See Direct Antiglobulin Test

Anti-Globulin Test, Indirect See Coombs, Indirect

Antiglomerular Basement Antibody See Glomerular Basement Membrane Antibody

Anti-HCV See Hepatitis C Antibody

Antihistone

Normal: <1 U
Elevated in: drug-induced lupus erythematosus

Antimitochondrial Antibody (AMA)

Normal range: <1:20 titer
Elevated in: primary biliary cirrhosis (85%–95%), chronic active hepatitis (25%–30%), cryptogenic cirrhosis (25%–30%)

Antineutrophil Cytoplasmic Antibody (ANCA)

Positive test:
 Cytoplasmic pattern (cANCA): positive in granulomatosis with polyangiitis (Wegener's granulomatosis)
 Perinuclear pattern (pANCA): positive in inflammatory bowel disease, primary biliary cirrhosis, primary sclerosing cholangitis, autoimmune chronic active hepatitis, crescentic glomerulonephritis

Antinuclear Antibody (ANA)

Normal: <1:20 titer

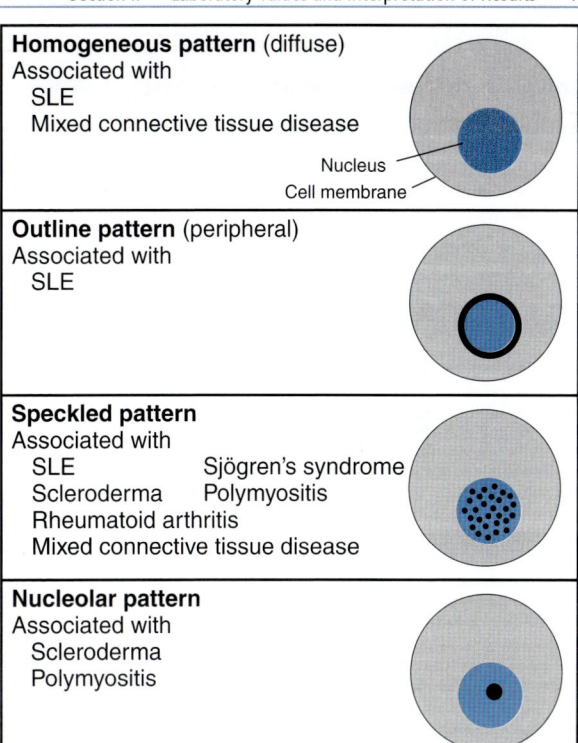

Homogeneous pattern (diffuse)
Associated with
SLE
Mixed connective tissue disease

Nucleus
Cell membrane

Outline pattern (peripheral)
Associated with
SLE

Speckled pattern
Associated with
SLE   Sjögren's syndrome
Scleroderma   Polymyositis
Rheumatoid arthritis
Mixed connective tissue disease

Nucleolar pattern
Associated with
Scleroderma
Polymyositis

Fig. 2.2 Patterns of immunofluorescent staining of antinuclear antibodies and the diseases with which they are associated. *SLE,* systemic lupus erythematosus. (From Pagana KD, Pagana TJ: *Mosby's diagnostic and laboratory test reference,* ed 8, St. Louis, Mosby, 2006.)

Positive test: systemic lupus erythematosus (more significant if titer > 1:160), drug therapy (phenytoin, ethosuximide, primidone, methyldopa, hydralazine, carbamazepine, penicillin, procainamide, chlorpromazine, griseofulvin, thiazides), chronic active hepatitis, age older than 60 years (particularly age older than 80), rheumatoid arthritis, scleroderma, mixed connective tissue disease, necrotizing vasculitis, Sjögren's syndrome Fig. 2.2 describes diagnostic tests and diagnoses to consider from ANA patterns.

Antiphospholipid Antibody See Lupus Anticoagulant (LA) Test

Anti-RNP Antibody See Extractable Nuclear Antigen

Anti-Scl-70

Normal: absent
Elevated in: scleroderma

Anti-Smith (Anti-Sm) Antibody See Extractable Nuclear Antigen

Anti-Smooth Muscle Antibody See Smooth Muscle Antibody

Antistreptolysin O Titer (Streptozyme, ASLO Titer)

Normal for adults: <160 Todd units
Elevated in: streptococcal upper airway infection, acute rheumatic fever, acute glomerulonephritis, increased levels of β-lipoprotein (false-positive ASLO test)
Note: A fourfold increase in titer between acute and convalescent specimens is diagnostic of streptococcal upper airway infection regardless of the initial titer.

Antithrombin III

Normal range: 81% to 120% of normal activity; 17 to 30 mg/dL
Elevated in: warfarin drug therapy, post–myocardial infarction
Decreased in: hereditary deficiency of antithrombin III, disseminated intravascular coagulation, pulmonary embolism, cirrhosis, thrombolytic therapy, chronic liver failure, postsurgery, third trimester of pregnancy, oral contraceptive use, nephrotic syndrome, intravenous heparin therapy >3 days, sepsis, acute leukemia, carcinoma, thrombophlebitis

Apolipoprotein A-1 (Apo A-1)

Normal: recommended >120 mg/dL
Elevated in: familial hyperalphalipoproteinemia, statins, niacin, estrogens, weight loss, familial cholesteryl ester transfer protein deficiency
Decreased in: familial hypoalphalipoproteinemia, Tangier disease, diuretic use, androgens, cigarette smoking, hepatocellular disorders, chronic renal failure, nephritic syndrome, coronary heart disease, cholestasis

Apolipoprotein B (Apo B)

Normal: desirable <100 mg/dL, high risk >120 mg/dL
Elevated in: high-saturated-fat diet, high-cholesterol diet, hyper-apobetalipoproteinemia, familial combined hyperlipidemia, anabolic steroids, diuretic use, beta blocker therapy, corticosteroid use, progestin use, diabetes, hypothyroidism, chronic renal failure, liver disease, Cushing's syndrome, coronary heart disease
Decreased in: statin therapy, niacin, low-cholesterol diet, malnutrition, abetalipoproteinemia, hypobetalipoproteinemia, hyperthyroidism

Arterial Blood Gases

Normal range:
  P_{O_2}: 75 to 100 mm Hg
  P_{CO_2}: 35 to 45 mm Hg
  HCO^{-3}: 24 to 28 mEq/L
  pH: 7.35-7.45
Abnormal values: Acid-base disturbances (see following). Table 2.3 summarizes acid-base abnormalities and appropriate compensatory responses.
1. Metabolic acidosis
  a. Metabolic acidosis with increased anion gap (AG acidosis)
    i. Lactic acidosis

TABLE 2.3   Acid-Base Abnormalities and Appropriate Compensatory Responses for Simple Disorders

PRIMARY ACID-BASE DISORDERS	PRIMARY DEFECT	EFFECT ON pH	COMPENSATORY RESPONSE	EXPECTED RANGE OF COMPENSATION	LIMITS OF COMPENSATION
Respiratory acidosis	Alveolar hypoventilation (↑ P_{CO_2})	↓	↑ Renal HCO_3^- reabsorption (HCO_3^- ↑)	**Acute** Δ $[HCO_3^-]$ = +1 mEq/L for each ↑ Δ P_{CO_2} of 10 mm Hg	$[HCO_3^-]$ = 38 mEq/L
				Chronic Δ $[HCO_3^-]$ = +4 mEq/L for each ↑ Δ P_{CO_2} of 10 mm Hg	$[HCO_3^-]$ = 45 mEq/L
Respiratory alkalosis	Alveolar hyperventilation (↓ P_{CO_2})	↑	↓ Renal HCO_3^- reabsorption (HCO_3^- ↓)	**Acute** Δ $[HCO_3^-]$ = −2 mEq/L for each ↓ Δ P_{CO_2} of 10 mm Hg	$[HCO_3^-]$ = 18 mEq/L
				Chronic Δ $[HCO_3^-]$ = −5 mEq/L for each ↓ Δ P_{CO_2} of 10 mm Hg	$[HCO_3^-]$ = 15 mEq/L
Metabolic acidosis	Loss of HCO_3^- or gain of H^+ (↓HCO_3^-)	↓	Alveolar hyperventilation to ↑ pulmonary CO_2 excretion (↓P_{CO_2})	P_{CO_2} = 1.5$[HCO_3^-]$ + 8 ± 2 P_{CO_2} = last 2 digits of pH × 100 P_{CO_2} = 15 + $[HCO_3^-]$	P_{CO_2} = 15 mm Hg
Metabolic alkalosis	Gain of HCO_3^- Δ or loss of H^+ (↑HCO_3^-)	↑	Alveolar hypoventilation to ↓ pulmonary CO_2 excretion (↑P_{CO_2})	P_{CO_2} = +0.6 mm Hg for Δ $[HCO_3^-]$ of 1 mEq/L P_{CO_2} = 15 + $[HCO_3^-]$	P_{CO_2} = 55 mm Hg

P_{CO_2}, carbon dioxide pressure.

Adapted from Bidani A, Tauzon DM, Heming TA: Regulation of whole body acid-base balance. In DuBose TD, Hamm LL, [eds]: *Acid-base and electrolyte disorders; a companion to Brenner and Rector's the kidney*, Philadelphia, 2002, Saunders, pp. 1-2. From Skorecki K, Chetow GM, Marsden PA, et al: *Brenner and Rector's the kidney*, ed 10, Philadelphia, Elsevier, 2016.

    ii. Ketoacidosis (diabetes mellitus, alcoholic ketoacidosis)

    iii. Uremia (chronic renal failure)

    iv. Ingestion of toxins (paraldehyde, methanol, salicylate, ethylene glycol)

    v. High-fat diet (mild acidosis)

    vi. Metabolic acidosis with normal AG (hyperchloremic acidosis)

    vii. Renal tubular acidosis (including acidosis of aldosterone deficiency)

    viii. Intestinal loss of HCO_3^- (diarrhea, pancreatic fistula)

    ix. Carbonic anhydrase inhibitors (e.g., acetazolamide)

    x. Dilutional acidosis (as a result of rapid infusion of bicarbonate-free isotonic saline)

    xi. Ingestion of exogenous acids (ammonium chloride, methionine, cystine, calcium chloride)

    xii. Ileostomy

    xiii. Ureterosigmoidostomy

    xiv. Drug therapy: amiloride, triamterene, spironolactone, beta blockers

2. Respiratory acidosis
   a. Pulmonary disease (chronic obstructive pulmonary disease, severe pneumonia, pulmonary edema, interstitial fibrosis)
   b. Airway obstruction (foreign body, severe bronchospasm, laryngospasm)
   c. Thoracic cage disorders (pneumothorax, flail chest, kyphoscoliosis)
   d. Defects in muscles of respiration (myasthenia gravis, hypokalemia, muscular dystrophy)
   e. Defects in peripheral nervous system (amyotrophic lateral sclerosis, poliomyelitis, Guillain-Barré syndrome, botulism, tetanus, organophosphate poisoning, spinal cord injury)
   f. Depression of respiratory center (anesthesia, narcotics, sedatives, vertebral artery embolism or thrombosis, increased intracranial pressure)
   g. Failure of mechanical ventilator

3. Metabolic alkalosis. Metabolic alkalosis is divided into chloride-responsive (urinary chloride <15 mEq/L) and chloride-resistant forms (urinary chloride level >15 mEq/L).
   a. Chloride-responsive
      i. Vomiting
      ii. Nasogastric suction
      iii. Diuretics
      iv. Posthypercapnic alkalosis
      v. Stool losses (laxative abuse, cystic fibrosis, villous adenoma)
      vi. Massive blood transfusion
      vii. Exogenous alkali administration
   b. Chloride-resistant
      i. Hyperadrenocorticoid states (Cushing's syndrome, primary hyperaldosteronism, secondary mineralocorticoidism [licorice ingestion, chewing tobacco use])
      ii. Hypomagnesemia
      iii. Hypokalemia
      iv. Bartter's syndrome

4. Respiratory alkalosis
   a. Hypoxemia (pneumonia, pulmonary embolism, atelectasis, high-altitude living)
   b. Drugs (salicylates, xanthines, progesterone, epinephrine, thyroxine, nicotine)
   c. Central nervous system disorders (tumor, cerebrovascular accident, trauma, infections)
   d. Psychogenic hyperventilation (anxiety, hysteria)
   e. Hepatic encephalopathy
   f. Gram-negative sepsis
   g. Hyponatremia
   h. Sudden recovery from metabolic acidosis
   i. Assisted ventilation

Arthrocentesis Fluid

Interpretation of results:
1. Color: Normally it is clear or pale yellow; cloudiness indicates inflammatory process or presence of crystals, cell debris, fibrin, or triglycerides.
2. Viscosity: Normally it has a high viscosity because of hyaluronate; when fluid is placed on a slide, it can be stretched to a string greater than 2 cm in length before separating (low viscosity indicates breakdown of hyaluronate [lysosomal enzymes from leukocytes] or the presence of edema fluid).
3. Mucin clot: Add 1 mL of fluid to 5 mL of a 5% acetic acid solution and allow 1 minute for the clot to form; a firm clot (does not fragment on shaking) is normal and indicates the presence of large molecules of hyaluronic acid (this test is non-specific and infrequently done).
4. Glucose: Normally glucose approximately equals serum glucose level; a difference of more than 40 mg/dL is suggestive of infection.
5. Protein: Total protein concentration is less than 2.5 g/dL in the normal synovial fluid; it is elevated in inflammatory and septic arthritis.
6. Microscopic examination for crystals
   a. Gout: monosodium urate crystals
   b. Pseudogout: calcium pyrophosphate dihydrate crystals
   Table 2.4 describes synovial fluid findings in common disorders.

TABLE 2.4	Knee Joint Synovial Fluid Findings in Common Forms of Arthritis			
	NORMAL	**OSTEOARTHRITIS**	**RHEUMATOID AND OTHER INFLAMMATORY ARTHRITIS**	**SEPTIC ARTHRITIS**
Gross appearance	Clear	Clear	Opaque	Opaque
Volume (mL)	0–1	0–10	5–50	5–50
Viscosity	High	High	Low	Low
Total white cell count/mm³	<200	200–10,000	500–75,000	>50,000
% Polymorphonuclear cells	<25	<50	>50	>75

From Hochberg MC, Silma AJ, Smolen JS, et al, [eds]: *Rheumatology*, ed 3, St. Louis, Mosby, 2003.

ASLO Titer See Antistreptolysin O Titer

Aspartate Aminotransferase (AST, Serum Glutamic Oxaloacetic Transaminase [SGOT])

Normal range: 0 to 35 U/L
Elevated in: liver disease (hepatitis, hemochromatosis, cirrhosis, Reye's syndrome, Wilson's disease), alcohol abuse, drug therapy (acetaminophen, statins, nonsteroidal antiinflammatory drugs, angiotensin-converting enzyme inhibitors, heparin, labetalol, phenytoin, amiodarone, chlorpromazine), hepatic congestion, infec-

tious mononucleosis, myocardial infarction, myocarditis, severe muscle trauma, dermatomyositis and polymyositis, muscular dystrophy, malignancy, renal and pulmonary infarction, convulsions, eclampsia

Decreased in: uremia, vitamin B_6 deficiency

Atrial Natriuretic Hormone (ANH)

Normal range: 20 to 77 pg/mL
Elevated in: congestive heart failure, volume overload, cardiovascular disease with high filling pressure
Decreased with: alpha blocker use

Basophil Count

Normal range: 0.4% to 1% of total white blood cells; 40 to 100/mm³
Elevated in: inflammatory processes, leukemia, polycythemia vera, Hodgkin's lymphoma, hemolytic anemia, after splenectomy, myeloid metaplasia, myxedema
Decreased in: stress, hypersensitivity reaction, steroids, pregnancy, hyperthyroidism

Bicarbonate

Normal range:
   Arterial: 21 to 28 mEq/L
   Venous: 22 to 29 mEq/L
Elevated in: metabolic alkalosis, compensated respiratory acidosis, diuretics, cortico-steroids, laxative abuse
Decreased in: metabolic acidosis; compensated respiratory alkalosis; acetazolamide, cyclosporine, or cholestyramine use; methanol or ethylene glycol poisoning

Bile Acid Breath Test

Normal: The test determines the radioactivity of $^{14}CO_2$ in breath samples at 2 and 4 hours.
   2 hours after dose: 0.11 ± 0.14
   4 hours after dose: 0.52 ± 0.09
Elevated in: gastrointestinal bacterial overgrowth, H_2 blockers use

Bile, Urine See Urine Bile

Bilirubin, Direct (Conjugated Bilirubin)

Normal range: 0 to 0.2 mg/dL
Elevated in: hepatocellular disease, biliary obstruction, drug-induced cholestasis, heredi-tary disorders (Dubin-Johnson syndrome, Rotor's syndrome), advanced neoplastic states

Bilirubin, Indirect (Unconjugated Bilirubin)

Normal range: 0 to 1.0 mg/dL
Elevated in: hemolysis, liver disease (hepatitis, cirrhosis, neoplasm), hepatic conges-tion caused by congestive heart failure, hereditary disorders (Gilbert's disease, Crigler-Najjar syndrome)

Bilirubin, Total

Normal range: 0 to 1 mg/dL

Elevated in: liver disease (hepatitis, cirrhosis, cholangitis, neoplasm, biliary obstruction, infectious mononucleosis), hereditary disorders (Gilbert's disease, Dubin-Johnson syndrome), drug therapy (steroids, diphenylhydantoin, phenothiazines, penicillin, erythromycin, clindamycin, captopril, amphotericin B, sulfonamides, azathioprine, isoniazid, 5-aminosalicylic acid, allopurinol, methyldopa, indomethacin, halothane, oral contraceptives, procainamide, tolbutamide, labetalol), hemolysis, pulmonary embolism or infarct, hepatic congestion resulting from congestive heart failure
Table 2.5 compares hereditary disorders of bilirubin metabolism and transport.

TABLE 2.5	Hereditary Disorders of Bilirubin Metabolism and Transport				
PARAMETER	**GILBERT'S**	**TYPE I CRIGLER-NAJJAR**	**TYPE II CRIGLER-NAJJAR**	**DUBIN-JOHNSON**	**ROTOR'S**
Incidence	6%–12%	Very rare	Uncommon	Uncommon	Rare
Gene affected	*UGT1A1*	*UGT1A1*	*UGT1A1*	*MRP2*	*OATP1B1* and *OATP1B3*
Metabolic defect	↓Bilirubin conjugation	No bilirubin conjugation	↓↓Bilirubin conjugation	Impaired canalicular export of conjugated bilirubin	Impaired canalicular export of conjugated bilirubin
Plasma bilirubin (mg/dL)	≤3 in absence of fasting or hemolysis, almost all unconjugated	Usually >20 (range, 17–50), all unconjugated	Usually <20 (range, 6–45), almost all unconjugated	Usually <7, about half conjugated	Usually <7, about half conjugated
Liver histology	Usually normal, occasional →lipofuscin	Normal	Normal	Coarse pigment in centrilobular hepatocytes	Normal
Other distinguishing features	↓Bilirubin concentration with phenobarbital	No response to phenobarbital	↓Bilirubin concentration with phenobarbital	→Bilirubin concentration with estrogens; →→→urinary coproporphyrin I/III ratio	Mild →urinary coproporphyrin I/III ratio
Prognosis	Normal (theoretical risk of selected drug toxicity)	Death in infancy if untreated	Usually normal	Normal (theoretical risk of selected drug toxicity)	Normal (theoretical risk of selected drug toxicity)
Treatment	None	Phototherapy as a bridge to liver transplantation	Phenobarbital for →→→bilirubin concentration	Avoid estrogens	None available

MRP2, multidrug resistance–associated protein-2 gene; *OATP*, organic anion transporter; *UGT1A1*, bilirubin uridine diphosphate-glucuronyl transferase gene.
From Feldman M, Friedman LS, Brandt LJ: *Sleisenger and Fordtran's gastrointestinal and liver disease*, ed 10, Philadelphia, Elsevier, 2016.

Bilirubin, Urine See Urine Bile

Bladder Tumor–Associated Antigen

Normal: ≤14 U/mL. The test is used to detect bladder cancer recurrence. Sensitivity is 57% to 83% and specificity 68% to 72%.

Elevated in: bladder cancer, renal stones, nephritis, urinary tract infection, hematuria, renal cancer, cystitis, recent bladder or urinary tract trauma

Bleeding Time (Modified IVY Method)

Normal range: 2 to 9.5 minutes

Elevated in: thrombocytopenia, capillary wall abnormalities, platelet abnormalities (Bernard-Soulier disease, Glanzmann's disease), drug therapy (aspirin, warfarin, antiinflammatory medications, streptokinase, urokinase, dextran, b-lactam antibiotics, moxalactam), disseminated intravascular coagulation, cirrhosis, uremia, myeloproliferative disorders, von Willebrand's disease

Comments: The bleeding time test as a method to evaluate suspected hemostatic incompetence has been replaced in many laboratories with the platelet function analysis–100 assay. The bleeding time test's ability to predict excessive bleeding in clinical situations such as surgery or invasive diagnostic procedures is poor. It may play a limited residual role in the evaluation of suspected hereditary disorders of hemostasis.

Blood Volume, Total

Normal: 60 to 80 mL/kg

Elevated in: polycythemia vera, pulmonary disease, congestive heart failure, renal insufficiency, pregnancy, acidosis, thyrotoxicosis

Decreased in: anemia, hemorrhage, vomiting, diarrhea, dehydration, burns, starvation

Bordetella pertussis Serology

Test description: Polymerase chain reaction of nasopharyngeal aspirates or secretions is used to identify *Bordetella pertussis,* the organism responsible for whooping cough.

BRCA-1, BRCA-2

This test involves the detection of carriers of mutations in the gene that are characterized by predisposition to breast and ovarian cancers. Women found to carry the mutation should undergo earlier and more intensive surveillance for breast cancer. Pretest counseling should be provided before genetic testing.

Breath Hydrogen Test

Normal: This test is for bacterial overgrowth. Fasting H_2 excretion is 4.6 ± 5.1, after lactulose challenge, with an early increase of less than 12. Lactulose usually results in a colonic response more than 30 minutes after ingestion.

Elevated in: A high fasting breath H_2 level and an increase of at least 12 ppm within 30 minutes after lactulose challenge are indicative of bacterial overgrowth in the small intestine. The increase must precede the colonic response.

False positives in: accelerated gastric emptying, laxative use

False negatives in: use of antibiotics and patients who are nonhydrogen producers

B-Type Natriuretic Peptide (BNP)

Normal range: up to 100 mg/L. Natriuretic peptides are secreted to regulate fluid volume, blood pressure, and electrolyte balance. They have activity in both the central and peripheral nervous system. In humans the main source of circulatory BNP is the heart ventricles.

Elevated in: heart failure. This test is useful in the emergency department setting to differentiate heart failure patients from those with chronic obstructive pulmonary disease presenting with dyspnea. Levels are also increased in asymptomatic left ventricular dysfunction, arterial and pulmonary hypertension, cardiac hypertrophy, valvular heart disease, arrhythmia, and acute coronary syndrome.

BUN See Urea Nitrogen

C3 See Complement (C3, C4)

C4 See Complement (C3, C4)

CA 15-3 See Cancer Antigen 15-3

CA 27-29 See Cancer Antigen 27-29

CA 72-4 See Cancer Antigen 72-4

CA 125 See Cancer Antigen 125

Calcitonin (Serum)

Normal: <100 pg/mL
Elevated in: medullary carcinoma of the thyroid (particularly if level >1500 pg/mL), carcinoma of the breast, apudomas, carcinoids, renal failure, thyroiditis

Calcium (Serum)

Normal range: 8.8-10.3 mg/dL
Elevated in:
1. Malignancy: increased bone resorption via osteoclast-activating factors, secretion of pituitary hormone (PTH)–like substances, prostaglandin E_2, direct erosion by tumor cells, transforming growth factors, colony-stimulating activity. Hypercalcemia is common in the following neoplasms:
   a. Solid tumors: breast, lung, pancreas, kidneys, ovary
   b. Hematologic cancers: myeloma, lymphosarcoma, adult T-cell lymphoma, Burkitt's lymphoma
2. Hyperparathyroidism: increased bone resorption, gastrointestinal (GI) absorption, and renal absorption. Hyperparathyroidism can be caused by the following conditions:
   a. Parathyroid hyperplasia, adenoma
   b. Hyperparathyroidism or renal failure with secondary hyperparathyroidism
3. Granulomatous disorders: increased GI absorption (e.g., sarcoidosis)
4. Paget's disease: increased bone resorption, seen only during periods of immobilization
5. Vitamin D intoxication, milk-alkali syndrome, increased GI absorption
6. Thiazides: increased renal absorption
7. Other causes: familial hypocalciuric hypercalcemia, thyrotoxicosis, adrenal insufficiency, prolonged immobilization, vitamin A intoxication, recovery from acute renal failure, lithium administration, pheochromocytoma, disseminated systemic lupus erythematosus

Decreased in:
1. Renal insufficiency: hypocalcemia caused by the following:
   a. Increased calcium deposits in bone and soft tissue secondary to increased serum PO_4^{-3} level
   b. Decreased production of 1,25-dihydroxyvitamin D
   c. Excessive loss of 25-OHD (nephrotic syndrome)
2. Hypoalbuminemia: Each decrease in serum albumin (g/L) will decrease serum calcium by 0.8 mg/dL but will not change free (ionized) calcium
3. Vitamin D deficiency
   a. Malabsorption (most common cause)
   b. Inadequate intake
   c. Decreased production of 1,25-dihydroxyvitamin D (vitamin D–dependent rickets, renal failure)
   d. Decreased production of 25-OHD (parenchymal liver disease)
   e. Accelerated 25-OHD catabolism (phenytoin, phenobarbital)
   f. End organ resistance to 1,25-dihydroxyvitamin D
4. Hypomagnesemia: hypocalcemia caused by the following:
   a. Decreased PTH secretion
   b. Inhibition of PTH effect on bone
5. Pancreatitis, hyperphosphatemia, osteoblastic metastases: Hypocalcemia is secondary to increased calcium deposits (bone, abdomen).
6. Pseudohypoparathyroidism: autosomal recessive disorder characterized by short stature, shortening of metacarpal bones, obesity, and mental retardation. Hypocalcemia is secondary to congenital end organ resistance to PTH.
7. Idiopathic hypoparathyroidism, surgical removal of parathyroids (e.g., neck surgery)
8. "Hungry bones syndrome": rapid transfer of calcium from plasma into bones after removal of a parathyroid tumor
9. Sepsis
10. Massive blood transfusion (as a result of EDTA in blood)
    Table 2.6 summarizes a differential diagnosis of hypocalcemia based on laboratory evaluation.

Calcium, Urine See Urine Calcium

Cancer Antigen 15-3 (CA 15-3)

Normal: <30 U/mL
Elevated in: approximately 80% of women with metastatic breast cancer. Clinical sensitivity is 0.60, specificity 0.87, positive predictive value 0.91. This test is generally used to predict recurrence of breast cancer and to evaluate response to therapy. May also be elevated in liver cancer, pancreatic cancer, ovarian cancer, and colorectal cancer. Elevations can also occur with benign breast and liver disease.

Cancer Antigen 27-29 (CA 27-29)

Normal: <38 U/mL
Elevated in: approximately 75% of women with metastatic breast cancer. Clinical sensitivity is 0.57, specificity 0.97, positive predictive value 0.83, negative predictive value 0.92. This test is generally used to predict recurrence of breast cancer and to evaluate response to therapy. May also be elevated in liver cancer, pancreatic cancer, ovarian cancer, and colorectal cancer. Elevations can also occur with benign breast and liver disease.

Cancer Antigen 72-4 (CA 72-4)

Normal: <4.0 ng/mL

TABLE 2.6 Laboratory Differential Diagnosis of Hypocalcemia

DIAGNOSIS	PLASMA TESTS						URINE TESTS				COMMENTS
	Ca	PO$_4$	PTH	25(OH)D	1,25(OH)$_2$D	cAMP	cAMP AFTER PTH	TmP/GFR	TmP/GFR AFTER PTH	Ca	
Hypoparathyroidism	↓	↑	N/↓	N	↓	↓	↑↑	↓	↓↓	N/↓	Deficiency of PTH
Pseudohypoparathyroidism type I	↓	↑	↑↑	N	↓	↓	NC	↑	↑	N/↓	Resistance to PTH; patients may have Albright's hereditary osteodystrophy and resistance to multiple hormones
Type II	↓	N	↑↑	N	↓	↑	↑	↑	↑	N/↓	Renal resistance to cAMP
Vitamin D deficiency	↓	N/↓	↑↑	↓↓	N/↓	↑	↑	↓	↓	↓↓	Deficient supply (e.g., nutrition) or absorption (e.g., pancreatic insufficiency) of vitamin D
Vitamin D–dependent rickets											
Type I	↓	N/↓	↑↑	N	↓	↑	↑	↓	↓↓	↓↓	Deficient activity of renal 25(OH)D-1α-hydroxylase
Type II	↓	N/↓	↑↑	N	↑↑	↑	↑	↓	↓↓	↓↓	Resistance to 1,25(OH)$_2$D

Ca, Calcium; cAMP, cyclic adenosine monophosphate; GFR, glomerular filtration rate; (OH)D, hydroxycholecalciferol D; N, normal; NC, no change; OH$_2$D, dihydroxycholecalciferol; PO$_4$, phosphate; PTH, parathyroid hormone; TmP, renal threshold for phosphorus.
From Moore WT, Eastman RC: Diagnostic endocrinology, ed 2, St. Louis, Mosby, 1996.

Elevated in: gastric cancer (elevated in >50% of patients). Often used in combination with CA 19-9 and carcinoembryonic antigen to monitor gastric cancer after treatment.

Cancer Antigen 125 (CA-125)

Normal range: <35 U/mL
Elevated in: epithelial ovarian cancer; carcinoma of fallopian tube and endometrium; nonovarian abdominal malignancies; all forms of liver disease, especially those with cirrhotic ascites

Captopril Stimulation Test

Normal: This test is performed by giving 25 mg captopril orally after overnight fast. The patient should be seated during the test. After captopril administration, aldosterone is less than 15 ng/dL, renin greater than 2 ng Al/mL/hr.
Interpretation: In patients with primary aldosteronism, plasma aldosterone remains high and plasma renin activity remains low after captopril administration.

Carbamazepine (Tegretol)

Normal therapeutic range: 4 to 12 mcg/mL

Carbohydrate Antigen 19-9

Normal: <37.0 U/mL
Elevated in: gastrointestinal cancer, most commonly pancreatic cancer. The amount of elevation has no relation to tumor mass. Elevations can also occur with cirrhosis, cholangitis, and chronic or acute pancreatitis.

Carbon Dioxide, Partial Pressure

Normal range:
   Male: 35 to 48 mm Hg
   Female: 32 to 45 mm Hg
Elevated in: respiratory acidosis
Decreased in: respiratory alkalosis

Carbon Monoxide See Carboxyhemoglobin

Carboxyhemoglobin (COHb)

Normal: saturation of hemoglobin <2%; smokers <9% (coma: 50%; death: 80%)
Elevated in: smoking, exposure to smoking, exposure to automobile exhaust fumes, malfunctioning gas-burning appliances
   Fig. 2.3 illustrates the effects of oxygen on the dissociation of CO from carboxyhemoglobin.

Cardiac Markers (Serum) Fig. 2.4 describes typical cardiac marker diagnostic window curves and serum levels after myocardial infarction.

Cardiac Troponins See Troponins, Serum

Carcinoembryonic Antigen (CEA)

Normal range: nonsmokers: 0 to 2.5 ng/mL; smokers: 0 to 5 ng/mL

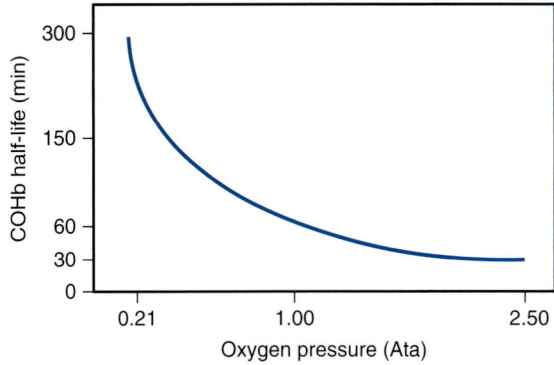

Fig. 2.3 The effects of oxygen on the dissociation of CO from carboxyhemoglobin (COHb). Oxygen breathing at 1 atm decreases the half-life of COHb to 60 minutes from approximately 300 minutes in air. Hyperbaric oxygen at 2.5 atm reduces the half-life to 30 minutes, allowing most of the COHb to be removed from the body within 90 minutes. *Ata,* atmosphere absolute. (From Auerbach P: *Field guide to wilderness medicine,* ed 4, St. Louis, Mosby, Elsevier, 2013.)

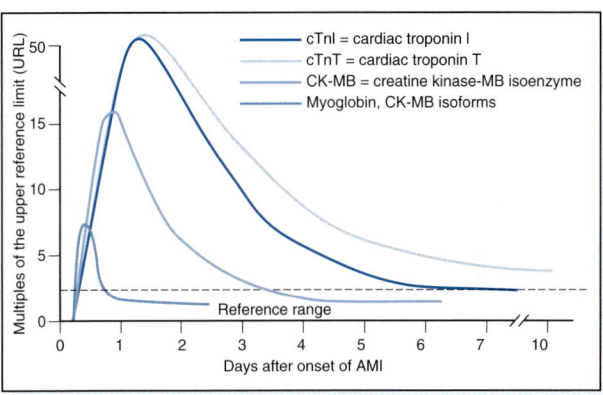

Fig. 2.4 Typical cardiac marker diagnostic window curves and serum levels after acute myocardial infarction (AMI). (From Lehmann CA: *Saunders manual of clinical laboratory science,* Philadelphia, WB Saunders, 1998.)

Elevated in: Colorectal carcinomas, pancreatic carcinomas, and metastatic disease usually produce higher elevations (>20 ng/mL); carcinomas of the esophagus, stomach, small intestine, liver, breast, ovary, lung, and thyroid usually produce lesser elevations; benign conditions (smoking, inflammatory bowel disease, hypothyroidism, cirrhosis, pancreatitis, infections) usually produce levels less than 10 ng/mL.

Cardio-CRP See C-Reactive Protein

Carotene (Serum)

Normal range: 50 to 250 mcg/dL
Elevated in: carotenemia, chronic nephritis, diabetes mellitus, hypothyroidism, nephrotic syndrome, hyperlipidemia
Decreased in: fat malabsorption, steatorrhea, pancreatic insufficiency, lack of carotenoids in diet, high fever, liver disease

Catecholamines, Urine See Urine Catecholamines

CBC See Complete Blood Cell (CBC) Count

CCK See Cholecystokinin-Pancreozymin

CCK-PZ See Cholecystokinin-Pancreozymin

CD4 T-Lymphocyte Count (CD4 T-Cells)

Calculated as total white blood cells × % lymphocytes × % lymphocytes stained with CD4. This test is used primarily to evaluate immune dysfunction in human immunodeficiency virus (HIV) infection and should be measured periodically in all HIV-infected persons. It is useful as a prognostic indicator and as a criterion for initiating prophylaxis for several opportunistic infections that are sequelae of HIV infection. Progressive depletion of CD4 T-lymphocytes is associated with an increased likelihood of clinical complications. Adolescents and adults with HIV are classified as having acquired immune deficiency syndrome (AIDS) if their CD4 lymphocyte count is less than 200/μL or if their CD4 T-lymphocyte percentage is less than 14%. Corticosteroids decrease CD4 T-cell percentage and absolute number.

CD40 Ligand

Normal: <5 mcg/L. CD40 ligand is a soluble protein that is shed from activated leukocytes and platelets and used in risk stratification for acute coronary syndrome.
Elevated in: acute coronary syndrome. Increased CD40 ligand is associated with higher incidence of death or nonfatal myocardial infarction.

CEA See Carcinoembryonic Antigen

Cerebrospinal Fluid (CSF)

Normal range:
    Appearance: clear
    Glucose: 40 to 70 mg/dL
    Protein: 20 to 45 mg/dL
    Chloride: 116 to 122 mEq/L
    Pressure: 100 to 200 mm H_2O
Cell count (cells/mm³) and cell type: <6 lymphocytes, no polymorphonucleocytes
Interpretation of results:
1. Appearance of the fluid:
    a. Clear fluid indicates that results are normal.
    b. Yellow (xanthochromia) in the supernatant of centrifuged CSF within 1 hour or less after collection is usually the result of previous bleeding (subarachnoid hemorrhage); it may also be caused by increased CSF protein, melanin from meningeal melanosarcomas, or carotenoids.

    c. Pinkish color is usually the result of a bloody tap; the color generally clears progressively from tubes 1 to 4 (the supernatant is usually crystal clear in traumatic taps).

    d. Turbidity usually indicates the presence of leukocytes (bleeding introduces approximately 1 WBC/500 red blood cells [RBCs] into the CSF).

2. CSF pressure: Elevated pressure can be seen with meningitis, meningoencephalitis, pseudotumor cerebri, mass lesions, and intracerebral bleeding.

3. Cell count: In adults the CSF is normally free of cells (although up to 5 mononuclear cells/mm^3 is considered normal); the presence of granulocytes is never normal.

    a. Neutrophils: These are seen in bacterial meningitis, early viral meningoencephalitis, and early tuberculosis (TB) meningitis.

    b. Increased lymphocytes are seen in TB meningitis, viral meningoencephalitis, syphilitic meningoencephalitis, and fungal meningitis.

4. Protein: Serum proteins are generally too large to cross the normal blood-CSF barrier; however, increased CSF protein is seen with meningeal inflammation, traumatic tap, increased CNS synthesis, tissue degeneration, obstruction to CSF circulation, and Guillain-Barré syndrome.

5. Glucose:

    a. Decreased glucose is seen with bacterial meningitis, TB meningitis, fungal meningitis, subarachnoid hemorrhage, and some cases of viral meningitis.

    b. A mild increase in CSF glucose can be seen in patients with very elevated serum glucose levels.

    Table 2.7 summarizes characteristic cerebrospinal fluid abnormalities.

Ceruloplasmin (Serum)

Normal range: 20 to 35 mg/dL

Elevated in: pregnancy, estrogen therapy, oral contraceptive use, neoplastic diseases (leukemias, Hodgkin's lymphoma, carcinomas), inflammatory states, systemic lupus erythematosus, primary biliary cirrhosis, rheumatoid arthritis

Decreased in: Wilson's disease (values often <10 mg/dL), nephrotic syndrome, advanced liver disease, malabsorption, total parenteral nutrition, Menkes' syndrome

Chlamydia Group Antibody Serologic Test

Test description: Acute and convalescent sera are drawn 2 to 4 weeks apart. A fourfold increase in titer between acute and convalescent sera is necessary for confirmation. A single titer 1:64 or higher is considered indicative of psittacosis or lymphogranuloma venereum.

Chlamydia Trachomatis Polymerase Chain Reaction (PCR)

Test description: performed on endocervical swab, urine, and intraurethral swab for detection of Chlamydia infection.

Chloride (Serum)

Normal range: 95 to 105 mEq/L

Elevated in: dehydration, sodium loss greater than chloride loss, respiratory alkalosis, excessive infusion of normal saline solution, cystic fibrosis, hyperparathyroidism, renal tubular disease, metabolic acidosis, prolonged diarrhea, acetazolamide administration, diabetes insipidus, ureterosigmoidostomy

Decreased in: vomiting, gastric suction, primary aldosteronism, congestive heart failure, syndrome of inappropriate antidiuretic hormone secretion, Addison's disease, salt-losing nephritis, continuous infusion of D$_5$W, thiazide diuretic administration, diaphoresis, diarrhea, burns, diabetic ketoacidosis

TABLE 2.7   Characteristic Cerebrospinal Fluid Abnormalities

	TURBIDITY AND COLOR	OPENING PRESSURE	WBC COUNT	DIFFERENTIAL CELLS	RBC COUNT	PROTEIN	GLUCOSE
Normal	Clear, colorless	70–180 mm H$_2$O	0–5 cells/μL^3	Mononuclear	0	<60 mg/dL	> serum
Bacterial meningitis	Cloudy, straw colored	↑	↑↑	PMNs	0	↑↑	↓
Viral meningitis	Clear or cloudy, colorless	↑	↑	Lymphocytes	0	↑	Normal
Fungal and tuberculous meningitis	Cloudy, straw colored	↑	↑	Lymphocytes	0	↑↑	↓↓
Viral encephalitis	Clear or cloudy, straw colored	Normal to ↑	↑	Lymphocytes	0 (herpes ↑)	Normal to ↑	Normal
Subarachnoid hemorrhage	Cloudy, pink	↑	↑	PMNs and lymphocytes	↑↑	↑	Normal (early); ↓ (late)
Guillain-Barré syndrome	Clear, yellow	Normal to ↑	0–5 cells/μL^3	Mononuclear	0	↑	Normal

PMN, polymorphonuclear leukocyte; *RBC,* red blood cell; *WBC,* white blood cell.
From Goldman L, Schafer AI: *Goldman's Cecil medicine,* ed 24, Philadelphia, Saunders, 2012.

Chloride (Sweat)

Normal range: 0 to 40 mmol/L
Borderline/indeterminate: 41 to 60 mmol/L
Consistent with cystic fibrosis: >60 mmol/L
False low results can occur with edema, excessive sweating, and hypoproteinemia.

Chloride, Urine See Urine Chloride

Cholecystokinin-Pancreozymin (CCK, CCK-PZ)

Normal: <80 pg/mL
Elevated in: pancreatic disease, celiac disease, gastric ulcer, postgastrectomy state, irritable bowel syndrome, fatty food intolerance

Cholesterol, Low-Density Lipoprotein See Low-Density Lipoprotein Cholesterol

Cholesterol, High-Density Lipoprotein See High-Density Lipoprotein Cholesterol

Cholesterol, Total

Normal: generally <200 mg/dL
Elevated in: primary hypercholesterolemia, biliary obstruction, diabetes mellitus, nephrotic syndrome, hypothyroidism, primary biliary cirrhosis, diet high in cholesterol and total and saturated fat, third trimester of pregnancy, drug therapy (steroids, phenothiazines, oral contraceptives)
Decreased in: use of lipid-lowering agents (statins, niacin, ezetimibe, cholestyramine, colesevelam), starvation, malabsorption, abetalipoproteinemia, hyperthyroidism, hepatic failure, carcinoma, infection, inflammation

Chorionic Gonadotropin (hCG), Human (Serum)

Normal, serum:
    Male: <0.7 IU/L
    Female premenopausal: <0.8 IU/L
    Female postmenopausal: <3.3 IU/L
Elevated in: pregnancy, choriocarcinoma, gestational trophoblastic neoplasia (including molar gestations), placental site trophoblastic tumors. Human antimouse antibodies can produce false serum assay for hCG.
The principal use of this test is to diagnose pregnancy. In pregnancy, the concentration of hCG increases significantly during the initial 6 weeks of pregnancy. Peak values approaching 100,000 IU/L occur 60 to 70 days after implantation.
hCG levels generally double every 1 to 3 days. In patients with concentration less than 2000 IU/L, an increase of serum hCG less than 66% after 2 days is suggestive of spontaneous abortion or ruptured ectopic gestation.

Chymotrypsin

Normal: <10 mc/L
Elevated in: acute pancreatitis, chronic renal failure, oral enzyme preparations, gastric cancer, pancreatic cancer
Decreased in: chronic pancreatitis, late cystic fibrosis

Circulating Anticoagulant (Antiphospholipid Antibody, Lupus Anticoagulant)

Normal: negative

Detected in: systemic lupus erythematosus, drug-induced lupus, long-term phenothiazine therapy, multiple myeloma, ulcerative colitis, rheumatoid arthritis, postpartum, hemophilia, neoplasms, chronic inflammatory states, acquired immune deficiency syndrome, nephrotic syndrome

Note: The name is a misnomer because these patients are prone to hypercoagulability and thrombosis.

CK See Creatine Kinase

Clonidine Suppression Test

Interpretation: Clonidine inhibits neurogenic catecholamine release and will cause a decrease in plasma norepinephrine into the reference interval in hypertensive subjects without pheochromocytoma. The test is performed by giving 4.3 mg clonidine/kg orally after overnight fast. Norepinephrine is measured at 3 hours. Result should be within established reference range and decrease to less than 50% of baseline concentration. Lack of decrease in norepinephrine is suggestive of pheochromocytoma.

Clostridium difficile Toxin Assay (Stool)

Normal: negative

Detected in: antibiotic-associated diarrhea (*Clostridium difficile* infection [CDI]) and pseudomembranous colitis

CO See Carboxyhemoglobin

Coagulation Factors

Factor reference ranges:
  V: >10%
  VII: >10%
  VIII: 50% to 170%
  IX: 60% to 136%
  X: >10%
  XI: 50% to 150%
  XII: >30%

Fig. 2.5 illustrates the blood coagulation pathways.

Cold Agglutinins Titer

Normal: <1:32

Elevated in: primary atypical pneumonia (Mycoplasma pneumonia), infectious mononucleosis, cytomegalovirus infection, others (hepatic cirrhosis, acquired hemolytic anemia, frostbite, multiple myeloma, lymphoma, malaria)

Complement (C3, C4)

Normal range: C3: 70 to 160 mg/dL; C4: 20 to 40 mg/dL

Abnormal values:

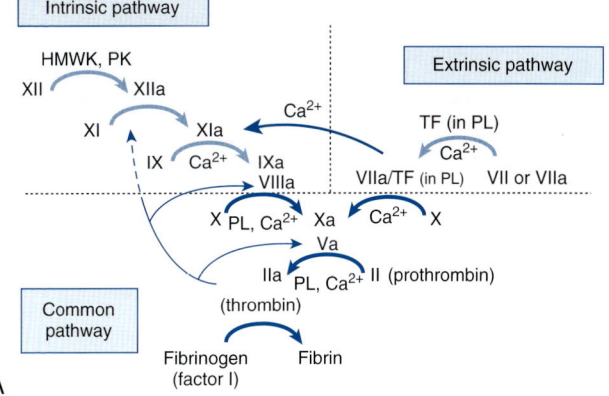

A

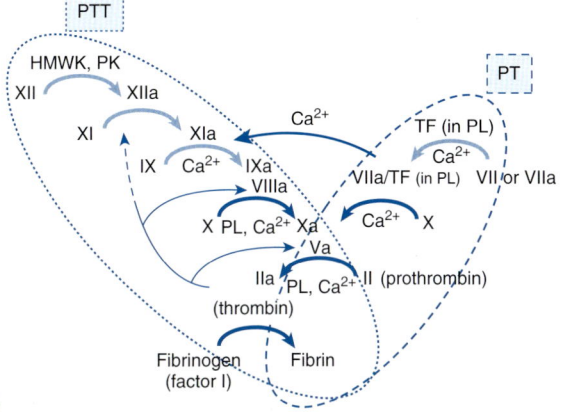

B

Fig. 2.5 Simplified coagulation cascade. A, The coagulation cascade has historically been divided into two main pathways—the intrinsic and extrinsic pathways—both of which culminate in the formation of fibrin through the common pathway. It is now believed that factor IX activation by the factor VIIa/tissue factor (TF) complex plays a major role in the initiation of normal hemostasis. Once coagulation is activated, factor Xa binds to the tissue factor pathway inhibitor, which then effectively inhibits factor VIIa/TF. The factor VIIIa/IXa complex becomes the dominant generator of factor Xa and thus thrombin and fibrin formation. This model is consistent with the observation that deficiencies in factors VIII, IX, and, to a lesser extent, XI cause a bleeding diathesis, whereas the absence of factor XII, prekallikrein (PK), or high-molecular-weight kininogen (HMWK) does not. In the intrinsic pathway, factor XIIa activates prekallikrein into kallikrein. Kallikrein then activates more factor XIIa from factor XII. HMWK acts as a cofactor in both these reactions. HMWK also acts as a cofactor in the activation of factor XI by factor XII. Kallikrein releases bradykinin from HMWK, which has vasoactive activities. B, The activated partial thromboplastin time measures the clotting time from factor XII through fibrin formation. The prothrombin time (PT) measures the clotting time from factor VII through fibrin formation. TF is a transmembrane protein; thus, it is associated with phospholipid in vivo. Ca^{2+}, calcium; *PL*, phospholipid, *PTT*, partial thromboplastin time. (From Henry JB, Davey FR, Herman CJ, et al, editors: *Clinical diagnosis and management by laboratory methods*, ed 20, Philadelphia, Saunders, 2001.)

Decreased C3: active systemic lupus erythematosus (SLE), immune complex disease, acute glomerulonephritis, inborn C3 deficiency, membranoproliferative glomerulonephritis, infective endocarditis, serum sickness, autoimmune or chronic active hepatitis

Decreased C4: immune complex disease, active SLE, infective endocarditis, inborn C4 deficiency, hereditary angioedema, hypergammaglobulinemic states, cryoglobulinemic vasculitis

Complete Blood Cell (CBC) Count

White blood cells: 3200 to 9800/mm^3
Red blood cells: 4.3 to 5.9 10^6/mm^3 (male); 3.5 to 5 10^6/mm^3 (female)
Hemoglobin: 13.6 to 17.7 g/dL (male); 12 to 15 g/dL (female)
Hematocrit: 39% to 49% (male); 33% to 43% (female)
Mean corpuscular volume: 76 to 100 μm^3
Mean corpuscular hemoglobin: 27 to 33 pg
Mean corpuscular hemoglobin concentration: 33 to 37 g/dL
Red cell distribution width: 11.5% to 14.5%
Platelet count: 130 to 400 × 10^3/mm^3
Differential: 2 to 6 bands (early mature neutrophils); 60 to 70 segs (mature neutrophils); 1 to 4 eosinophils; 0 to 1 basophils; 2 to 8 monocytes; 25 to 40 lymphocytes

Conjugated Bilirubin See Bilirubin, Direct

Coombs, Direct See Direct Antiglobulin Test

Coombs, Indirect (Antiglobulin, Indirect)

Normal: negative
Positive: (Fig. 2.6) acquired hemolytic anemia, incompatible cross-matched blood, anti-Rh antibodies, drug therapy (methyldopa, mefenamic acid, levodopa)

Copper (Serum)

Normal range: 70 to 140 mg/dL
Elevated in: aplastic anemia, biliary cirrhosis, systemic lupus erythematosus, hemochromatosis, hyperthyroidism, hypothyroidism, infection, iron deficiency anemia, leukemia, lymphoma, oral contraceptive use, pernicious anemia, rheumatoid arthritis
Decreased in: Wilson's disease, malabsorption, malnutrition, nephrosis, total parenteral nutrition, acute leukemia in remission

Copper, Urine See Urine Copper

Corticotropin-Releasing Hormone (CRH) Stimulation Test

Normal: A dose of 0.5 mg of dexamethasone is given every 6 hours for 2 days; 2 hours after last dose, 1 mcg/kg CRH is given intravenously. Samples are drawn after 15 minutes. Normally there is a twofold to fourfold increase in mean baseline concentration of adrenocorticotropic hormone or cortisol. Cortisol greater than 1.4 mcg/L is virtually 100% specific and 100% diagnostic.
Interpretation: normal or exaggerated response: pituitary Cushing's disease
No response: ectopic ACTH-secreting tumor

INDIRECT COOMBS TEST

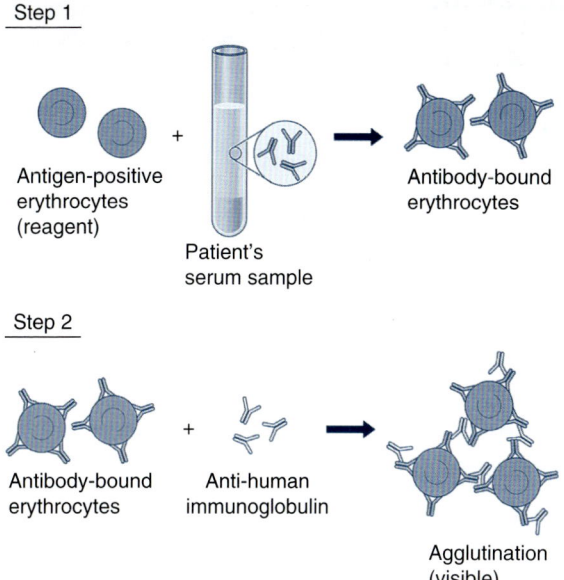

Fig. 2.6 Positive indirect Coombs test (indirect antiglobulin). In step one, reagent red blood cells coated with antigen are added to the patient's serum, which contains antibody. In the presence of antigen-antibody specificity, the antibody from the patient's serum coats the reagent red blood cells (in vitro); this does not result in visible agglutination. In step two, reagent antihuman immunoglobulin is added to the antibody-bound reagent red blood cells. The reagent antihuman immunoglobulin attaches to the antibodies that are coating the reagent red blood cells, causing visible agglutination. (From Young NS, Gerson SL, High KA: *Clinical hematology*, St. Louis, Mosby, 2006.)

A positive response to CRH or a suppressed response to high-dose dexamethasone has a 97% positive predictive value for Cushing's disease. However, a lack of response to either test excludes Cushing's disease in only 64% to 78% of patients. When the tests are considered together, negative responses from both have a 100% predictive value for ectopic ACTH secretion.

Cortisol (Plasma)

Normal range: varies with time of collection (circadian variation):
  8 AM: 4 to 19 mcg/dL
  4 PM: 2 to 15 mcg/dL
Elevated in: ectopic adrenocorticotropic hormone production (i.e., oat cell carcinoma of lung), loss of normal diurnal variation, pregnancy, chronic renal failure, iatrogenic, stress, adrenal or pituitary hyperplasia or adenomas
Decreased in: primary adrenocortical insufficiency, anterior pituitary hypofunction, secondary adrenocortical insufficiency, adrenogenital syndromes

TABLE 2.8	C-Reactive Protein and Cardiac Risk
CARDIO-CRP LEVEL (mg/L)	**RISK**
<0.6	Lowest risk
0.7–1.1	Low risk
1.2–1.9	Moderate risk
2.0–3.8	High risk
3.9–4.9	Highest risk
≥5.0	Results may be confounded by acute inflammatory disease. If clinically indicated, a repeat test should be performed in 2 or more weeks.

CRP, C-reactive protein.

C-Peptide

Normal range (serum): 0.51 to 2.70 ng/mL
Elevated in: insulinoma, sulfonylurea administration, type 2 diabetes mellitus (DM), renal failure
Decreased in: type 1 DM, factitious insulin administration

CPK See Creatine Kinase

C-Reactive Protein (CRP)

Normal: <1 mg/dL. CRP levels are valuable in the clinical assessment of chronic in-flammatory disorders such as rheumatoid arthritis, systemic lupus erythematosus, vasculitis syndromes, and inflammatory bowel disease.
Elevated in: inflammatory and neoplastic diseases, myocardial infarction, third trimester of pregnancy (acute-phase reactant), oral contraceptive use. Moderately high CRP concentrations (3–10 mg/L) predict increased risk of myocardial infarction and stroke. Markedly high levels (>10 mg/L) have been shown to predict cardiovascular risk.
Note: High-sensitivity C-reactive protein (hs-CRP, cardio-CRP) is used as a cardiac risk marker. It is increased in patients with silent atherosclerosis for a prolonged period before a cardiovascular event and is independent of cholesterol level and other lipoproteins. It can be used to help stratify cardiac risk (Table 2.8).

Creatinine Clearance

Normal range: 75 to 124 mL/min
Elevated in: pregnancy, exercise
Decreased in: renal insufficiency, drug therapy (e.g., cimetidine, procainamide, antibiotics, quinidine)
   Fig. 2.7 illustrates the relationship between creatinine clearance and serum creatinine.

Creatine Kinase (CK), Creatine Phosphokinase (CPK)

Normal range: 0 to 130 U/L
Elevated in: vigorous exercise, intramuscular injections, myocardial infarction, myocarditis, rhabdomyolysis, myositis, crush injury or trauma, polymyositis, der-matomyositis, muscular dystrophy, myxedema, seizures, malignant hyperthermia syndrome, cerebrovascular accident, pulmonary embolism and infarction, acute dissection of aorta

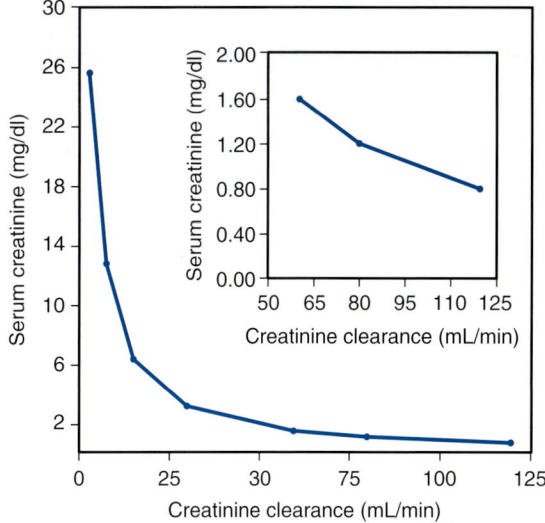

Fig. 2.7 Relationship between creatinine clearance and serum creatinine. In steady state, serum creatinine should increase twofold for each 50% reduction in creatinine clearance. Inset represents enlarged view of changes in serum creatinine as creatinine clearance decreases from 120 to 60 mL/min. If serum creatinine is 0.8 mg/dL when creatinine clearance is 120 mL/min, creatinine clearance can decrease by 33% such that increased serum creatinine is still within normal range. (From Vincent JL, Abraham E, Moore FA, et al: *Textbook of critical care*, ed 6, Philadelphia, Saunders, 2011.)

Decreased in: steroids, decreased muscle mass, connective tissue disorders, alcoholic liver disease, metastatic neoplasms

Creatine Kinase Isoenzymes

CK-MB
Elevated in: myocardial infarction (MI), myocarditis, pericarditis, muscular dystrophy, cardiac defibrillation, cardiac surgery, extensive rhabdomyolysis, strenuous exercise (e.g., marathon runners), mixed connective tissue disease, cardiomyopathy, hypothermia

Note: CK-MB exists in the blood in two subforms; MB_2 is released from cardiac cells and converted in the blood in MB_1. Rapid assay of CK-MB subforms can detect MI ($CK\text{-}MB_2 \geq 1$ U/L, with a ratio of $CK\text{-}MB_2/CK\text{-}MB_1 \geq 1.5$) within the first 6 hours of onset of symptoms.

CK-MM
Elevated in: crush injury, seizures, malignant hyperthermia syndrome, rhabdomyolysis, myositis, polymyositis, dermatomyositis, vigorous exercise, muscular dystrophy, intramuscular injections, acute dissection of aorta

CK-BB

Elevated in: cerebrovascular accident, subarachnoid hemorrhage, neoplasms (prostate, gastrointestinal tract, brain, ovary, breast, lung), severe shock, bowel infarction, hypothermia, meningitis

Creatinine (Serum)

Normal range: 0.6 to 1.2 mg/dL

Elevated in: renal insufficiency (acute and chronic), decreased renal perfusion (hypotension, dehydration, congestive heart failure), rhabdomyolysis, administration of contrast dyes, ketonemia, drug therapy (antibiotics [aminoglycosides, cephalosporins], angiotensin-converting enzyme inhibitors [in patients with renal artery stenosis], diuretics)

Falsely elevated in: diabetic ketoacidosis, administration of some cephalosporins (e.g., cefoxitin, cephalothin)

Decreased in: decreased muscle mass (including amputees and older adults), pregnancy, prolonged debilitation

Creatinine, Urine See Urine Creatinine

Cryoglobulins (Serum)

Normal: not detectable

Present in: collagen-vascular diseases, chronic active hepatitis, chronic lymphocytic leukemia, hemolytic anemias, multiple myeloma, Waldenström's macroglobulinemia, Hodgkin's disease

Cryptosporidium Antigen by Enzyme Immunoassay (EIA) (Stool)

Normal: not detected
Present in: cryptosporidiosis

CSF See Cerebrospinal Fluid

Cystatin C

Normal: Cystatin C is a cysteine protease inhibitor that is produced at a constant rate by all nucleated cells. It is freely filtered by the glomerulus and reabsorbed (but not secreted) by the renal tubules with no extrarenal excretion. Its concentration is not affected by diet, muscle mass, or acute inflammation. Normal range when measured by particle-enhanced nephelometric immunoassay is <0.28 mg/L.

Elevated in: renal disorders. Good predictor of the severity of acute tubular necrosis. Cystatin C increases more rapidly than creatinine in the early stages of glomerular filtration rate impairment. The cystatin C concentration is an independent risk factor for heart failure in older adults and appears to provide a better measure of risk assessment than the serum creatinine concentration.

Cystic Fibrosis Polymerase Chain Reaction (PCR)

Test description: Can be performed on whole blood or tissue. Common mutations in the cystic fibrosis transmembrane regulator gene can be used to detect 75% to 80% of mutant alleles.

Cytomegalovirus by Polymerase Chain Reaction (PCR)

Test description: Can be performed on whole blood, plasma, or tissue. Qualitative PCR is highly sensitive but may not be able to differentiate between latent and active infection.

DAT See Direct Antiglobulin Test

D-Dimer

Normal: <0.5 mcg/mL
Elevated in: deep-vein thrombosis (DVT), pulmonary embolism, high levels of rheumatoid factor, activation of coagulation and fibrolytic system from any cause

D-Dimer assay by enzyme-linked immunosorbent assay assists in the diagnosis of DVT and pulmonary embolism. This test has significant limitations because it can be elevated whenever the coagulation and fibrinolytic systems are activated and can also be falsely elevated with high rheumatoid factor levels.

Dehydroepiandrosterone Sulfate

Normal range:
  Males: age 19 to 30: 125 to 619 mg/dL
    age 31 to 50: 59 to 452 mg/dL
    age 51 to 60: 20 to 413 mg/dL
    age 61 to 83: 10 to 285 mg/dL
  Females: age 19 to 30: 29 to 781 mg/dL
    age 31 to 50: 12 to 379 mg/dL
    Postmenopausal: 30 to 260 mg/dL
Elevated in: hirsutism, congenital adrenal hyperplasia, adrenal carcinomas, adrenal adenomas, polycystic ovarian syndrome, ectopic adrenocorticotropic hormone–producing tumors, Cushing's disease, spironolactone

Deoxycorticosterone (11-Deoxycorticosterone, DOC), Serum

Normal range: 2 to 19 ng/dL. Normal secretion depends on adrenocorticotropic hormone and is suppressible by dexamethasone.
Elevated in: androgenital syndromes caused by 17- and 11-hydroxylase deficiencies, pregnancy
Decreased in: preeclampsia

Dexamethasone Suppression Test, Overnight

Normal: This test is performed by giving 1 mg dexamethasone orally at 11 PM and measuring serum cortisol at 8 AM on the following morning. Normal response is cortisol suppression to less than 3 mcg/dL. If a dose of 4 mg dexamethasone is given, cortisol suppression will be less than 50% of baseline.
Interpretation: Cushing's syndrome (>10 mcg/dL), endogenous depression (half of patients suppress test values >5 mcg/dL). Most patients with pituitary Cushing's disease demonstrate suppression, whereas patients with adrenal adenoma, carcinoma, and ectopic adrenocorticotropic hormone–producing tumors do not.

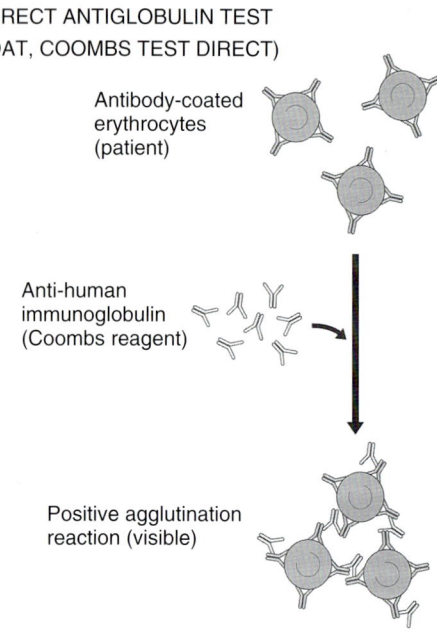

DIRECT ANTIGLOBULIN TEST
(DAT, COOMBS TEST DIRECT)

Antibody-coated
erythrocytes
(patient)

Anti-human
immunoglobulin
(Coombs reagent)

Positive agglutination
reaction (visible)

Fig. 2.8 Positive direct Coombs test (direct antiglobulin test). Antihuman immunoglobulin (reagent) is added to the patient's red blood cells, which have been coated with antibody (in vivo). The reagent antihuman immunoglobulin attaches to the antibodies coating the patient's red blood cells, causing visible agglutination. (From Young NS, Gerson SL, High KA: *Clinical hematology,* St. Louis, Mosby, 2006.)

Dihydrotestosterone, Serum, Urine

Normal range:
  Serum:
    Males: 30 to 85 ng/dL
    Females: 4 to 22 ng/dL
  Urine, 24 h:
    Males: 20 to 50 mcg/day
    Females: <8 mcg/day
Elevated in: hirsutism
Decreased in: 5-alpha-reductase deficiency, hypogonadism

Direct Antiglobulin Test (DAT, Coombs Test Direct)

Normal: negative
Positive: autoimmune hemolytic anemia, erythroblastosis fetalis, transfusion reactions, drug therapy (methyldopa, penicillins, tetracycline, sulfonamides, levodopa, cephalosporins, quinidine, insulin) (see Fig. 2.8)
False positive: may be seen with cold agglutinins

Disaccharide Absorption Tests

Normal: This test is used to diagnose malabsorption caused by disaccharide deficiency. It is performed by giving disaccharide orally 1 g/kg body weight to a total of 25 g. Blood is drawn at 0, 30, 60, 90, and 120 minutes. Normal response is a change in glucose from fasting value more than 30 mg/dL; results are inconclusive when the increase is 20 to 30 mg/dL and abnormal when the increase is less than 20 mg/dL. The test can also be performed by measuring air at 0, 30, 60, 90, and 120 minutes. Normal is H_2 more than 20 ppm above baseline level before a colonic response.

Decreased in: disaccharide deficiency (lactose, fructose, sorbitol), celiac disease, sprue, acute gastroenteritis

DOC See Deoxycorticosterone

Donath-Landsteiner (D-L) Test for Paroxysmal Cold Hemoglobinuria

Normal: no hemolysis

Interpretation: hemolysis indicates presence of bithermic cold hemolysins or Donath-Landsteiner antibodies (D-L Abs).

Digoxin (Lanoxin)

Normal therapeutic range: 0.5 to 2 ng/mL

Elevated in: impaired renal function; excessive dosing; concomitant use of quinidine, amiodarone, verapamil, fluoxetine, nifedipine

Dilantin See Phenytoin

Dopamine

Normal range: 0 to 175 pg/mL

Elevated in: pheochromocytomas, neuroblastomas, stress, vigorous exercise, ingestion of certain foods (bananas, chocolate, coffee, tea, vanilla)

d-Xylose Absorption Test

Normal:
   Urine: ≥4 g/5 h (5-hour urine collection in adults ≥12 years [25-g dose])
   Serum: ≥25 mg/dL (adult, 1 h, 25-g dose, normal renal function)
   Normal results: In patients with malabsorption, normal results suggest pancreatic disease as a cause of the malabsorption.
   Abnormal results: celiac disease, Crohn's disease, tropical sprue, surgical bowel resection, acquired immune deficiency syndrome. False positives can occur with decreased renal function, dehydration/hypovolemia, surgical blind loops, decreased gastric emptying, and vomiting.

Electrolytes, Urine See specific Urine Electrolytes (e.g., urine potassium, urine sodium, etc.)

Electrophoresis, Hemoglobin See Hemoglobin Electrophoresis

Electrophoresis, Protein See Protein Electrophoresis

ENA Complex See Extractable Nuclear Antigen

Endomysial Antibodies

Normal: not detected
Present in: celiac disease, dermatitis herpetiformis

Eosinophil Count

Normal range: 1% to 4% eosinophils (0–440/mm³)
Elevated in: allergy, parasitic infestations (trichinosis, aspergillosis, hydatidosis), angioneurotic edema, drug reactions, warfarin sensitivity, collagen-vascular diseases, acute hypereosinophilic syndrome, eosinophilic nonallergic rhinitis, myeloproliferative disorders, Hodgkin's lymphoma, non-Hodgkin's lymphoma, radiation therapy, L-tryptophan ingestion, urticaria, pernicious anemia, pemphigus, inflammatory bowel disease, bronchial asthma

Epinephrine, Plasma

Normal range: 0 to 90 pg/mL
Elevated in: pheochromocytomas, neuroblastomas, stress, vigorous exercise, ingestion of certain foods (bananas, chocolate, coffee, tea, vanilla), hypoglycemia

Epstein-Barr Virus (EBV) Serology

Normal: immunoglobulin (Ig) G anti–viral capsid antigen (VCA) <1:10 or negative
IgM anti-VCA <1:10 or negative
Anti–EBV nuclear antigen (EBNA) <1.5 or negative
Abnormal: IgG anti-VCA >1:10 or positive indicates either current or previous infection
IgM anti-VCA >1:10 or positive indicates current or recent infection
Anti-EBNA ≥1.5 or positive indicates previous infection
 Fig. 2.9 illustrates the pattern of EBV serology during acute infection.

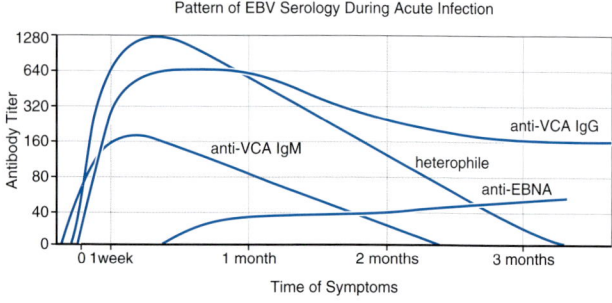

Fɪɢ. 2.9 Patterns of Epstein-Barr virus (EBV) serology during acute infection. (*EBNA,* Epstein-Barr virus nuclear antigen; *Ig,* immunoglobulin; *VCA,* viral capsid antigen.) (From Young NS, Gerson SL, High KA: *Clinical hematology,* St. Louis, Mosby, 2006.)

Erythrocyte Sedimentation Rate (ESR) (Westergren)

Normal range:
   Male: 0 to 15 mm/h

Female: 0 to 20 mm/h

Elevated in: inflammatory states (acute-phase reactant), collagen-vascular diseases, infections, myocardial infarction, neoplasms, hyperthyroidism, hypothyroidism, rouleaux formation, older adults, pregnancy

Note: Sedimentation rates greater than 100 mm/h are strongly associated with serious underlying disease (collagen-vascular, infection, malignancy). Some clinicians use ESR as a "sickness index"; high rates encountered without obvious reason should be repeated rather than pursuing extensive search for occult disease.

Decreased in: sickle cell disease, polycythemia, corticosteroids, spherocytosis, anisocytosis, hypofibrinogenemia, increased serum viscosity, microcytosis

Erythropoietin (EP)

Normal range: 3.7 to 16.0 IU/L by radioimmunoassay

Erythropoietin is a glycoprotein secreted by the kidneys that stimulates RBC production by acting on erythroid committed stem cells.

Elevated in: patients with severe anemia (generally extremely high; hematocrit [Hct] <25, hemoglobin [Hb] <7), such as in cases of aplastic anemia, severe hemolytic anemia, hematologic cancers. Very high in patients with mild to moderate anemia (Hct 25-35, Hb 7-10); high in patients with mild anemia (e.g., acquired immune deficiency syndrome, myelodysplasia).

Erythropoietin can be inappropriately elevated in patients with malignant neoplasms, renal cysts, meningioma, hemangioblastoma, and leiomyoma and after renal transplant.

Decreased in: renal failure, polycythemia vera, autonomic neuropathy

Estradiol (Serum)

Normal range:
   Male, adult: 10 to 50 pg/mL
   Female, premenopausal: 30 to 400 pg/mL, depending on phase of menstrual cycle
   Female, postmenopausal: 0 to 30 pg/mL

Elevated in: tumors of ovary, testis, adrenal glands, or nonendocrine sites (rare)

Decreased in: ovarian failure

Estrogens, Total

Normal range:
   Male: 20 to 80 pg/mL
   Female follicular phase: 60 to 200 pg/mL
   Female luteal phase: 160 to 400 pg/mL
   Female postmenopausal: <130 pg/mL

Elevated in: ovarian tumor–producing estrogens, testicular tumors, tumors or hyperplasia of adrenal cortex, chorioepithelioma

Decreased in: menopause, primary ovarian failure, hypopituitarism, anorexia nervosa, gonadotropin-releasing hormone deficiency, psychogenic stress

Ethanol (Blood)

Normal range:
   Negative (values <10 mg/dL are considered negative)
   Ethanol is metabolized at 10 to 25 mg/dL/hr. Levels 80 mg/dL or higher are considered evidence of impairment for driving. Fatal blood concentration is considered to be more than 400 mg/dL.

Extractable Nuclear Antigen (ENA Complex, Anti-RNP Antibody, Anti-SM, Anti-Smith)

Normal: negative
Present in: systemic lupus erythematosus, rheumatoid arthritis, Sjögren's syndrome, mixed connective tissue disease

Factor V Leiden

Test description: Polymerase chain reaction (PCR) test performed on whole blood or tissue. This single mutation, found in 2% to 8% of the general white population, is the single most common cause of hereditary thrombophilia.

FDP See Fibrin Degradation Product

FENA See Fractional Excretion of Sodium

Fecal FAT, Qualitative See Sudan III Stain

Fecal FAT, Quantitative (72-Hour Collection)

Normal range: 2 to 6 g/24 h
Elevated in: malabsorption syndrome

Fecal Globin Immunochemical Test

Normal: negative. This test is performed by immunochromatography on a cellulose strip impregnated with various antibodies. The test uses a small amount of toilet water as the specimen and is placed onto absorbent pads of a card similar to a traditional occult blood card. There is no direct handling of stool. This test is specific for the globin portion of the hemoglobin molecule, which confers lower gastrointestinal (GI) bleeding specificity. It specifically detects blood from the lower GI tract, whereas guaiac tests are not lower GI–specific. It is more sensitive than a typical Hemoccult test (detection limit 50 mcg Hb/g feces versus >500 mcg Hb/g feces for Hemoccult). It has no dietary restrictions and gives no false positives resulting from plant peroxidases and red meats. It has no medication restrictions. Iron supplements and nonsteroidal antiinflammatory drugs do not cause false positives. Vitamin C does not cause false negatives.
Positive in: lower GI bleeding

Ferritin (Serum)

Normal range: 18 to 300 ng/mL
Elevated in: inflammatory states, liver disease (ferritin elevated from necrotic hepatocytes), hyperthyroidism, neoplasms (neuroblastomas, lymphomas, leukemia, breast carcinoma), iron replacement therapy, hemochromatosis, hemosiderosis
Decreased in: iron deficiency anemia

Fibrin Degradation Product (FDP)

Normal: <10 mcg/mL
Elevated in: disseminated intravascular coagulation, primary fibrinolysis, pulmonary embolism, severe liver disease
Note: The presence of rheumatoid factor may cause falsely elevated FDP.

Fibrinogen

Normal range: 200 to 400 mg/dL
Elevated in: tissue inflammation or damage (acute-phase protein reactant), oral contraceptive use, pregnancy, acute infection, myocardial infarction
Decreased in: disseminated intravascular coagulation, hereditary afibrinogenemia, liver disease, primary or secondary fibrinolysis, cachexia

5-Hydroxyindole-Acetic Acid, Urine See Urine
5-Hydroxyindole-Acetic Acid

5′ Nucleotidase

Normal range: 2 to 16 IU/L
Elevated in: biliary obstruction, metastatic neoplasms to liver, primary biliary cirrhosis, renal failure, pancreatic carcinoma, chronic active hepatitis

Fluorescent Treponemal Antibody See FTA-ABS

Folate (Folic Acid)

Normal:
    Plasma: Low: <3.4 ng/mL
    Normal: >5.4 ng/mL
    Red blood cell: >280 ng/mL
Elevated in: folic acid therapy
Decreased in: folic acid deficiency (inadequate intake, malabsorption), alcoholism, drug therapy (methotrexate, trimethoprim, phenytoin, oral contraceptives, ÿzulfidine), vitamin B_{12} deficiency (defective red cell folate absorption), hemolytic anemia

Follicle-Stimulating Hormone (FSH)

Normal range:
    Male, adult: <22 IU/L
    Female adult, midcycle: <40 IU/L
    Female nonmidcycle: <20 IU/L
    Female postmenopausal: 40 to 160 IU/L
Elevated in: primary hypogonadism, gonadal failure, alcoholism, Klinefelter's syndrome, testicular feminization, anorchia, castration
Decreased in: precocious puberty related to adrenal tumors, congenital adrenal hyperplasia. Normal FSH in an adult nonovulating female is indicative of hypothalamic or pituitary dysfunction.

Fractional Excretion of Sodium (FENA)

Fractional excretion of sodium $(FE_{Na}) = [(U_{Na} \times P_{Cr}) / (P_{Na} \times U_{Cr})] \times 100\%)$.
    Urine indices used in the differential diagnosis of prerenal acute kidney injury and acute tubular necrosis are summarized in Table 2.9.

Free T₄ See T_4, Free

Free Thyroxine Index

Normal range: 1.1 to 4.3

TABLE 2.9	Urine Indices Used in the Differential Diagnosis of Prerenal Acute Kidney Injury and Acute Tubular Necrosis	
DIAGNOSTIC INDEX	**PRERENAL ACUTE KIDNEY INJURY**	**ACUTE TUBULAR NECROSIS**
Fractional excretion of sodium (%)	<1*	>2*
U_{Na} (mmol/L)	<20	>40
Urine creatinine/plasma creatinine ratio	>40	<20
Urine urea nitrogen/plasma urea nitrogen ratio	>8	<3
Urine specific gravity	>1.018	≈1.010
Urine osmolality (mOsm/kg H_2O)	>500	≈300
Plasma BUN/creatinine ratio	>20	<10–15
Renal failure index, $U_{Na}/(U_{Cr}/P_{Cr})$	<1	>1
Urine sediment	Hyaline casts	Muddy-brown granular casts

*Fractional excretion of sodium (FE_{Na}) may be >1% in prerenal acute kidney injury associated diuretic use and/or in the setting of bicarbonaturia or chronic disease; FE_{Na} often <1% in acute tubular necrosis caused by radiocontrast media or rhabdomyolysis. *BUN*, blood urea nitrogen; P_{Cr}, plasma creatinine concentration; U_{Cr}, urine creatinine concentration; U_{Na}, urinary sodium concentration.
From Skorecki K, Chertow GM, Marsden PA, et al: *Brenner and Rector's the kidney*, ed 10, Philadelphia, Elsevier, 2016.

Serum free T_4 directly measures unbound thyroxine. Free T_4 can be measured by equilibrium dialysis (gold standard of free T_4 assays) or by immunometric techniques (influenced by serum levels of lipids, proteins, and certain drugs); the free thyroxine index (FTI) can also be easily calculated by multiplying T_4 times T_3RU and dividing the result by 100; the FTI corrects for any abnormal T_4 values secondary to protein binding:

$$FTI = T_4 \times T_3RU/100$$

Normal values equal 1.1 to 4.3.

FSH See Follicle-Stimulating Hormone

FTA-ABS (Serum)

Normal: nonreactive
Reactive in: syphilis, other treponemal diseases (yaws, pinta, bejel), systemic lupus erythematosus, pregnancy

---

Furosemide Stimulation Test

Normal: This test is performed by giving 60 mg furosemide orally after overnight fast. The patient should be on a normal diet without medications the week before the test. Normal results: renin 1 to 6 ng Al/mL/h.
Elevated in: renovascular hypertension, Bartter's syndrome, high-renin essential hypertension, pheochromocytoma
No response in: primary aldosteronism, low-renin essential hypertension, hyporeninemic hypoaldosteronism

---

Gamma-Glutamyl Transferase (GGT) See γ-Glutamyl Transferase

Gastrin (Serum)

Normal range: 0 to 180 pg/mL
Elevated in: Zollinger-Ellison syndrome (gastrinoma), use of proton pump inhibitors, chronic renal failure, gastric ulcer, chronic atrophic gastritis, pyloric obstruction, malignant neoplasms of the stomach, H_2 blockers, calcium therapy, ulcerative colitis, rheumatoid arthritis

Gastrin Stimulation Test

Normal: Gastrin stimulation test after calcium infusion is performed by giving a calcium infusion (15 mg Ca/kg in 500 mL normal saline over 4 hours). Serum is drawn in fasting state before infusion and at 1, 2, 3, and 4 hours. Normal response is little or no increase over baseline gastrin level.
Elevated in: gastrinoma (gastrin >400 pg/mL), duodenal ulcer (gastrin level increase <400 ng/L)
Decreased in: pernicious anemia, atrophic gastritis

Gliadin Antibodies, Immunoglobulin (Ig) A and IgG

Normal: <25 U, equivocal 20 to 25 U, positive >25 U. This test is useful to monitor compliance with gluten-free diet in patients with celiac disease.
Elevated in: celiac disease with dietary noncompliance

Glomerular Basement Membrane Antibody

Normal: negative
Present in: Goodpasture's syndrome

Glomerular Filtration Rate (GFR)

Normal:
  Age 20 to 29: 116 mL/min/1.73 m^2
  Age 30 to 39: 107 mL/min/1.73 m^2
  Age 40 to 49: 99 mL/min/1.73 m^2
  Age 50 to 59: 93 mL/min/1.73 m^2
  Age 60 to 69: 85 mL/min/1.73 m^2
  Age ≥70: 75 mL/min/1.73 m^2
Decreased in: renal insufficiency, decreased renal blood flow
  Table 2.10 summarizes common equations for estimating glomerular filtration rate.

Glucagon

Normal range: 20 to 100 pg/mL
Elevated in: glucagonoma (900–7800 pg/mL), chronic renal failure, diabetes mellitus, drug therapy (glucocorticoids, insulin, nifedipine, danazol, sympathomimetic amines)
Decreased in: hyperlipoproteinemia (types III, IV), beta blocker use, secretin therapy

Glucose, Fasting

Normal range: 60 to 99 mg/dL
Elevated in: diabetes mellitus, impaired glucose tolerance, stress, infections, myocardial infarction, cerebrovascular accident, Cushing's syndrome, acromegaly, acute pancreatitis, glucagonoma, hemochromatosis, drug therapy (glucocorticoids, diuretics [thiazides, loop diuretics])

TABLE 2.10	Common Equations for Estimating Glomerular Filtration Rate or Creatinine Clearance

Cockcroft-Gault (C_{Cr} • BSA/1.73 m^2)
For men: $C_{Cr} = [(140 - age) • weight (kg)]/S_{Cr} • 72$
For women: $C_{Cr} = ([(140 - age) • weight (kg)]/S_{Cr} • 72) • 0.85$

MDRD (1)
GFR = 170 • $[S_{Cr}]^{-0.999}$ • $[age]^{-0.176}$ • [0.762 if patient is female] • [1.18 if patient is black] • $[BUN]^{-0170}$ • $[Alb]^{0.318}$

MDRD (2)
GFR = 186 • $[S_{Cr}]^{-1.154}$ • $[age]^{-0.203}$ • [0.742 if patient is female] • [1.212 if patient is black]

Jellife (1) (C_{Cr} • BSA/1.73 m^2)
For men: $(98 - [0.8 • (age - 20)])/S_{Cr}$
For women: $(98 - [0.8 • (age - 20)])S_{Cr} • 0.90$

Jellife (2)
For men: $(100/S_{Cr}) - 12$
For women: $(80/S_{Cr}) - 7$

Mawer
For men: weight • $[29.3 - (0.203 • age)] • [1 - (0.03 • S_{Cr})]$
For women: weight • $[25.3 - (0.175 • age)] • [1 - (0.03 • S_{Cr})]$

Bjornsson
For men: $[27 - (0.173 • age)] • weight • 0/S_{Cr}$
For women: $[25 - (0.175 • age)] • weight • 0.07/S_{Cr}$

Gates
For men: $(89.4 • S_{Cr}^{-1.2}) + (55 - age) • (0.447 • S_{Cr}^{-1.1})$
For women: $(89.4 • S_{Cr}^{-1.2}) + (55 - age) • (0.447 • S_{Cr}^{-1.1})$

Salazar-Corcoran
For men: $[137 - age] • [(0.285 • weight) + (12.1 • height^2)]/(51 • S_{Cr})$
For women: $[146 - age] • [(0.287 • weight) + (9.74 • height^2)]/(60 • S_{Cr})$

Alb, albumin; *BSA,* body surface area; *BUN,* blood urea nitrogen; C_{Cr}, creatinine clearance rate; *GFR,* glomerular filtration rate; *MDRD,* modification of diet in renal disease; S_{Cr}, serum creatinine.
From Vincent JL, Abraham E, Moore FA, et al: *Textbook of critical care,* ed 6, Philadelphia, Saunders, 2011.

Decreased in: prolonged fasting, excessive dose of insulin or hypoglycemic agents, insulinoma

Glucose, Postprandial

Normal: <140 mg/dL
Elevated in: diabetes mellitus, impaired glucose tolerance
Decreased in: postgastrointestinal resection, reactive hypoglycemia, hereditary fructose intolerance, galactosemia, leucine sensitivity

Glucose Tolerance Test

Normal values above fasting:
    30 minutes: 30 to 60 mg/dL
    60 minutes: 20 to 50 mg/dL
    120 minutes: 5 to 15 mg/dL
    180 minutes: fasting level or below

Elevated in: impaired glucose tolerance, diabetes mellitus, Cushing's syndrome, acromegaly, pheochromocytoma, gestational diabetes

Glucose-6-Phosphate Dehydrogenase (G_6PD) Screen (Blood)

Normal: G_6PD enzyme activity detected
Elevated in: If a deficiency is detected, quantitation of G_6PD is necessary; a G_6PD screen may be falsely interpreted as "normal" after an episode of hemolysis because most G_6PD-deficient cells have been destroyed.

γ-Glutamyl Transferase (GGT)

Normal range: 0 to 30 U/L
Elevated in: chronic alcoholic liver disease, neoplasms (hepatoma, metastatic disease to the liver, carcinoma of the pancreas), nephrotic syndrome, sepsis, cholestasis, drug therapy (phenytoin, barbiturates)

Glycated (Glycosylated) Hemoglobin (HbA_{1c})

Normal range: 4% to 5.6%
Note: HbA_{1c} greater than 9% correlates with a mean glucose higher than 200 mg/dL. The goal of therapy should be an HbA_{1c} less than 7%.
Elevated in: uncontrolled diabetes mellitus (glycated hemoglobin levels reflect the level of glucose control over the preceding 120 days), lead toxicity, alcoholism, iron deficiency anemia, hypertriglyceridemia
Decreased in: hemolytic anemias; decreased red blood cell survival; pregnancy; acute or chronic blood loss; chronic renal failure; insulinoma; congenital spherocytosis; Hb S, Hb C, Hb D diseases

Glycohemoglobin See Glycated (Glycosylated) Hemoglobin

Growth Hormone

Normal range:
    Male: 1 to 9 ng/mL
    Female: 1 to 16 ng/mL
Elevated in: pituitary gigantism, acromegaly, ectopic growth hormone secretion, cirrhosis, renal failure, anorexia nervosa, stress, exercise, prolonged fasting, drug therapy (amphetamines, beta blockers, insulin, levodopa, metoclopramide, clonidine, vasopressin)
Decreased in: hypopituitarism, pituitary dwarfism, adrenocortical hyperfunction, drug therapy (bromocriptine, corticosteroids, glucose)

Growth Hormone–Releasing Hormone (GHRH)

Normal: <50 pg/mL
Elevated in: acromegaly caused by GHRH secretion by neoplasms

Growth Hormone Suppression Test (After Glucose)

Normal: This test is done by giving 1.75 g glucose/kg orally after overnight fast. Blood is drawn at baseline, after 60 minutes, and after 120 minutes of glucose load. Normal response is growth hormone suppression to less than 2 ng/mL or undetectable levels.
Abnormal: There is no or incomplete suppression from the high basal level in gigantism or acromegaly.

HAM Test (Acid Serum Test)

Normal: negative
Positive in: paroxysmal nocturnal hemoglobinuria
False positive in: hereditary or acquired spherocytosis, recent transfusion with aged red blood cells, aplastic anemia, myeloproliferative syndromes, leukemia, hereditary dyserythropoietic anemia type II

Haptoglobin (Serum)

Normal range: 50 to 220 mg/dL
Elevated in: inflammation (acute-phase reactant), collagen-vascular diseases, infections (acute-phase reactant), drug therapy (androgens), obstructive liver disease
Decreased in: hemolysis (intravascular more than extravascular), megaloblastic anemia, severe liver disease, large tissue hematomas, infectious mononucleosis, drug therapy (oral contraceptives)

HBA₁C See Glycated (Glycosylated) Hemoglobin

HDL See High-Density Lipoprotein Cholesterol

Helicobacter pylori (Serology, Stool Antigen)

Normal range: not detected
Detected in: *H. pylori* infection. Positive serology can indicate current or past infection. Positive stool antigen test indicates acute infection (sensitivity and specificity >90%). Stool testing should be delayed at least 4 weeks after eradication therapy. Table 2.11 describes diagnostic tests for *H. pylori.*

Hematocrit

Normal range:
  Male: 39% to 49%
  Female: 33% to 43%
Elevated in: polycythemia vera, smoking, chronic obstructive pulmonary disease, high altitudes, dehydration, hypovolemia
Decreased in: blood loss (gastrointestinal, genitourinary), repeated phlebotomy, pregnancy, prolonged medical illness, renal failure, trauma with blood loss, recent blood donation, hemolysis

Hemoglobin

Normal range:
  Male: 13.6 to 17.7 g/dL
  Female: 12.0 to 15.0 g/dL
Elevated in: hemoconcentration, dehydration, polycythemia vera, chronic obstructive pulmonary disease, high altitudes, false elevations (hyperlipemic plasma, white blood cells >50,000 mm³), stress
Decreased in: hemorrhage (gastrointestinal, genitourinary), anemia, prolonged medical illness, renal failure

Hemoglobin (Hb) Electrophoresis

Normal range:
  HbA₁: 95% to 98%

TABLE 2.11	Diagnostic Tests for *Helicobacter pylori**		
METHODS	**ADVANTAGES**	**DISADVANTAGES**	**USEFULNESS**
Noninvasive			
Serologic testing	Noninvasive, relatively cheap	Requires validation in local patient	Initial diagnosis, no follow-up after therapy
^{14}C urea breath test	Rapid, allows distinction between current and past infection	Involves ingestion of radioactivity Reduced sensitivity with acid suppression or antibiotics	Initial diagnosis, follow-up of treatment regimens
^{14}C urea breath test	No radioactivity, as for ^{14}C	Complex equipment, expensive Reduced sensitivity with acid suppression or antibiotics	Initial diagnosis, follow-up of treatment regimens
Invasive			
Rapid urease test	Rapid, inexpensive	Invasive Reduced sensitivity in those on acid suppression or with recent or active bleeding	Initial diagnosis
Histologic examination	Allows assessment of mucosa	Invasive, costly	Assess gastritis, metaplasia, atrophy, etc., initial diagnosis
Culture	Specificity: 100%	Invasive, costly, slow, less sensitive	Initial diagnosis, antimicrobial sensitivities, strain typing (macrolide resistance 4%-12%); metronidazole resistance is common

*Testing should be performed only if treatment is planned.
From Talley NJ, Martin CJ: *Clinical gastroenterology*, ed 2, Sidney, Churchill Livingstone, 2006.

HbA$_2$: 1.5% to 3.5%
HbF: <2%
HbC: absent
HbS: absent
    Table 2.12 summarizes the various neonatal hemoglobin electrophoresis patterns.

Hemoglobin, Glycated See Glycated (Glycosylated) Hemoglobin

Hemoglobin, Glycosylated See Glycated (Glycosylated) Hemoglobin

Hemoglobin H

Normal: negative
Present in: hemoglobin H disease, alpha thalassemia trait, unstable hemoglobin disorders

Hemoglobin, Urine See Urine Hemoglobin, Free

TABLE 2.12	Neonatal Hemoglobin Electrophoresis Patterns*
FA	Fetal Hb and adult normal Hb; the normal newborn pattern.
FAV	Indicates the presence of both HbF and HbA. However, an anomalous band (V) is present, which does not appear to be any of the common Hb variants.
FAS	Indicates fetal Hb, adult normal HbA, and HbS, consistent with benign sickle cell trait.
FS	Fetal and sickle HbS without detectable adult normal HbA. Consistent with clinically significant homozygous sickle Hb genotype (S/S) or sickle β-thalassemia, with manifestations of sickle cell anemia during childhood.
FC†	Designates the presence of HbC without adult normal HbA. Consistent with clinically significant homozygous HbC genotype (C/C), resulting in a mild hematologic disorder presenting during childhood.
FSC	HbS and HbC present. This heterozygous condition could lead to the manifestations of sickle cell disease during childhood.
FAC	HbC and adult normal HbA present, consistent with benign HbC trait.
FSA	Heterozygous HbS/β-thalassemia, a clinically significant sickling disorder.
F†	Fetal HbF is present without adult normal HbA. Although this may indicate a delayed appearance of HbA, it is also consistent with homozygous β-thalassemia major or homozygous hereditary persistence of fetal HbF.
FV†	Fetal HbF and an anomalous Hb variant (V) are present.
AF	May indicate prior blood transfusion. Submit another filter paper blood specimen when the infant is 4 mo of age, at which time the transfused blood cells should have been cleared.

NOTE: HbA: $\alpha_2\beta_2$; HbF: $\alpha_2\gamma_2$; HbA$_2$: $\alpha_2\delta_2$.
*Hemoglobin variants are reported in order of decreasing abundance; for example, FA indicates more fetal than adult hemoglobin.
†Repeat blood specimen should be submitted to confirm the original interpretation.
Johns Hopkins Hospital, Arcara KM, Tschudy MM: *The Harriet Lane handbook*, ed 19, Philadelphia, Mosby, 2012.

Hemosiderin, Urine See Urine Hemosiderin

Heparin-Induced Thrombocytopenia Antibodies

Normal: antigen assay: negative <0.45, weak 0.45 to 1, strong >1
Elevated in: heparin-induced thrombocytopenia

Hepatitis A Antibody

Normal: negative
Present in: viral hepatitis A; can be immunoglobulin (Ig) M or IgG (if IgM, acute hepatitis A; if IgG, previous infection with hepatitis A)

Hepatitis B Core Antibody

Normal: negative
Present in: hepatitis B. Anti-HBc assay is the first antibody test to become positive with exposure to hepatitis B virus and persists the longest after resolution of acute infection.
    Fig. 2.10 describes viral antigens and antibodies in hepatitis B infection.

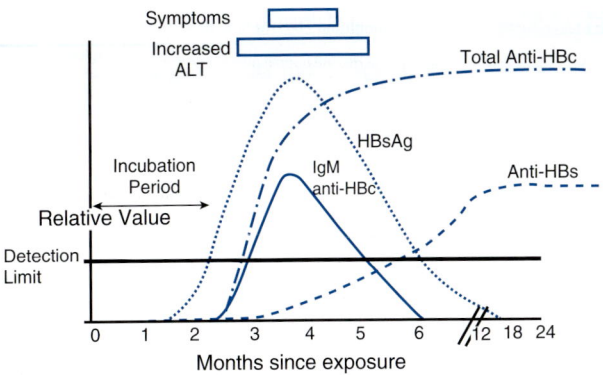

Fig. 2.10 Typical time course for appearance of viral antigens and antiviral antibodies in acute hepatitis B infection. After an incubation period of 1 to 3 months, hepatitis B surface antigen (HBsAg) is the first viral marker to appear. During the incubation period and while no detectable antibody is present, the patient remains asymptomatic. After 3 to 6 months total, antibody to the core antigen (anti-HBc) appears, typically first as an immunoglobulin (Ig) M antibody (IgM anti-HBc). At the time of antibody development, symptoms of acute infection begin, accompanied by increased cytoplasmic enzymes and, in many cases, by jaundice. At the time of development of jaundice, most patients still have measurable HBsAg. In a few patients, neither surface antigen nor its antibody is detectable, leaving IgM anti-HBc as the only marker of acute infection ("core window"). Development of anti-HBc indicates clearance of infectious virus and recovery from infection. IgM anti-HBc persists for approximately 3 to 6 months, but total anti-HBc is typically present for life. (From Henry JB, Davey FR, Herman CJ, et al, editors: *Clinical diagnosis and management by laboratory methods*, ed 20, Philadelphia, Saunders, 2001.)

Hepatitis B DNA

Normal: negative
Present in: active hepatitis B infection. It implies infectivity of the serum. This test is currently used to assess the response of hepatitis B to therapy.

Hepatitis Be Antigen (HbeAg) and Antibody

Normal: negative. These tests are ordered together and should be used only in patients who are chronically HbsAg positive. The main utility of these tests is to assess response of hepatitis B infection to therapy.
Present in: The presence of HbeAg implies that infective hepatitis B virus is present in serum. However, its absence on conversion to anti-Hbe does not rule out infection, especially in persons infected with genotypes other than A. Measurement of HBV-DNA is useful in persons with increased alanine aminotransferase (but negative HbeAg).

Hepatitis B Surface Antibody

Normal: negative
Present in: after vaccination for hepatitis B (a level >10 U/L for postvaccine testing is the accepted concentration that indicates protection), after infection with hepatitis B (it generally appears several weeks after disappearance of HbsAg)

TABLE 2.13	Serologic Markers of Hepatitis B Infection			
	HBsAg	**ANTI-HBc**	**ANTI-HBs**	**IgM ANTI-HBs**
Susceptible to infection	Negative	Negative	Negative	Negative
Immune because of natural infection	Negative	Positive	Positive	Negative
Immune because of hepatitis B vaccination	Negative	Negative	Positive	Negative
Acutely infected	Positive	Positive	Negative	Positive
Chronically infected	Positive	Positive	Negative	Negative

HBc, hepatitis B core; *HBsAg,* hepatitis B surface antigen; *Ig,* immunoglobulin.
From Ballinger A: *Essentials of Kumar and Clark's clinical medicine,* ed 5, Edinburgh, Saunders, 2012.

Hepatitis B Surface Antigen (HbsAg)

Normal: not detected
Detected in: acute viral hepatitis type B, chronic hepatitis B
    Table 2.13 summarizes serologic markers of hepatis B infection.

Hepatitis C Antibody (Anti-HCV)

Normal: negative
Present in: hepatitis C. Centers for Disease Control and Prevention guidelines recommend confirmation with recombinant immunoblot assay (RIBA) before reporting anti-HCV as positive. HCV-RNA can also be obtained if there is a high clinical suspicion of HCV despite a negative anti-HVC, especially in immunosuppressed individuals or in the setting of acute hepatitis. Anti-HCV and the RIBA often do not become positive during an acute infection; thus repeat testing several months later is required if HCV-RNA is negative.
    Fig. 2.11 describes antibody and antigen patterns in hepatitis C infection.

Hepatitis C RNA

Normal: negative
Elevated in: hepatitis C. Detection of hepatitis C RNA is used to confirm current infection and to monitor treatment. Quantitative assays (viral load) are needed before treatment to assess response (<2 log decrease after 12-week treatment indicates lack of response).

Hepatitis D Antigen and Antibody

Normal: negative
Elevated in: hepatitis D. Hepatitis D is a replication-defective RNA virus that requires the surface coat of hepatitis B (hepatitis B surface antigen [HbsAg]) to become an infectious virus. Testing for hepatitis D is therefore done only in patients positive for HbsAg. It is useful in patients with chronic hepatitis B if there is an exacerbation of stable hepatitis.

Her-2/nue

Normal: negative

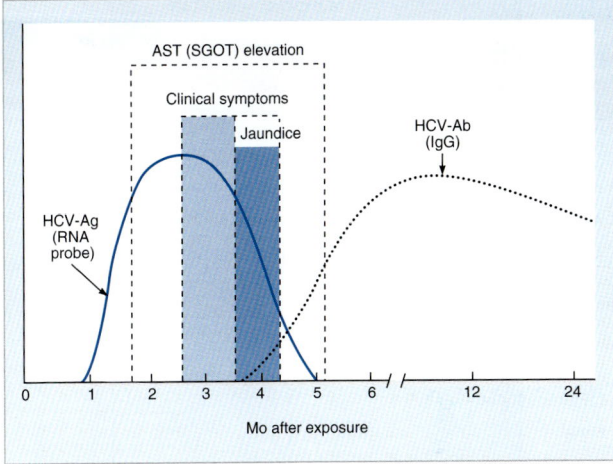

FIG. 2.11 Hepatitis C virus (HCV) antigen and antibody. (*AST,* aspartate aminotransferase; *Ig,* immunoglobulin.) (From Ravel R: *Clinical laboratory medicine,* ed 6, St. Louis, Mosby, 1995.)

Present in: 25% to 30% of primary breast cancers. It can also be found in other epithelial tumors, including lung, hepatocellular, pancreatic, colon, stomach, ovarian, cervical, and bladder cancers. Trastuzumab (Herceptin) is a humanized monoclonal antibody against Her-2/nue. Test is useful to identify patients with metastatic, recurrent, or treatment-refractory unresectable locally advanced breast cancer for trastuzumab treatment.

Herpes Simplex Virus (HSV)

Test description: Polymerase chain reaction test can be performed on serum biopsy samples, cerebrospinal fluid, and vitreous humor.

Table 2.14 describes laboratory diagnosis of herpes virus infections.

Heterophil Antibody

Normal: negative
Positive in: infectious mononucleosis

HFE Screen for Hereditary Hemochromatosis

Test description: Polymerase chain reaction test can be performed on whole blood or tissue. One mutation (C282Y) and two polymorphisms (H63D, S65C) account for the majority of alleles associated with this disease.

High-Density Lipoprotein (HDL) Cholesterol

Normal range:
Male: 45 to 70 mg/dL
Female: 50 to 90 mg/dL

TABLE 2.14 Laboratory Diagnosis of Herpes Virus Infections

VIRUS	DISEASE MANIFESTATION	VIRUS CULTURE	SEROLOGY	ANTIGEN DETECTION	DNA AMPLIFICATION
HSV-1	Skin lesions	+++	+++	+-	+++
	CNS infection				+++
HSV-2	Genital lesions	+++-	++	+-	+
	CNS infection				+++
VZV	Skin lesions	++-	+	++-	+++
	CNS infection		++		+++
CMV	Mononucleosis-like illness	-	+++	-	-
	Neonatal disease	+++	++	-	+++
	Systemic infection in immunocompromised	+	+	++	+++
	CNS disease	-	+	-	+++
EBV	Mononucleosis-like illness	-	+++	-	-
	Systemic infection in immunocompromised	-	+	+	+++
	CNS disease	-	+	-	+++
HHV-6	Exanthema subitum	+-	+++	+-	-
	CNS disease		++		+++
HHV-8	Kaposi's sarcoma	-	+	-	+++

CMV, cytomegalovirus; *CNS,* central nervous system; *EBV,* Epstein-Barr virus; *HHV,* human herpesvirus; *HSV,* herpes simplex virus; *VZV,* varicella zoster virus.
From Cohen J, Powderly WG: *Infectious diseases,* ed 2, St. Louis, Mosby, 2003.

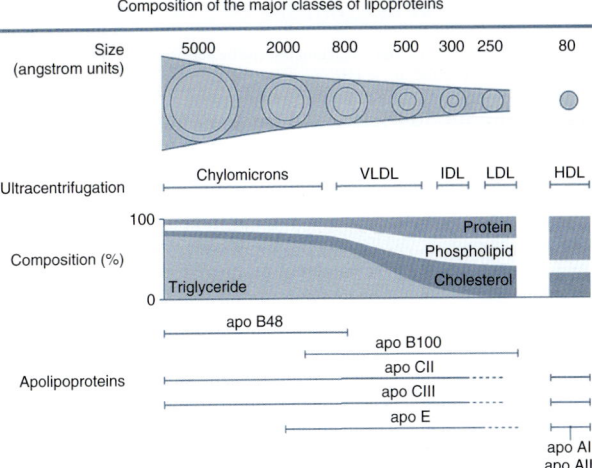

Fɪɢ. 2.12 Composition of the major classes of lipoproteins. Although each of the lipoproteins is distinct with respect to its relative proportions of cholesterol, triglycerides, phospholipids, and apolipoproteins, considerable heterogeneity exists within each lipoprotein class. Lipoproteins are graded according to their density: very-low-density lipoprotein (VLDL), low-density lipoprotein (LDL), intermediate-density lipoprotein (IDL), or high-density lipoprotein (HDL). Clinical measures of LDL include both LDL and IDL. (From Besser GM, Thorner MO: *Comprehensive clinical endocrinology*, ed 3, St. Louis, Mosby, 2002.)

Elevated in: use of fenofibrate, gemfibrozil, nicotinic acid, estrogens; regular aerobic exercise; mild to moderate (1 oz) daily alcohol intake, omega-3 fatty acids

Decreased in: familial deficiency of apoproteins, liver disease, probucol ingestion, sedentary lifestyle, acute myocardial infarction, stroke, starvation

Note: A cholesterol/HDL ratio 4.5 or more is associated with increased risk of coronary artery disease.

Fig. 2.12 describes the composition of the major classes of lipoproteins.

Homocysteine, Plasma

Normal range:
  0 to 30 years: 4.6 to 8.1 μmol/L
  30 to 59 years:
    Men: 6.3 to 11.2 μmol/L
    Women: 4.5 to 7.9 μmol/L
  >59 years: 5.8 to 11.9 μmol/L

Elevated in: thrombophilic states; B_6, B_{12}, folic acid, riboflavin deficiency; pregnancy; homocystinuria

Note: An increased homocysteine level is an independent risk factor for atherosclerosis.

Hs-CRP See C-reactive Protein

HSV See Herpes Simplex Virus

Human Herpes Virus 8 (HHV8)

Test description: Polymerase chain reaction test can be performed on whole blood, tissue, bone marrow, and urine. HHV8 is found in all forms of Kaposi's sarcoma.

Human Immunodeficiency Virus Antibody, Type 1 (HIV-1)

Normal: not detected

Abnormal result: HIV antibodies usually appear in the blood 1 to 4 months after infection.

Testing sequence:

1. Enzyme-linked immunosorbent assay (ELISA) is the recommended initial screening test. Sensitivity and specificity is greater than 99%. False-positive ELISA may occur with autoimmune disorders, administration of immune globulin manufactured before 1985 within 6 weeks of testing, presence of rheumatoid factor, presence of DLA-DR antibodies in multigravida females, administration of influenza vaccine within 3 months of testing, hemodialysis, positive plasma reagin test, and certain medical disorders (hemophilia, hypergammaglobulinemia, alcoholic hepatitis).

2. A positive ELISA is confirmed with Western blot (Fig. 2.13). False-positive Western blot may be caused by connective tissue disorders, human leukocyte antigen antibodies, polyclonal gammopathies, hyperbilirubinemia, presence of antibody to another human retrovirus, or cross-reaction with other non–virus-derived proteins in healthy persons. Undetermined Western blot may occur in patients with acquired immune deficiency syndrome with advanced immunodeficiency (from loss of antibodies) and in recent HIV infections.

3. Polymerase chain reaction is used to confirm indeterminate Western blot results or negative results in persons with suspected HIV infection.

   Fig. 2.14 illustrates the immune response to HIV and relationship to clinical symptoms.

Human Papilloma Virus (HPV)

Test description: Polymerase chain reaction test can be performed on cervical smears, biopsies, scrapings, liquid cytology specimen, and anogenital tissues.

Huntington's Disease Polymerase Chain Reaction (PCR)

Test description: PCR test can be performed on whole blood. Huntington's disease is caused by the expansion of the trinucleotide repeat CAG within IT 15. Genetic counseling should be performed pre and post testing

Immune Complex Assay

Normal: negative

Detected in: collagen-vascular disorders, glomerulonephritis, neoplastic diseases, malaria, primary biliary cirrhosis, chronic acute hepatitis, bacterial endocarditis, vasculitis

Immunoglobulin (Ig)

Normal range:
  IgA: 50 to 350 mg/dL
  IgD: <6 mg/dL

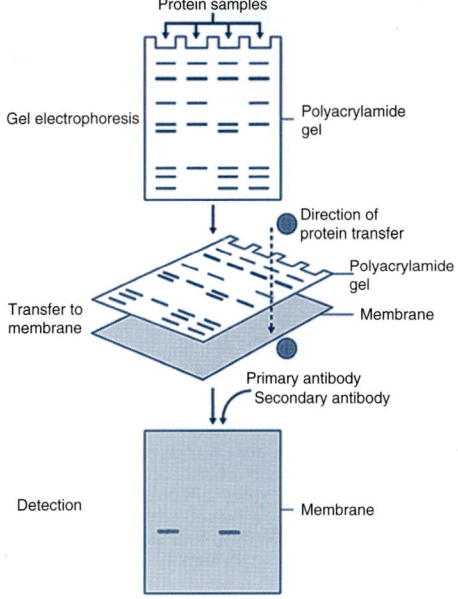

Protein samples

Gel electrophoresis — Polyacrylamide gel

Direction of protein transfer

Polyacrylamide gel

Transfer to membrane — Membrane

Primary antibody
Secondary antibody

Detection — Membrane

Fɪɢ. 2.13 Western blot technique. The solubilized protein mix is separated on a polyacrylamide gel and transferred electrophoretically to a membrane. The membrane is soaked in a buffer-containing antibody. The bound antibody is detected by a chromogenic or chemiluminescent assay. (From Bolognia JL, Jorizzo JL, Rapini RP: *Dermatology,* St. Louis, Mosby, 2003.)

IgE: <25 mg/dL
IgG: 800 to 1500 mg/dL
IgM: 45 to 150 mg/dL

Elevated in:
IgA: lymphoproliferative disorders, Berger's nephropathy, chronic infections, autoimmune disorders, liver disease
IgE: allergic disorders, parasitic infections, immunologic disorders, IgE myeloma, acquired immune deficiency syndrome, pemphigoid
IgG: chronic granulomatous infections, infectious diseases, inflammation, myeloma, liver disease
IgM: primary biliary cirrhosis, infectious diseases (brucellosis, malaria), Waldenström's macroglobulinemia, liver disease

Decreased in:
IgA: nephrotic syndrome, protein-losing enteropathy, congenital deficiency, lymphocytic leukemia, ataxia-telangiectasia, chronic sinopulmonary disease
IgE: hypogammaglobulinemia, neoplasm (breast, bronchial, cervical), ataxia-telangiectasia
IgG: congenital or acquired deficiency, lymphocytic leukemia, phenytoin, methylprednisolone, nephrotic syndrome, protein-losing enteropathy
IgM: congenital deficiency, lymphocytic leukemia, nephrotic syndrome

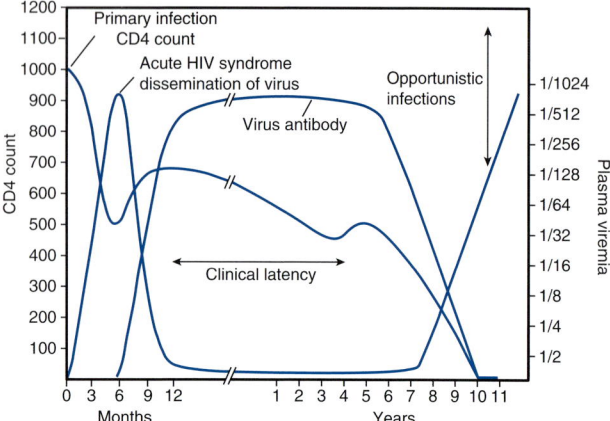

Fig. 2.14 Immune response to human immunodeficiency virus (HIV) and relationship to clinical symptoms and development of acquired immune deficiency syndrome. After primary HIV infection, a period of viremia is followed by development of immunoglobulin (Ig) M and the IgG antibody 4 to 7 weeks later. Development of antibody is accompanied by disappearance of the virus from the circulation. Eventually CD4 counts decrease, viral antibody titers decline, and viremia recurs. (From Young NS, Gerson SL, High KA: *Clinical hematology*, St. Louis, Mosby, 2006.)

Influenza A and B Tests

Test description: Polymerase chain reaction can be performed on nasopharyngeal swab, wash, or aspirate.

INR See International Normalized Ratio

Insulin Autoantibodies

Normal: negative
Present in: exogenous insulin from insulin therapy. The presence of islet cell antibodies indicates ongoing beta cell destruction. This test is useful in the early diagnosis of type 1a diabetes mellitus and in the identification of patients at high risk for type 1a diabetes.

Insulin, Free

Normal: <17 microU/mL
Elevated in: insulin overdose, insulin resistance syndromes, endogenous hyperinsulinemia
Decreased in: inadequately treated type 1 diabetes mellitus

Insulin-Like Growth Factor 1 (IGF-1), Serum

Normal range:
    Age 16 to 24: 182 to 780 ng/mL
    Age 25 to 39: 114 to 492 ng/mL

Age 40 to 54: 90 to 360 ng/mL
Age >55: 71 to 290 ng/mL
Elevated in: adolescence, acromegaly, pregnancy, precocious puberty, obesity
Decreased in: malnutrition, delayed puberty, diabetes mellitus, hypopituitarism, cirrhosis, old age

Insulin-Like Growth Factor II

Normal range: 288 to 736 ng/mL
Elevated in: hypoglycemia associated with non–islet cell tumors, hepatoma, Wilms' tumor
Decreased in: growth hormone deficiency

International Normalized Ratio (INR)

The INR is a comparative rating of prothrombin time (PT) ratios. The INR represents the observed PT ratio adjusted by the International Reference Sensitivity Index: INR = PT patient/PT mean. The INR provides a universal result indicative of what the patient's PT result would have been if measured by using the primary World Health Organization International Reference reagent. For proper interpretation of INR values, the patient should be on stable anticoagulant therapy. Therapeutic INR range is 2 to 3 in most patients.

Box 2.1 summarizes recommended INR ranges.

BOX 2.1	Recommended International Normalized Ratio Ranges
DISORDER	**INR RANGE**
Proximal deep vein thrombosis	2-3
Pulmonary embolism	2-3
Transient ischemic attacks	2-3
Atrial fibrillation	2-3
Mechanical prosthetic valves	2.5-3.5
Recurrent venous thromboembolic disease	2.5-3.5

INR, international normalized ratio.

Intrinsic Factor Antibodies

Normal: negative
Present in: pernicious anemia (>50% of patients). Cyanocobalamin may give false-positive results.

Iron-Binding Capacity (Total Iron-Binding Capacity [TIBC])

Normal range: 250 to 460 mcg/dL
Elevated in: iron deficiency anemia, pregnancy, polycythemia, hepatitis, weight loss
Decreased in: anemia of chronic disease, hemochromatosis, chronic liver disease, hemolytic anemias, malnutrition (protein depletion)

Iron Saturation (% Transferrin Saturation)

Normal range:
Male: 20% to 50%
Female: 15% to 50%

Elevated in: hemochromatosis, excessive iron intake, aplastic anemia, thalassemia, vitamin B^6 deficiency

Decreased in: hypochromic anemias, gastrointestinal malignancy

Iron, Serum

Normal range: Male: 65 to 175 mcg/dL; female: 50 to 1170 mcg/dL

Elevated in: hemochromatosis, excessive iron therapy, repeated transfusions, lead poisoning, hemolytic anemia, aplastic anemia, pernicious anemia

Decreased in: iron deficiency anemia, hypothyroidism, chronic infection, pregnancy uremia

Lactate (Blood)

Normal range: 0.5 to 2.0 mEq/L

Elevated in: tissue hypoxia (shock, respiratory failure, severe congestive heart failure, severe anemia, carbon monoxide or cyanide poisoning), systemic disorders (liver or renal failure, seizures), abnormal intestinal flora (D-lactic acidosis), drugs or toxins (salicylates, ethanol, methanol, ethylene glycol), glucose-6-phosphate dehydrogenase deficiency

Lactate Dehydrogenase (LDH)

Normal range: 50 to 150 U/L

Elevated in: infarction of myocardium, lung, kidney; diseases of cardiopulmonary system, liver, collagen, central nervous system; hemolytic anemias, megaloblastic anemias, transfusions, seizures, muscle trauma, muscular dystrophy, acute pancreatitis, hypotension, shock, infectious mononucleosis, inflammation, neoplasia, intestinal obstruction, hypothyroidism

Lactate Dehydrogenase (LDH) Isoenzymes

Normal range:
  LDH_1: 22% to 36% (cardiac, red blood cells [RBCs])
  LDH_2: 35% to 46% (cardiac, RBCs)
  LDH_3: 13% to 26% (pulmonary)
  LDH_4: 3% to 10% (striated muscle, liver)
  LDH_5: 2% to 9% (striated muscle, liver)

Normal range:
  $LDH_1 < LDH_2$
  $LDH_5 < LDH_4$

Abnormal values:
  $LDH_1 > LDH_2$: myocardial infarction (can also be seen with hemolytic anemias, pernicious anemia, folate deficiency, renal infarct)
  $LDH_5 > LDH_4$: liver disease (cirrhosis, hepatitis, hepatic congestion)

Lactose Tolerance Test (Serum)

Normal: This test is performed by giving 2 g/kg body weight lactose orally and drawing glucose level at 0, 30, 45, 60, and 90 minutes. A normal response is a change in glucose from fasting value to more than 30 mg/dL. An inconclusive response is an increase of 20 to 30 mg/dL; an abnormal response is an increase less than 20 mg/dL.

Elevated in: lactase deficiency

Lanoxin See Digoxin

Lap Score See Leukocyte Alkaline Phosphatase

Lead

Normal: <10 mcg/dL
Elevated in: lead exposure, lead poisoning

LDH See Lactate Dehydrogenase

LDL See Low-Density Lipoprotein Cholesterol

Legionella pneumophila Polymerase Chain Reaction (PCR)

Test description: PCR can be performed on lung tissue, water sputum, bronchoalveolar lavage, and other respiratory fluids. Table 2.15 summarizes specific diagnostic tests for Legionnaire's disease.

TABLE 2.15	Specific Diagnostic Tests for Legionnaires' Disease Caused by *Legionella pneumophila**			
TEST	**SPECIMEN TYPES**	**SENSITIVITY**	**SPECIFICITY**	**NOTES**
Culture	Sputum, other lower respiratory tract secretions; lung; pleural fluid; blood; extrapulmonary tissues, fluids	20%–95%	100%	May be positive up to several days after treatment; requires special media and expertise
Antigenuria	Urine	60%–95%	>99%	Highest sensitivity for *L. pneumophila* serogroup 1, Pontiac type; may remain positive for days to months
Immunofluorescent microscopy	Same as culture	20%–50%	99%	Highest specificity with monoclonal antibody; requires high level of technical expertise
Antibody	Paired serum	20%–70%	95%–99%	Highest specificity for *L. pneumophila* serogroup 1
Molecular amplification	Sputum, other lower respiratory tract secretions; urine	70%–95%	90%–95%	Not well standardized; good performance in some reference laboratories

*Pertains only to *Legionella pneumophila* infections. The yield of diagnostic tests is lower for infection caused by other species, especially those tests based on immunoassay.
From Bennett JE, Dolin R, Blaser MJ: Mandell, Douglas, and Bennett's principles and practice of infectious diseases, ed 8, Philadelphia, Saunders, 2015.

Legionella Titer

Normal: negative
Positive in: Legionnaire's disease (presumptive: ≥1:256 titer; definitive: fourfold titer increase to ≥1:128)

Leukocyte Alkaline Phosphatase (LAP)

Normal range: 13 to 100
Elevated in: leukemoid reactions, neutrophilia resulting from infections (except in sickle cell crisis—no significant increase in LAP score), Hodgkin's disease, polycythemia vera, hairy cell leukemia, aplastic anemia, Down syndrome, myelofibrosis
Decreased in: acute and chronic granulocytic leukemia, thrombocytopenic purpura, paroxysmal nocturnal hemoglobinuria, hypophosphatemia, collagen disorders

LH See Luteinizing Hormone (LH), Blood

Lipase

Normal range: 0 to 160 U/L
Elevated in: acute pancreatitis, perforated peptic ulcer, carcinoma of pancreas (early stage), pancreatic duct obstruction, bowel infarction, intestinal obstruction

Lipoprotein (A)

Normal range:
    Male: 1.35 to 19.6 mg/dL
    Female: 1.24 to 20.1 mg/dL
Elevated in: coronary artery disease, uncontrolled diabetes, hypothyroidism, chronic renal failure, pregnancy, tobacco use, infections, nephritic syndrome
Decreased in: niacin, estrogens, tamoxifen therapy, omega-3 fatty acid use

Lipoprotein Cholesterol, Low Density See Low-Density Lipoprotein Cholesterol

Lipoprotein Cholesterol, High Density See High-Density Lipoprotein Cholesterol

Liver Kidney Microsome Type 1 (LKM1) Antibodies

Normal: <20 U
Elevated in: autoimmune hepatitis type 2

Low-Density Lipoprotein (LDL) Cholesterol

Normal: <130 mg/dL (<70 mg/dL in diabetics and patients with cardiovascular risk factors)
Elevated in: diet high in saturated fat, familial hyperlipidemia, sedentary lifestyle, poorly controlled diabetes mellitus, nephritic syndrome, hypothyroidism
Decreased in: use of lipid-lowering agents (statins, niacin, ezetimibe, cholestyramine, colesevelam), starvation, malabsorption, abetalipoproteinemia, hyperthyroidism, hepatic failure, carcinoma, infection, inflammation

Lupus Anticoagulant (LA) Test

Normal: negative
Present in: antiphospholipid antibody syndrome. False positives may occur with oral anticoagulant therapy, factor deficiency, and specific factor inhibitors.

Luteinizing Hormone (LH), Blood

Normal range:
    Female, adult: Follicular phase: 1 to 18 IU/L
    Midcycle phase: 20 to 80 IU/L
    Luteal phase: 0.5 to 18 IU/L
    Postmenopausal: 12 to 55 IU/L
    Male, adult: 1 to 9 IU/L
Elevated in: gonadal failure, anorchia, menopause, testicular feminization syndrome
Decreased in: primary pituitary or hypothalamic failure

Lymphocytes

Normal range: 15% to 40%
    Total lymphocyte count: 800 to 2600/mm³
    Total T-lymphocytes: 800 to 2200/mm³
    CD4 lymphocytes ≥400/mm³
    CD8 lymphocytes = 200 to 800/mm³
    Normal CD4/CD8 ratio is 2.
Elevated in: chronic infections, infectious mononucleosis and other viral infections, chronic lymphocytic leukemia, Hodgkin's disease, ulcerative colitis, hypoadrenalism, idiopathic thrombocytopenic purpura
Decreased in: human immunodeficiency virus infection, bone marrow suppression from chemotherapeutic agents or chemotherapy, aplastic anemia, neoplasms, steroids, adrenocortical hyperfunction, neurologic disorders (multiple sclerosis, myasthenia gravis, Guillain-Barré syndrome)
CD4 lymphocytes are calculated as total white blood cells × % lymphocytes × % lymphocytes stained with CD4. They are decreased in acquired immune deficiency syndrome and other forms of immune dysfunction.

Magnesium (Serum)

Normal range: 1.8 to 3 mg/dL
Elevated in:
    Renal failure (decreased glomerular filtration rate)
    Decreased renal excretion secondary to salt depletion
    Abuse of antacids and laxatives containing magnesium in patients with renal insufficiency
    Endocrinopathies (deficiency of mineralocorticoid or thyroid hormone)
    Increased tissue breakdown (rhabdomyolysis)
    Redistribution: acute diabetic ketoacidosis (DKA), pheochromocytoma
    Other: lithium therapy, volume depletion, familial hypocalciuric hypercalcemia
Decreased in:
1. Gastrointestinal (GI) and nutritional
    a. Defective GI absorption (malabsorption)
    b. Inadequate dietary intake (e.g., alcoholics)
    c. Parenteral therapy without magnesium
    d. Chronic diarrhea, villous adenoma, prolonged nasogastric suction, fistulas (small bowel, biliary)

2. Excessive renal losses
   a. Diuretic use
   b. Renal tubular acidosis
   c. Diuretic phase of acute tubular necrosis
   d. Endocrine disturbances (DKA, hyperaldosteronism, hyperthyroidism, hyperparathyroidism), syndrome of inappropriate antidiuretic hormone, Bartter's syndrome, hypercalciuria, hypokalemia
   e. Cisplatin therapy; alcohol use; cyclosporine, digoxin, pentamidine, mannitol, amphotericin B, foscarnet, methotrexate therapy
   f. Antibiotic therapy (gentamicin, ticarcillin, carbenicillin)
   g. Redistribution: hypoalbuminemia, cirrhosis, administration of insulin and glucose, theophylline use, epinephrine use, acute pancreatitis, cardiopulmonary bypass
   h. Miscellaneous: sweating, burns, prolonged exercise, lactation, "hungry-bones" syndrome

Mean Corpuscular Volume (MCV)

Normal range: 76 to 100 μm^3
Elevated in: alcohol abuse, reticulocytosis, vitamin B_{12} deficiency, folic acid deficiency, liver disease, hypothyroidism, marrow aplasia, myelofibrosis
Decreased in: iron deficiency, anemia of chronic disease, thalassemia trait or syndrome, other hemoglobinopathies, sideroblastic anemia, chronic renal failure, lead poisoning. Table 2.16 summarizes MCV and red cell distribution width (RBCDW) usefulness in the diagnosis of anemia.

TABLE 2.16	**Usefulness of the Mean Corpuscular Value and Red Blood Cell Distribution Width in the Diagnosis of Anemia**		
	LOW MCV (<80 fL)	**NORMAL MCV (80–99 fL)**	**HIGH MCV (≥100 fL)**
Normal RDW	Anemia of chronic disease α- or β-thalassemia trait	Acute blood loss Anemia of chronic disease	Aplastic anemia Chronic liver disease
	Hemoglobin E trait	Anemia of renal disease	Chemotherapy, antivirals, or alcohol
Elevated RDW	Iron deficiency Sickle cell–β-thalassemia	Early iron, folate, or vitamin B_{12} deficiency Dimorphic anemia (for example, iron + folate deficiency)	Folate or vitamin B_{12} deficiency Immune hemolytic anemia
		Sickle cell anemia	Cytotoxic chemotherapy
		Sickle cell disease	Chronic liver disease
		Chronic liver disease	Myelodysplasia
		Myelodysplasia	

MCW, mean corpuscular value; *RDW,* red blood cell distribution width.
Hoffmann R, Benz EJ, Silberstein LE, et al: *Hematology: basic principles and practice,* ed 6, Philadelphia, Saunders, 2013.

Metanephrines, Urine See Urine Metanephrines

Methylmalonic Acid, Serum

Normal: <0.2 $\mu mol/L$
Elevated in: vitamin B_{12} deficiency, pregnancy, methylmalonic acidemia

Mitochondrial Antibody (Antimitochondrial Antibody [AMA])

Normal: negative
Present in: primary biliary cirrhosis (>90% of patients)

Monocyte Count

Normal range: 2% to 8%
Elevated in: viral diseases, parasites, infections, neoplasms, inflammatory bowel disease, monocytic leukemia, lymphomas, myeloma, sarcoidosis
Decreased in: viral syndrome, glucocorticoid administration, aplastic anemia, lymphocytic leukemia

Mycoplasma pneumoniae Polymerase Chain Reaction (PCR)

Test description: PCR can be performed on sputum, bronchoalveolar lavage, nasopharyngeal and throat swabs, other respiratory fluids, and lung tissue.

Myelin Basic Protein, Cerebrospinal Fluid

Normal: <2.5 ng/mL
Elevated in: multiple sclerosis, central nervous system trauma, stroke, encephalitis

Myoglobin, Urine See Urine Myoglobin

Natriuretic Peptide See B-type Natriuretic Peptide

Neisseria gonorrhoeae Polymerase Chain Reaction (PCR)

Test description: Test can be performed on endocervical swab, urine, and intraurethral swab.

Neutrophil Count

Normal range: 50% to 70%
    Subsets: Bands (early mature neutrophils): 2% to 6%
    Segs (mature neutrophils): 60% to 70%
Elevated in: acute bacterial infections, acute myocardial infarction, stress, neoplasms, myelocytic leukemia
Decreased in: viral infections, aplastic anemias, immunosuppressive drugs, radiation therapy to bone marrow, agranulocytosis, drug therapy (antibiotics, antithyroidals), lymphocytic and monocytic leukemias

Norepinephrine

Normal range: 0 to 600 pg/mL
Elevated in: pheochromocytomas, neuroblastomas, stress, vigorous exercise, certain foods (bananas, chocolate, coffee, tea, vanilla)

Osmolality, Serum

Normal range: 280 to 300 mOsm/kg

It can also be estimated by the following formula: 2([Na] + [K]) + Glucose/18 + BUN/2.8.

Elevated in: dehydration, hypernatremia, diabetes insipidus, uremia, hyperglycemia, mannitol therapy, ingestion of toxins (ethylene glycol, methanol, ethanol), hypercalcemia, diuretic use

Decreased in: syndrome of inappropriate antidiuretic hormone, hyponatremia, overhydration, Addison's disease, hypothyroidism

Osmolality, Urine See Urine Osmolality

Osmotic Fragility Test

Normal: Hemolysis begins at 0.50, w/v (5.0 g/L) and is complete at 0.30, w/v (3.0 g/L) NaCl.

Elevated in: hereditary spherocytosis, hereditary stomatocytosis, spherocytosis associated with acquired immune hemolytic anemia

Decreased in: iron deficiency anemia, thalassemias, liver disease, leptocytosis associated with asplenia

Paracentesis Fluid Testing and evaluation of results:

1. Process the fluid as follows:
   a. Tube 1: lactate dehydrogenase (LDH), glucose, albumin
   b. Tube 2: protein, specific gravity
   c. Tube 3: cell count and differential
   d. Tube 4: save until further notice
2. Draw serum LDH, protein, albumin.
3. Gram stain, acid-fast bacteria stain, bacterial and fungal cultures, amylase, and triglycerides should be ordered only when clearly indicated; bedside inoculation of blood culture bottles with ascitic fluid improves sensitivity in detecting bacterial growth.
4. If malignant ascites is suspected, consider a carcinoembryonic antigen level on the paracentesis fluid and cytologic evaluation.
5. In suspected spontaneous bacterial peritonitis (SBP), the incidence of positive cultures can be increased by injecting 10 to 20 mL of ascitic fluid into blood culture bottles.
6. Peritoneal effusion can be subdivided as exudative or transudative based on its characteristics.
7. The serum-ascites albumin gradient (serum albumin level–ascitic fluid albumin level) correlates directly with portal pressure and can also be used to classify ascites. Patients with gradients 1.1 g/dL or more have portal hypertension, and those with gradients less than 1.1 g/dL do not; the accuracy of this method is greater than 95%.
8. An ascitic fluid polymorphonuclear leukocyte count greater than 500/μL is suggestive of SBP.
9. A blood-ascitic fluid albumin gradient less than 1.1 g/dL is suggestive of malignant ascites.
   Table 2.17 describes characteristics of paracentesis fluid in various disorders.

Parathyroid Hormone

Normal range: serum, intact molecule: 10 to 65 pg/mL; plasma 1 to 5 pmol/L

Elevated in: hyperparathyroidism (primary or secondary), pseudohypoparathyroidism, drug therapy (anticonvulsants, corticosteroids, lithium, isoniazid, rifampin, phosphates), Zollinger-Ellison syndrome, hereditary vitamin D deficiency

Decreased in: hypoparathyroidism, sarcoidosis, drug therapy (cimetidine, beta blockers), hyperthyroidism, hypomagnesemia

TABLE 2.17	**Characteristics of Paracentesis Fluid**					
ETIOLOGIC FACTORS	**COLOR**	**SAAG (g/L)**	**RBC (10⁶/L)**	**WBC (10⁶/L)**	**CYTOLOGIC FINDINGS**	**OTHER**
Cirrhosis	Straw	≥11	Few	<250	—	—
Infected ascites	Straw	≥11	Few	≥250 poly-morphs or ≥500 cells	—	+ culture
Neoplastic	Straw/hem-orrhagic/mucinous	<11	Variable	Variable	Malignant cells	—
Tuberculosis	Clear/tur-bid/hemor-rhagic	<11	Many	>1000 70% lympho-cytes	—	Acid-fast bacilli + culture
Cardiac failure	Straw	≥11		<250	—	—
Pancreatic	Turbid/hemor-rhagic	<11	Variable	Variable	—	Amylase increased
Lymphatic obstruction or disruption	White	<11	0	0	—	Fat glob-ules on staining

RBC, red blood cells; *SAAG,* serum ascites albumin gradient; *WBC,* white blood cells.
From Talley NJ, Martin CJ: *Clinical gastroenterology,* ed 2, Sidney, Churchill Livingstone, 2006.

Parietal Cell Antibodies

Normal: negative
Present in: pernicious anemia (>90%), atrophic gastritis (up to 50%), thyroiditis (30%), Addison's disease, myasthenia gravis, Sjögren's syndrome, type 1 diabetes mellitus

Partial Thromboplastin Time (PTT), Activated Partial Thromboplastin Time (aPTT)

Normal range: 25 to 41 seconds
Elevated in: heparin therapy, coagulation factor deficiency (I, II, V, VIII, IX, X, XI, XII), liver disease, vitamin K deficiency, disseminated intravascular coagulation, circulating anticoagulant, warfarin therapy, specific factor inhibition (penicillin reaction, rheumatoid arthritis), thrombolytic therapy, nephrotic syndrome
Note: This test is useful to evaluate the intrinsic coagulation system.
    Table 2.18 describes clinical peculiarities of coagulation protein screening tests.

Pepsinogen I

Normal range: 124 to 142 ng/mL
Elevated in: Zollinger-Ellison syndrome, duodenal ulcer, acute gastritis
Decreased in: atrophic gastritis, gastric carcinoma, myxedema, pernicious anemia, Addison's disease

TABLE 2.18	Clinical Peculiarities of Coagulation Protein Screening Tests
LONG aPTT, NORMAL OR LONG PT, NO BLEEDING	**NORMAL aPTT, PT, WITH BLEEDING**
Long aPTT only	
Factor XII deficiency	Factor XIII deficiency or inhibitor
Prekallikrein deficiency	α_2-Antiplasmin deficiency or defect
High-molecular-weight kininogen	Plasminogen activator inhibitor deficiency or defect
Lupus anticoagulant	α_1-Antitrypsin Pittsburgh defect
Long aPTT and PT	
Dysfibrinogenemia with	
fibrinopeptide B release defect	
Lupus anticoagulant	

aPTT, activated partial thromboplastin time; *PT,* prothrombin time.
From Hoffman R, Benz EJ, Shattil SJ, et al: *Hematology: basic principles and practice,* ed 5, Philadelphia, Churchill Livingston, 2009.

PFA See Platelet Function Analysis (PFA) 100 Assay

pH, Blood

Normal values:
    Arterial: 7.35 to 7.45
    Venous: 7.32 to 7.42
For abnormal values, refer to Arterial Blood Gases.

Phenobarbital

Normal therapeutic range: 15 to 30 mcg/mL for epilepsy control

Phenytoin (Dilantin)

Normal therapeutic range: 10 to 20 mcg/mL

Phosphatase, Acid See Acid Phosphatase

Phosphatase, Alkaline See Alkaline Phosphatase

Phosphate (Serum)

Normal range: 2.5 to 5 mg/dL
Elevated in:
1. Excessive phosphate administration
    a. Excessive oral intake or intravenous administration
    b. Laxatives containing phosphate (phosphate tablets, phosphate enemas)
2. Decreased renal phosphate excretion
    a. Acute or chronic renal failure
    b. Hypoparathyroidism or pseudohypoparathyroidism

   c. Acromegaly, thyrotoxicosis
   d. Bisphosphonate therapy
   e. Tumor calcinosis
   f. Sickle cell anemia
3. Transcellular shift out of cells
   a. Chemotherapy of lymphoma or leukemia, tumor lysis syndrome, hemolysis
   b. Acidosis
   c. Rhabdomyolysis, malignant hyperthermia
   d. Artifact: in vitro hemolysis
   e. Pseudohyperphosphatemia: hyperlipidemia, paraproteinemia, hyperbilirubinemia

Decreased in:
1. Decreased intake (prolonged starvation, alcoholics), hyperalimentation, intravenous infusion without phosphate
2. Malabsorption
3. Phosphate-binding antacids
4. Renal loss
   a. Renal tubular acidosis
   b. Fanconi's syndrome, vitamin D–resistant rickets
   c. Acute tubular necrosis (diuretic phase)
   d. Hyperparathyroidism (primary or secondary)
   e. Familial hypophosphatemia
   f. Hypokalemia, hypomagnesemia
   g. Acute volume expansion
   h. Glycosuria, idiopathic hypercalciuria
   i. Acetazolamide
5. Transcellular shift into cells
   a. Alcohol withdrawal
   b. Diabetic ketoacidosis (recovery phase)
   c. Glucose-insulin or catecholamine infusion
   d. Anabolic steroids
   e. Total parenteral nutrition
   f. Theophylline overdose
   g. Severe hyperthermia; recovery from hypothermia
   h. "Hungry bones" syndrome

pH, Urine See Urine pH

Plasminogen

Normal: Immunoassay (antigen): <20 mg/dL
Elevated in: infection, trauma, neoplasm, myocardial infarction (acute-phase reactant), pregnancy, bilirubinemia
Decreased in: disseminated intravascular coagulation, severe liver disease, thrombolytic therapy with streptokinase or urokinase, alteplase therapy

Platelet Aggregation

Normal: full aggregation (generally >60%) in response to epinephrine, thrombin, ristocetin, adenosine diphosphate (ADP), collagen
Elevated in: heparin therapy, hemolysis, lipemia, nicotine use, hereditary and acquired disorders of platelet adhesion, activation, aggregation
Decreased in: drug therapy (aspirin, some penicillins, chloroquine, chlorpromazine, clofibrate, captopril), Glanzmann's thrombasthenia, Bernard-Soulier syndrome, Wiskott-Aldrich syndrome, cyclooxygenase deficiency. In von Willebrand's disease there is normal aggregation with ADP, collagen, and epinephrine use but abnormal agglutination with ristocetin use.

Platelet Antibodies

Normal: absent
Present in: idiopathic thrombocytopenic purpura (ITP) (>90% of patients with chronic ITP). Patients with nonimmune thrombocytopenias may have false-positive results.

Platelet Count

Normal range: 130 to 400 × 10^3/mm^3
Elevated in: iron deficiency, posthemorrhage, neoplasms (gastrointestinal tract), chronic myelogenous leukemia, polycythemia vera, myelofibrosis with myeloid metaplasia, infections, postsplenectomy, postpartum, hemophilia, pancreatitis, cirrhosis
Decreased in:
1. Increased destruction
   a. Immunologic
   b. Drug therapy: quinine, quinidine, digitalis, procainamide, thiazide diuretics, sulfonamides, phenytoin, aspirin, penicillin, heparin, gold, meprobamate, sulfa drugs, phenylbutazone, nonsteroidal antiinflammatory drugs, methyldopa, cimetidine, furosemide, isoniazid, cephalosporins, chlorpropamide, organic arsenicals, chloroquine, platelet glycoprotein Iib/IIIa receptor inhibitors, ranitidine, indomethacin, carboplatin, ticlopidine, clopidogrel
   c. Idiopathic thrombocytopenic purpura
   d. Transfusion reaction: transfusion of platelets with plasminogen activator (PLA) in recipients without PLA-1
   e. Fetal-maternal incompatibility
   f. Collagen-vascular diseases (e.g., systemic lupus erythematosus)
   g. Autoimmune hemolytic anemia
   h. Lymphoreticular disorders (e.g., chronic lymphocytic leukemia)
   i. Nonimmunologic
   j. Prosthetic heart valves
   k. Thrombotic thrombocytopenic purpura
   l. Sepsis
   m. Disseminated intravascular coagulation
   n. Hemolytic-uremic syndrome
   o. Giant cavernous hemangioma
2. Decreased production
   a. Abnormal marrow
   b. Marrow infiltration (e.g., leukemia, lymphoma, fibrosis)
   c. Marrow suppression (e.g., chemotherapy, alcohol, radiation)
   d. Hereditary disorders
   e. Wiskott-Aldrich syndrome: X-linked disorder characterized by thrombocytopenia, eczema, and repeated infections
   f. May-Hegglin anomaly: increased megakaryocytes but ineffective thrombopoiesis
   g. Vitamin deficiencies (e.g., vitamin B_{12}, folic acid)
3. Splenic sequestration, hypersplenism
4. Dilutional, as a result of massive transfusion

Platelet Function Analysis (PFA) 100 Assay

Normal: This test is a two-component assay in which blood is aspirated through two capillary tubes, one of which is coated with collagen (COL) and adenosine diphosphate (ADP) and the other with COL and epinephrine (EPI). The test measures the ability of platelets to occlude an aperture in a biologically active membrane treated with COL/ADP and COL/EPI. During the test the platelets adhere to the surface of the tube and cause blood flow to cease. The *closing time* refers to the cessation of blood flow and is reported in conjunction with the hematocrit and platelet count. Hematocrit count must be more than 25% and platelet count less than 50,000/μL for the test to be performed.

COL/ADP: 70 to 120 sec
COL/EP: 75 to 120 sec

Elevated in: acquired platelet dysfunction, von Willebrand's disease, anemia, thrombocytopenia, use of aspirin and nonsteroidal antiinflammatory drugs

Potassium (Serum)

Normal range: 3.5 to 5 mEq/L
Elevated in:
1. Pseudohyperkalemia
   a. Hemolyzed specimen
   b. Severe thrombocytosis (platelet count $>10^6$ mL)
   c. Severe leukocytosis (white blood cell count $>10^5$ mL)
   d. Fist clenching during phlebotomy
2. Excessive potassium intake (often in setting of impaired excretion)
   a. Potassium replacement therapy
   b. High-potassium diet
   c. Salt substitutes with potassium
   d. Potassium salts of antibiotics
3. Decreased renal excretion
   a. Potassium-sparing diuretics (e.g., spironolactone, triamterene, amiloride)
   b. Renal insufficiency
   c. Mineralocorticoid deficiency
   d. Hyporeninemic hypoaldosteronism (diabetes mellitus)
   e. Tubular unresponsiveness to aldosterone (e.g., systemic lupus erythematosus, multiple myeloma, sickle cell disease)
   f. Type 4 renal tubular acidosis (RTA)
   g. Angiotensin-converting enzyme inhibitors
   h. Heparin administration
   i. Nonsteroidal antiinflammatory drugs
   j. Trimethoprim-sulfamethoxazole
   k. Beta blockers
   l. Pentamidine
4. Redistribution (excessive cellular release)
   a. Acidemia (each 0.1 decrease in pH increases the serum potassium by 0.4–0.6 mEq/L). Lactic acidosis and ketoacidosis cause minimal redistribution.
   b. Insulin deficiency
   c. Drug therapy (e.g., succinylcholine, markedly increased digitalis level, arginine, β-adrenergic blockers)
   d. Hypertonicity
   e. Hemolysis
   f. Tissue necrosis, rhabdomyolysis, burns
   g. Hyperkalemic periodic paralysis

Decreased in:
1. Cellular shift (redistribution) and undetermined mechanisms
   a. Alkalosis (each 0.1 increase in pH decreases serum potassium by 0.4–0.6 mEq/L)
   b. Insulin administration
   c. Vitamin B_{12} therapy for megaloblastic anemias, acute leukemias
   d. Hypokalemic periodic paralysis: rare familial disorder manifested by recurrent attacks of flaccid paralysis and hypokalemia
   e. β-adrenergic agonists (e.g., terbutaline), decongestants, bronchodilators, theophylline, caffeine

    f. Barium poisoning, toluene intoxication, verapamil intoxication, chloroquine intoxication

    g. Correction of digoxin intoxication with digoxin antibody fragments (Digibind)

2. Increased renal excretion
   a. Drugs
      i. Diuretics, including carbonic anhydrase inhibitors (e.g., acetazolamide)
      ii. Amphotericin B
      iii. High-dose sodium penicillin, nafcillin, ampicillin, carbenicillin
      iv. Cisplatin
      v. Aminoglycosides
      vi. Corticosteroids, mineralocorticoids
      vii. Foscarnet sodium
   b. RTA: distal (type 1) or proximal (type 2)
   c. Diabetic ketoacidosis, ureteroenterostomy
   d. Magnesium deficiency
   e. Postobstruction diuresis, diuretic phase of acute tubular necrosis
   f. Osmotic diuresis (e.g., mannitol)
   g. Bartter's syndrome: hyperplasia of juxtaglomerular cells leading to increased renin and aldosterone, metabolic alkalosis, hypokalemia, muscle weakness, tetany (seen in young adults)
   h. Increased mineralocorticoid activity (primary or secondary aldosteronism), Cushing's syndrome
   i. Chronic metabolic alkalosis from loss of gastric fluid (increased renal potassium secretion)

3. Gastrointestinal loss
   a. Vomiting, nasogastric suction
   b. Diarrhea
   c. Laxative abuse
   d. Villous adenoma
   e. Fistulas

4. Inadequate dietary intake (e.g., anorexia nervosa)

5. Cutaneous loss (excessive sweating)

6. High dietary sodium intake, excessive use of licorice

Potassium, Urine See Urine Potassium

Procainamide

Normal therapeutic range: 4 to 10 mcg/mL

Progesterone, Serum

Normal range:
    Male: 15 to 70 ng/dL
    Female follicular phase: 15 to 70 ng/dL
    Female luteal phase: 200 to 2500 ng/dL

Elevated in: congenital adrenal hyperplasia, drug therapy (clomiphene, corticosterone, 11-deoxycortisol, dihydroprogesterone), molar pregnancy, lipoid ovarian tumor

Decreased in: primary or secondary hypogonadism, oral contraceptive use, ampicillin therapy, threatened abortion

Prolactin

Normal range: <20 ng/mL

Elevated in: prolactinomas (level >200 highly suggestive), drug therapy (phenothiazines, cimetidine, tricyclic antidepressants, metoclopramide, estrogens, antihypertensives [methyldopa, verapamil], haloperidol), postpartum, stress, hypoglycemia, hypothyroidism

Prostate-Specific Antigen (PSA)

Normal range: 0 to 4 ng/mL. It is important to remember that there is no PSA level below which prostate cancer can be ruled out and no level above which prostate cancer is certain. The individual's PSA is only part of the equation. Other risk factors need to be considered, such as age, race, family history, findings on digital rectal examination, percent free PSA ratio, and PSA velocity (rate of change from prior PSA measurement).

Elevated in: benign prostatic hypertrophy, carcinoma of prostate, postrectal examination, prostate trauma, androgen therapy, prostatitis, urethral instrumentation, drug therapy (finasteride, dutasteride, antiandrogens)

Note: Measurement of "free PSA" is useful to assess the probability of prostate cancer in patients with normal digital rectal examination and total PSA between 4 and 10 ng/mL. In these patients the global risk of prostate cancer is 25% to 40%; however, if the free PSA is greater than 25%, the risk of prostate cancer decreases to 8%, whereas if the free PSA is less than 10%, the risk of cancer increases to 56%. Free PSA is also useful to evaluate the aggressiveness of prostate cancer. A low free PSA percentage generally indicates a high-grade cancer, whereas a high free PSA percentage is generally associated with a slower-growing tumor.

Prostatic Acid Phosphatase

Normal range: 0 to 0.8 U/L

Elevated in: prostate cancer (especially in metastatic prostate cancer), benign prostatic hyperplasia, prostatitis, post–prostate surgery or manipulation, hemolysis, androgen use, clofibrate therapy

Decreased in: ketoconazole use

Protein (Serum)

Normal range: 6 to 8 g/dL

Elevated in: dehydration, sarcoidosis, collagen-vascular diseases, multiple myeloma, Waldenström's macroglobulinemia

Decreased in: malnutrition, cirrhosis, nephrosis, low-protein diet, overhydration, malabsorption, pregnancy, severe burns, neoplasms, chronic diseases

Protein C Assay

Normal range: 70% to 140%

Elevated in: oral contraceptive use, stanozolol therapy

Decreased in: congenital protein C deficiency, warfarin therapy, vitamin K deficiency, renal insufficiency, consumptive coagulopathies

Protein Electrophoresis (Serum)

Fig. 2.15 illustrates clinicopathologic correlations with serum protein electrophoresis.

Normal range:
Albumin: 60% to 75%
Alpha$_1$: 1.7% to 5%
Alpha$_2$: 6.7% to 12.5%
Beta: 8.3% to 16.3%
Gamma: 10.7% to 20%
Albumin: 3.6 to 5.2 g/dL
Alpha$_1$: 0.1 to 0.4 g/dL

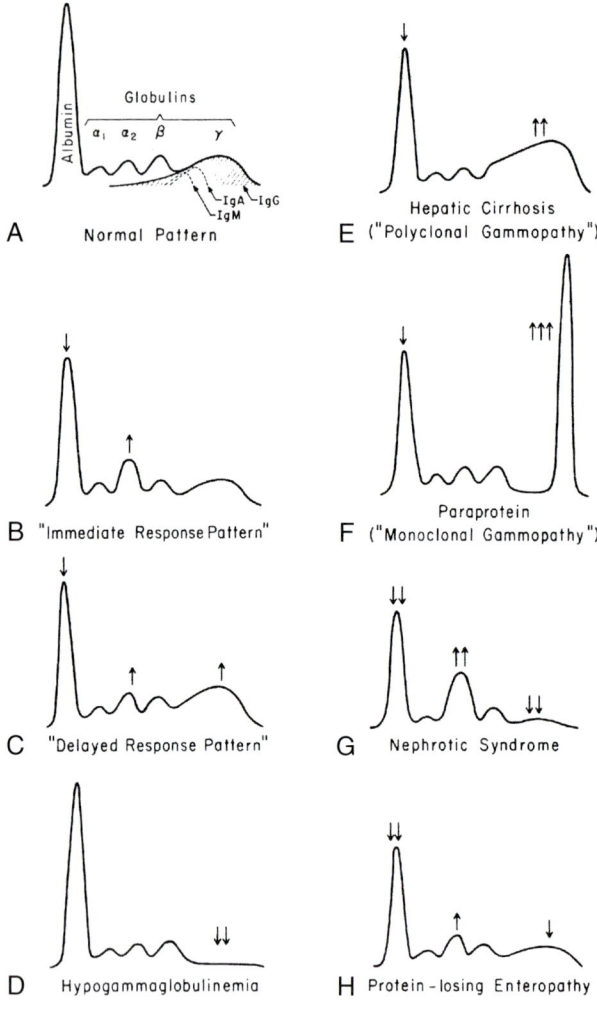

Fıɢ. 2.15 Serum protein electrophoresis: clinicopathologic correlations. (From Henry JB, Davey FR, Herman CJ, et al, editors: *Clinical diagnosis and management by laboratory methods*, ed 20, Philadelphia, Saunders, 2001.)

Alpha$_2$: 0.4 to 1.0 g/dL
Beta: 0.5 to 1.2 g/dL
Gamma: 0.6 to 1.6 g/dL
Elevated in: Albumin: dehydration
Alpha$_1$: neoplastic diseases, inflammation
Alpha$_2$: neoplasms, inflammation, infection, nephrotic syndrome
Beta: hypothyroidism, biliary cirrhosis, diabetes mellitus
Gamma: see Immunoglobulins
Decreased in: Albumin: malnutrition, chronic liver disease, malabsorption, nephrotic syndrome, burns, systemic lupus erythematosus
Alpha$_1$: emphysema (alpha$_1$-antitrypsin deficiency), nephrosis
Alpha$_2$: hemolytic anemias (decreased haptoglobin), severe hepatocellular damage
Beta: hypocholesterolemia, nephrosis
Gamma: see Immunoglobulins

Protein S Assay

Normal range: 65% to 140%
Elevated in: presence of lupus anticoagulant
Decreased in: hereditary deficiency, acute thrombotic events, disseminated intravascular coagulation, surgery, oral contraceptive use, pregnancy, hormone replacement therapy, L-asparaginase treatment

Prothrombin Time (PT)

Normal range: 11 to 13.2 seconds
Note: The prothrombin time is reported as absolute clotting time in seconds and also as a derivative number called the *International Normalized Ratio (INR)*. This ratio is derived from the actual PT of the patient divided by the mean PT of a group of healthy subjects. INR should always be used when interpreting prothrombin time.
Elevated in: liver disease, factor deficiency (I, II, V, VII, X), disseminated intravascular coagulation, vitamin K deficiency, afibrinogenemia, dysfibrinogenemia, drug therapy (oral anticoagulant [warfarin], heparin, salicylate, chloral hydrate, diphenylhydantoin, estrogens, antacids, phenylbutazone, quinidine, antibiotics, allopurinol, anabolic steroids)
Decreased in: vitamin K supplementation, thrombophlebitis, drug therapy (glutethimide, estrogens, griseofulvin, diphenhydramine)

Protoporphyrin (Free Erythrocyte)

Normal range: 16 to 36 mcg/dL of RBC
Elevated in: iron deficiency, lead poisoning, sideroblastic anemias, anemia of chronic disease, hemolytic anemias, erythropoietic protoporphyria

PSA See Prostate-Specific Antigen

PT See Prothrombin Time

PTT See Partial Thromboplastin Time (PTT), Activated Partial Thromboplastin Time (aPTT)

Rapid Plasma Reagin

Description: Nontreponemal test traditionally used as a screening test for syphilis. It is a quantitative test, and antibody titers can be monitored to assess treatment response.

Normal: Negative
Positive: Syphilis. False-positive results may occur with pregnancy, autoimmune diseases, tuberculosis, and other inflammatory conditions. Positive results should be confirmed with treponemal serologic tests (e.g., *Treponema pallidum* enzyme immunoassay [TP-EIA]).

RDW See Red Blood Cell Distribution Width

Red Blood Cell Count

Normal range:
   Male: 4.3 to $5.9 \times 10^6/mm^3$
   Female: 3.5 to $5 \times 10^6/mm^3$
Elevated in: hemoconcentration and dehydration, stress, polycythemia vera, smokers, high altitude, cardiovascular disease, renal cell carcinoma and other erythropoietin-producing neoplasms
Decreased in: anemias, hemolysis, chronic renal failure, hemorrhage, failure of marrow production

Red Blood Cell Distribution Width (RDW)

Test description: This test measures the variability of red blood cell size (anisocytosis).
Normal range: 11.5 to 14.5%
Normal RDW and elevated mean corpuscular volume (MCV): aplastic anemia, preleukemia
   Normal MCV: normal, anemia of chronic disease, acute blood loss or hemolysis, chronic lymphocytic leukemia (CLL), chronic myelogenous leukemia, nonanemic enzymopathy or hemoglobinopathy
   Decreased MCV: anemia of chronic disease, heterozygous thalassemia
Elevated RDW and elevated MCV: vitamin B_{12} deficiency, folate deficiency, immune hemolytic anemia, cold agglutinins, CLL with high count, liver disease
   Normal MCV: early iron deficiency, early vitamin B_{12} deficiency, early folate deficiency, anemic globinopathy
   Decreased MCV: iron deficiency, red blood cell fragmentation, hemoglobin H disease, thalassemia intermedia

Red Blood Cell Folate See Folate (Folic Acid)

Red Blood Cell Mass (Volume)

Normal range:
   Male: 20 to 36 mL/kg of body weight (1.15–1.21 L/m^2 body surface area [BSA])
   Female: 19 to 31 mL/kg of body weight (0.95–1 L/m^2 BSA)
Elevated in: polycythemia vera, hypoxia (smokers, high altitude, cardiovascular disease), hemoglobinopathies with high oxygen affinity, erythropoietin-producing tumors (renal cell carcinoma)
Decreased in: hemorrhage, chronic disease, failure of marrow production, anemias, hemolysis

Renin (Serum)

Elevated in: renal hypertension, reduced plasma volume, secondary aldosteronism, drug therapy (thiazides, estrogen, minoxidil), chronic renal failure, Bartter's syndrome, pregnancy (normal), pheochromocytoma

TABLE 2.19	Factors That May Cause False-Positive or False-Negative Results of the Aldosterone/Renin Ratio	
	FALSE POSITIVES	**FALSE NEGATIVES**
Aldosterone relatively high	Potassium loading	
Renin relatively low	β-Blockers Central antiadrenergics Direct renin inhibitors Nonsteroidal antiinflammatory drugs Chronic kidney disease Sodium loading	
Aldosterone relatively low		Hypokalemia
Renin relatively high		Diuretics ACE inhibitors, angiotensin receptor blockers Calcium channel blockers (dihydropyridines) Acute sodium depletion

ACE, angiotensin-converting enzyme.
From Skorecki K, Chertow GM, Marsden PA, et al: *Brenner and Rector's the kidney*, ed 10, Philadelphia, Elsevier, 2016.

Decreased in: primary aldosteronism, adrenocortical hypertension, increased plasma volume, drug therapy (propranolol, reserpine, clonidine)
Comments: Factors that may cause false-positive or false-negative results of aldosterone/renin ratio are summarized in Table 2.19.

Respiratory Syncytial Virus (RSV) Screen

Test description: Polymerase chain reaction test can be performed on nasopharyngeal swab, wash, or aspirate.

Reticulocyte Count

Normal range: 0.5% to 1.5%
Elevated in: hemolytic anemia (sickle cell crisis, thalassemia major, autoimmune hemolysis), hemorrhage, postanemia therapy (folic acid, ferrous sulfate, vitamin B_{12}), chronic renal failure
Decreased in: aplastic anemia, marrow suppression (sepsis, chemotherapeutic agents, radiation), hepatic cirrhosis, blood transfusion, anemias of disordered maturation (iron deficiency anemia, megaloblastic anemia, sideroblastic anemia, anemia of chronic disease)
    Table 2.20 describes combining the reticulocyte count and red blood cell parameters for diagnosis.

Rheumatoid Factor

Normal: negative
Present in titer >1:20: rheumatoid arthritis, systemic lupus erythematosus, chronic inflammatory processes, old age, infections, liver disease, multiple myeloma, sarcoidosis, pulmonary fibrosis, Sjögren's syndrome
    Table 2.21 describes the sensitivity and specificity of rheumatoid factor.

TABLE 2.20	Combining the Reticulocyte Count and Red Blood Cell Parameters for Diagnosis	
MCV, RDW	**RETICULOCYTE COUNT <75,000/μL**	**RETICULOCYTE COUNT >100,000/μL**
Low, Normal	Anemia of chronic disease	
Normal, Normal	Anemia of chronic disease	
High, Normal	Chemotherapy, antivirals, alcohol Aplastic anemia	Chronic liver disease
Low, High	Iron deficiency anemia	Sickle cell–β-thalassemia
Normal, High	Early iron, folate, vitamin B_{12} deficiency Myelodysplasia	Sickle cell anemia, sickle cell disease
High, High	Folate or vitamin B_{12} deficiency	Immune hemolytic anemia
	Myelodysplasia	Chronic liver disease

MCV, mean corpuscular volume; *RDW*, red blood cell distribution width.
From Hoffman R, Benz EJ, Shattil SJ, et al: *Hematology: basic principles and practice*, ed 5, Philadelphia, Churchill Livingston, 2009.

TABLE 2.21	Sensitivity and Specificity of Rheumatoid Factor		
DIAGNOSIS	**≥150 U/ML**	**≥50 U/ML**	**≥100 U/ML**
Rheumatoid arthritis	66*	46	26
Sjögren's syndrome	62	52	33
Systemic lupus erythematosus	27	10	3
Mixed connective tissue disease	23	13	6
Scleroderma	44	18	2
Polymyositis	18	0	0
Reactive arthritis	0	0	0
Osteoarthritis	25	4	4
Healthy controls	13	0	0
Sensitivity (%)	66	46	26
Specificity (%)	74	88 (92†)	95 (98†)

Rheumatoid factors were determined by nephelometry in 100 patients with rheumatoid arthritis, in more than 200 patients with other rheumatic diseases, and in 30 healthy control persons.
*Percentage of positive patients.
†Specificity when a diagnosis of Sjögren's syndrome can be excluded.
Hochberg MC, Silma AJ, Smolen JS, et al, [eds]: *Rheumatology*, ed 3, St. Louis, Mosby, 2003.

RNP See Extractable Nuclear Antigen

RPR See Rapid Plasma Reagin

Rotavirus Serology

Test description: Polymerase chain reaction test is performed on a stool specimen.

Schilling Test See Fig. 2.16.

Sedimentation Rate See Erythrocyte Sedimentation Rate

Semen Analysis

Normal: Volume: 2 to 6 mL
Sperm density: >20 million/mL
Total number of spermatozoa: >80 million/ejaculate
Progressive motility score evaluated 2 to 4 hours after ejaculate: 3 to 4
Live spermatozoa: ≥50% of total
Normal spermatozoa: ≥0% of total
Immature forms: <4%
Decreased in: cryptorchidism, testicular failure, obstruction of ejaculatory system, postvasectomy, drug therapy (cimetidine, ketoconazole, nitrofurantoin, cancer chemotherapy agents, sulfasalazine), testicular radiation

SGOT See Aspartate Aminotransferase

SGPT See Alanine Aminotransferase

Sickle Cell Test

Normal: negative
Positive in: sickle cell anemia, sickle cell trait, combination of hemoglobin S gene with other disorders such as alpha thalassemia, beta thalassemia

Smooth Muscle Antibody

Normal: negative
Present in: chronic active hepatitis (≥1:80), primary biliary cirrhosis (≥1:80), infectious mononucleosis

Sodium (Serum)

Normal range: 135 to 147 mEq/L
Elevated in:
1. Isovolemic hypernatremia (decreased total body water [TBW], normal total body sodium [TBNa], and extracellular fluid [ECF])
   a. Diabetes insipidus (neurogenic and nephrogenic)
   b. Skin loss (hyperhemia), iatrogenic, reset osmostat
2. Hypervolemic hypernatremia (increased TBW, markedly increased TBNa and ECF)
   a. Iatrogenic (administration of hypernatremic solutions)
   b. Mineralocorticoid excess (Conn's syndrome, Cushing's syndrome)
   c. Salt ingestion
3. Hypovolemic hypernatremia: loss of H_2O and Na^+ (H_2O loss > Na^+)
   a. Renal losses (e.g., diuretics, glycosuria)
   b. Gastrointestinal (GI), respiratory, skin losses
   c. Adrenal deficiencies
Decreased in:
1. Hypotonic hyponatremia
2. Isovolemic hyponatremia
   a. Syndrome of inappropriate diuretic hormone secretion

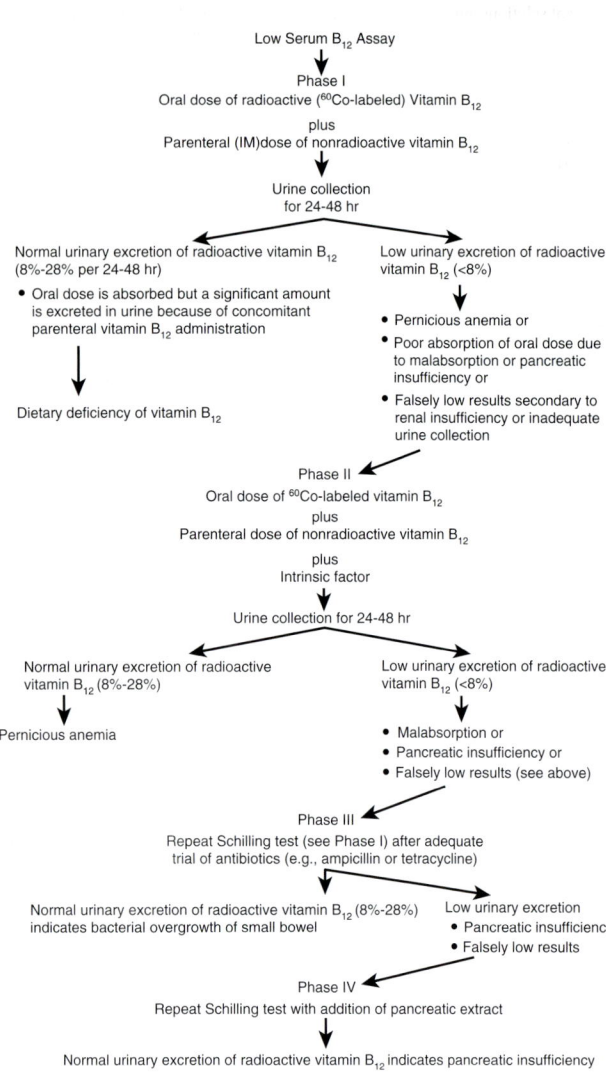

Fig. 2.16 Schilling test. (*IM,* intramuscular.) (From Ferri FF: *Practical guide to the care of the medical patient*, ed 7, St. Louis, Mosby, 2007.)

    b. Water intoxication (e.g., schizophrenia, primary polydipsia, sodium-free irrigant solutions, multiple tap-water enemas, dilute infant formulas). These entities are rare and often associated with a deranged antidiuretic hormone axis.

    c. Renal failure

    d. Reset osmostat (e.g., chronic active tuberculosis, carcinomatosis)

    e. Glucocorticoid deficiency (hypopituitarism)

    f. Hypothyroidism

    g. Thiazide diuretics, nonsteroidal antiinflammatory drugs, carbamazepine, amitriptyline, thioridazine, vincristine, cyclophosphamide, colchicine, tolbutamide, chlorpropamide, angiotensin-converting enzyme inhibitors, clofibrate, oxytocin, selective serotonin reuptake inhibitors, amiodarone. With these medications, various drug-induced mechanisms are involved.

3. Hypovolemic hyponatremia

    a. Renal losses (diuretics, partial urinary tract obstruction, salt-losing renal disease)

    b. Extrarenal losses: GI (vomiting, diarrhea), extensive burns, third spacing (peritonitis, pancreatitis)

    c. Adrenal insufficiency

4. Hypervolemic hyponatremia

    a. Congestive heart failure

    b. Nephrotic syndrome

    c. Cirrhosis

    d. Pregnancy

5. Isotonic hyponatremia (normal serum osmolality)

    a. Pseudohyponatremia (increased serum lipids and serum proteins). Newer sodium assays eliminate this problem.

    b. Isotonic infusion (e.g., glucose, mannitol)

6. Hypertonic hyponatremia (increased serum osmolality)

    a. Hyperglycemia: Each 100 mL/dL increment in blood sugar above normal decreases plasma sodium concentration by 1.6 mEq/L.

    b. Hypertonic infusions (e.g., glucose, mannitol)

Streptozyme See Antistreptolysin O Titer

Sucrose Hemolysis Test (Sugar Water Test)

Normal: absence of hemolysis
Positive in: paroxysmal nocturnal hemoglobinuria
False positive: autoimmune hemolytic anemia, megaloblastic anemias
False negative: may occur with use of heparin or ethylenediaminetetraacetic acid

Sudan III Stain (Qualitative Screening for Fecal Fat)

Normal: negative. This test should be preceded by diet containing 100 to 150 g of dietary fat per day for 1 week, avoidance of a high-fiber diet, and avoidance of suppositories or oily material before specimen collection.
Positive in: steatorrhea, use of castor oil or mineral oil droplets

T₃ (Triiodothyronine)

Normal range: 75 to 220 ng/dL
Abnormal values: elevated in hyperthyroidism (usually earlier and to a greater extent than serum T_4)
Useful in diagnosing:

1. T_3 hyperthyroidism (thyrotoxicosis): increased T3, normal free thyroxine index (FTI)
2. Toxic nodular goiter: increased T3, normal or increased T4
3. Iodine deficiency: normal T3, possibly decreased T4

4. Thyroid replacement therapy with liothyronine (Cytomel): normal T4, increased T3 if patient is symptomatically hyperthyroid

T_3 is not ordered routinely but is indicated when hyperthyroidism is suspected and serum free T_4 or FTI inconclusive.

T_3 Resin Uptake (T_3RU)

Normal range: 25% to 35%

Abnormal values: increased in hyperthyroidism. T_3 resin uptake (T_3RU or RT_3U) measures the percentage of free T_4 (not bound to protein); it does not measure serum T_3 concentration; T_3RU and other tests that reflect thyroid hormone binding to plasma protein are also known as thyroid hormone-binding ratios.

T_4, Serum T_4, and Free (Free Thyroxine)

Normal range: 0.8 to 2.8 ng/dL

Abnormal values: serum thyroxine (T_4)

Elevated in:

1. Graves' disease
2. Toxic multinodular goiter
3. Toxic adenoma
4. Iatrogenic and factitious
5. Transient hyperthyroidism
   a. Subacute thyroiditis
   b. Hashimoto's thyroiditis
   c. Silent thyroiditis
6. Rare causes: hypersecretion of thyroid-stimulating hormone (e.g., pituitary neoplasms), struma ovarii, ingestion of large amounts of iodine in a patient with preexisting thyroid hyperplasia or adenoma (Jod-Basedow phenomenon), hydatidiform mole, carcinoma of thyroid, amiodarone therapy of arrhythmias

Serum thyroxine test measures both circulating thyroxine bound to protein (represents > 99% of circulating T_4) and unbound (free) thyroxine. Values vary with protein binding. To eliminate the suspected influence of protein binding on thyroxine values, two additional tests are available: T_3 resin uptake and serum free thyroxine.

Serum Free T_4

Elevated in: Graves' disease, toxic multinodular goiter, toxic adenoma, iatrogenic and factitious causes, transient hyperthyroidism

Serum free T_4 directly measures unbound thyroxine. Free T_4 can be measured by equilibrium dialysis (gold standard of free T_4 assays) or by immunometric techniques (influenced by serum levels of lipids, proteins, and certain drugs). The free thyroxine index can also be easily calculated by multiplying T_4 times T_3RU and dividing the result by 100; the free thyroxine index (FTI) corrects for any abnormal T_4 values secondary to protein binding:

$$FTI = T_4 \times T_3RU/100$$

Normal values equal 1.1 to 4.3.

Tegretol See Carbamazepine

Testosterone

Elevated in: anabolic steroid abuse, testosterone replacement therapy, adrenogenital syndrome, polycystic ovarian syndrome

Decreased in: Klinefelter's syndrome, male hypogonadism

Fig. 2.17 illustrates hypogonadism in aging men.

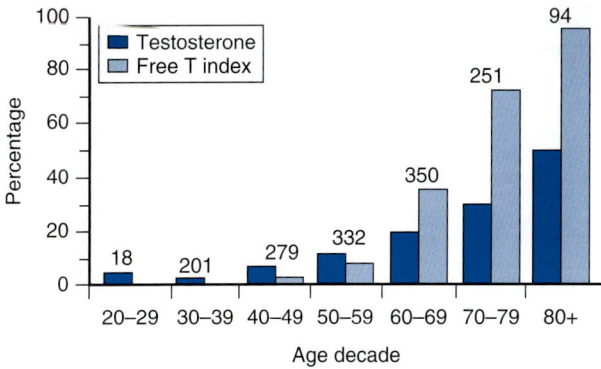

Fɪɢ. **2.17** Hypogonadism in aging men. Bar height indicates the percentage of men in each 10-year interval, from the third to the ninth decades, with at least one testosterone value in the hypogonadal range. The criteria used for these determinations are total testosterone less than 11.3 nmol/L (325 ng/dL) and testosterone and sex hormone–binding globulin (free T index) less than 0.153 nmol/nmol. The numbers above each pair of bars indicate the number of men studied in the corresponding decade. The fraction of men who are hypogonadal increases progressively after age 50 years by either criterion. More men are hypogonadal by free T index than by total testosterone after 50 years, and there seems to be a progressively greater difference between the two criteria with increasing age. (From Goldman L, Schafer AI: *Goldman's Cecil medicine,* ed 24, Philadelphia, Saunders, 2012.)

Theophylline

Normal therapeutic range: 10 to 20 µg/mL

Thiamine

Normal range: 275 to 675 ng/g

Elevated in: polycythemia vera, leukemia, Hodgkin's disease

Decreased in: alcoholism, dietary deficiency, excessive consumption of tea (contains antithiamine factor) or raw fish (contains a microbial thiaminase), chronic illness, prolonged illness, barbiturates

Thoracentesis Fluid

Testing and evaluation of results:

1. Pleural effusion fluid should be differentiated in exudate or transudate. The initial laboratory studies should be aimed only at distinguishing an exudate from a transudate.

   a. Tube 1: protein, lactate dehydrogenase (LDH), albumin

   b. Tubes 2, 3, 4: Save the fluid until further notice. In selected patients with suspected empyema, a pH level may be useful (generally ≤7). See the following for proper procedure to obtain a pH level from pleural fluid.

   Note: Do not order further tests until the presence of an exudate is confirmed on the basis of protein and LDH determinations (see Section III, Pleural Effusion);

however, if the results of protein and LDH determinations cannot be obtained within a reasonable time (resulting in unnecessary delay), additional laboratory tests should be ordered at the time of thoracentesis.

2. A serum/effusion albumin gradient of 1.2 g/dL or less is indicative of exudative effusions, especially in patients with congestive heart failure (CHF) treated with diuretics.
3. Note the appearance of the fluid:
   a. A grossly hemorrhagic effusion can be a result of a traumatic tap, neoplasm, or an embolus with infarction.
   b. A milky appearance indicates either of the following:
      i. Chylous effusion: caused by trauma or tumor invasion of the thoracic duct; lipoprotein electrophoresis of the effusion reveals chylomicrons and triglyceride levels greater than 115 mg/dL
      ii. Pseudochylous effusion: often seen with chronic inflammation of the pleural space (e.g., tuberculosis, connective tissue diseases)
4. If transudate, consider CHF, cirrhosis, chronic renal failure, and other hypoproteinemic states and perform subsequent workup accordingly
5. If exudate, consider ordering these tests on the pleural fluid:
   a. Cytologic examination for malignant cells (for suspected neoplasm)
   b. Gram stain, cultures (aerobic and anaerobic), and sensitivities (for suspected infectious process)
   c. Acid-fast bacilli stain and cultures (for suspected TB)
   d. pH: A value less than 7 suggests parapneumonic effusion or empyema; a pleural fluid pH must be drawn anaerobically and iced immediately; the syringe should be prerinsed with 0.2 mL of 1:1000 heparin.
   e. Glucose: A low glucose level suggests parapneumonic effusions and rheumatoid arthritis.
   f. Amylase: A high amylase level suggests pancreatitis or ruptured esophagus.
   g. Perplexing pleural effusions are often a result of malignancy (e.g., lymphoma, malignant mesothelioma, ovarian carcinoma), TB, subdiaphragmatic processes, prior asbestos exposure, or postcardiac injury syndrome.

Thrombin Time (TT)

Normal range: 11.3 to 18.5 seconds
Elevated in: thrombolytic and heparin therapy, disseminated intravascular coagulation, hypofibrinogenemia, dysfibrinogenemia

Thyroglobulin

Normal range: 3 to 40 ng/mL. Thyroglobulin is a tumor marker for monitoring the status of patients with papillary or follicular thyroid cancer after resection.
Elevated in: papillary or follicular thyroid cancer, Hashimoto's thyroiditis, Graves' disease, subacute thyroiditis

Thyroid Binding Globulin (TBG)

TBG can result in changes in the concentration of T4. Factors that can change TBG are summarized in Box 2.2.

Thyroid Microsomal Antibodies

Normal: undetectable. Low titers may be present in 5% to 10% of normal individuals.
Elevated in: Hashimoto's disease, thyroid carcinoma, early hypothyroidism, pernicious anemia

BOX 2.2	Changes in Thyroid Binding Globulin

INCREASED TBG (↑ T₄)	DECREASED TBG (↓ T₄)
Pregnancy	Androgens, glucocorticoids
Estrogens	Nephrotic syndrome, cirrhosis
Acute infectious hepatitis	Acromegaly
Oral contraceptives	Hypoproteinemia
Familial	Familial
Fluorouracil, clofibrate, heroin, methadone	Phenytoin, ASA and other NSAIDs, high-dose penicillin, asparaginase
Chronic debilitating illness	

ASA, acetylsalicylic acid; *NSAIDs*, nonsteroidal antiinflammatory drugs; *TBG*, thyroid binding globulin.

Thyroid-Stimulating Hormone (TSH)

Normal range: 2 to 11.0 µU/mL

Elevated in:

1. Primary hypothyroidism (thyroid gland dysfunction): cause of more than 90% of cases of hypothyroidism
   a. Hashimoto's thyroiditis (chronic lymphocytic thyroiditis); most common cause of hypothyroidism after 8 years of age
   b. Idiopathic myxedema (possibly a nongoitrous form of Hashimoto's thyroiditis)
   c. Previous treatment of hyperthyroidism (¹³¹I therapy, subtotal thyroidectomy)
   d. Subacute thyroiditis
   e. Radiation therapy of the neck (usually for malignant disease)
   f. Iodine deficiency or excess
   g. Drug therapy (lithium, polyglandular autoimmune syndrome, sulfonamides, phenylbutazone, amiodarone, thiourea)
   h. Congenital (approximately 1:4000 live births)
   i. Prolonged treatment of iodides
2. Tissue resistance to thyroid hormone (rare)

TSH is used primarily to diagnose hypothyroidism (the increased TSH level is the earliest thyroid abnormality detected); conventional TSH radioimmunoassays have been replaced by new third-generation TSH radioimmunoassays, which are useful to detect either clinical or subclinical thyroid hormone excess or deficiency. Various factors can influence TSH levels (recovery from severe illness and metoclopramide, chlorpromazine, haloperidol, and amiodarone use all elevate TSH; dopamine and corticosteroid therapies lower it). Apparently healthy ambulatory patients with subnormal TSH levels should be checked with measurement of free T_4 and total T_3. If they are normal, a T_3 level (by trace equilibrium dialysis) should be obtained to distinguish subclinical hyperthyroidism from overt free T_3 toxicosis.

Decreased in: hyperthyroidism, secondary hypothyroidism (pituitary dysfunction, postpartum necrosis, neoplasm, infiltrative disease causing deficiency of TSH), tertiary hypothyroidism (hypothalamic disease [granuloma, neoplasm, or irradiation causing deficiency of TSH])

Table 2.22 summarizes findings in thyroid function tests in various clinical conditions.

TABLE 2.22			Findings in Thyroid Function Tests in Various Clinical Conditions				
CONDITION	T$_4$	FT$_4$I	T$_3$	FT$_3$I	TSH	TSI	TRH STIMULATION
Hyperthyroidism							
Graves' disease	↑	↑	↑	↑	↓	+	↓
Toxic nodular goiter	↑	↑	↑	↑	↓	−	↓
Pituitary TSH-secreting tumors	↑	↑	↑	↑	↑	−	↓
T$_3$ thyrotoxicosis	N	N	↑	↑	↓	+, −	↓
T$_4$ thyrotoxicosis	↑	↑	N	N	↓	+, −	↓
Hypothyroidism							
Primary	↓	↓	↓	↓	↑	+, −	↑
Secondary	↓	↓	↓	↓	↓, N	−	↓
Tertiary	↓	↓	↓	↓	↓, N	−	N
Peripheral unresponsiveness	↑, N	↑, N	↑, N	↑	↑, N	−	N, ↑

−, Variable; *FT$_3$I,* free T$_3$ index; *FT$_4$I,* free T$_4$ index; *T$_3$,* triiodothyronine; *T$_4$,* thyroxine; *TRH,* thyrotropin-releasing hormone; *TSH,* thyroid-stimulating hormone; *TSI,* thyroid-stimulating immunoglobulin.
From Ferri FF: *Practical guide to the care of the medical patient,* ed 8, St. Louis, Mosby, 2011.

Thyrotropin (Thyroid-Stimulating Hormone [TSH]) Receptor Antibodies

Normal: <130% of basal activity
Elevated in: Values between 1.3 and 2 are found in 10% of patients with thyroid disease other than Graves' disease. Values greater than 2.8 have been found only in patients with Graves' disease.

Thyrotropin-Releasing Hormone (TRH) Stimulation Test

Normal: baseline thyroid-stimulating hormone (TSH) <11 microU/mL
Stimulated TSH: more than double the baseline.
In primary hypothyroidism, the TSH increases two to three times the normal value. In secondary hypothyroidism, no TSH response occurs. In tertiary hypothyroidism (hypothalamic failure), there is a delayed rise in the TSH level.

TIBC See Iron-binding Capacity

Tissue Transglutaminase Antibody

Normal: negative
Present in: celiac disease (specificity 94%–97%, sensitivity 90%–98%), dermatitis herpetiformis

Transferrin

Normal range: 170 to 370 mg/dL
Elevated in: iron deficiency anemia, oral contraceptive administration, viral hepatitis, late pregnancy

Decreased in: nephrotic syndrome, liver disease, hereditary deficiency, protein malnutrition, neoplasms, chronic inflammatory states, chronic illness, thalassemia, hemochromatosis, hemolytic anemia

Triglycerides

Normal range: <160 mg/dL
Elevated in: hyperlipoproteinemias (types I, IIb, III, IV, V), diet high in saturated fats, hypothyroidism, pregnancy, estrogen therapy, pancreatitis, alcohol intake, nephrotic syndrome, poorly controlled diabetes mellitus, sedentary lifestyle, glycogen storage disease
Decreased in: malnutrition, vigorous exercise, congenital abetalipoproteinemias, drug therapy (gemfibrozil, fenofibrate, nicotinic acid, metformin, clofibrate)

Triiodothyronine See T$_3$

Troponins, Serum

Normal range: 0 to 0.4 ng/mL (negative). If there is clinical suspicion of evolving acute myocardial infarction (MI) or ischemic episode, repeat testing in 5 to 6 hours is recommended.
Indeterminate: 0.05 to 0.49 ng/mL. Suggest further tests. In a patient with unstable angina and this troponin I level, there is an increased risk of a cardiac event in the near future.
Strong probability of acute MI: ≥0.50 ng/mL
  Cardiac troponin T (cTnT): highly sensitive marker for myocardial injury for the first 48 hours after MI and for up to 5 to 7 days. It may be also elevated in renal failure, chronic muscle disease, and trauma.
  Cardiac troponin I (cTnI): highly sensitive and specific for myocardial injury (≥creatine kinase MB) in the initial 8 hours, peaks within 24 hours, and lasts up to 7 days. With progressively higher levels of cTnI, the risk of mortality increases because the amount of necrosis increases.
Elevated in: In addition to acute coronary syndrome, many diseases such as sepsis, hypovolemia, atrial fibrillation, congestive heart failure, pulmonary embolism, myocarditis, myocardial contusion, and renal failure can be associated with an increase in troponin level.
  Box 2.3 summarizes causes of serum troponin T and I elevations.

TSH See Thyroid-Stimulating Hormone

TT See Thrombin Time

Unconjugated Bilirubin See Bilirubin, Indirect

Urea Nitrogen

Normal range: 8 to 18 mg/dL
Elevated in: dehydration, renal disease (glomerulonephritis, pyelonephritis, diabetic nephropathy), urinary tract obstruction (prostatic hypertrophy), drug therapy (aminoglycosides and other antibiotics, diuretics, lithium, corticosteroids), gastrointestinal bleeding, decreased renal blood flow (shock, congestive heart failure, myocardial infarction)
Decreased in: liver disease, malnutrition, third trimester of pregnancy

Uric Acid (Serum)

Normal range: 2 to 7 mg/dL

| **BOX 2.3** | **Causes of Serum Troponin T and I Elevations, Including Both Acute Coronary Syndromes, Noncoronary Cardiac Events, and Noncardiac Ailments** |

Acute coronary syndrome and acute myocardial infarction
Shock of any form (cardiogenic, obstructive, distributive)
Myocarditis and myopericarditis
Cardiomyopathies
Acute congestive heart failure (pulmonary edema)
Sepsis
Pulmonary embolism
Renal failure
Sympathomimetic ingestions
Polytrauma
Burns
Acute central nervous system event
Rhabdomyolysis
Cardiac neoplasm, inflammatory syndromes, and infiltrative diseases
Congenital coronary anomalies
Extreme physical exertion

(From Vincent JL, Abraham E, Moore FA, et al: *Textbook of critical care,* ed 6, Philadelphia, Saunders, 2011.)

Elevated in: hereditary enzyme deficiency (hypoxanthine-guanine-phosphoribosyl transferase), renal failure, gout, excessive cell lysis (chemotherapeutic agents, radiation therapy, leukemia, lymphoma, hemolytic anemia), acidosis, myeloproliferative disorders, diet high in purines or protein, drug therapy (diuretics, low doses of acetylsalicylic acid [ASA], ethambutol, nicotinic acid), lead poisoning, hypothyroidism

Decreased in: drug therapy (allopurinol, high doses of ASA, probenecid, warfarin, corticosteroid), deficiency of xanthine oxidase, syndrome of inappropriate diuretic hormone, renal tubular deficits (Fanconi's syndrome), alcoholism, liver disease, diet deficient in protein or purines, Wilson's disease, hemochromatosis

Urinalysis

Normal range:
  Color: light straw
  Appearance: clear
  pH: 4.5-8 (average, 6)
  Specific gravity: 1.005 to 1.030
  Protein: absent
  Ketones: absent
  Glucose: absent
  Occult blood: absent
Microscopic examination:
  Red blood cells: 0 to 5 (high-power field)
  White blood cells: 0 to 5 (high-power field)
  Bacteria (spun specimen): absent
  Casts: 0 to 4 hyaline (low-power field)

Urine Amylase

Normal range: 35 to 260 U Somogyi/h
Elevated in: pancreatitis, carcinoma of the pancreas

Urine Bile

Normal: absent
Abnormal: Urine bilirubin: hepatitis (viral, toxic, drug-induced), biliary obstruction
Urine urobilinogen: hepatitis (viral, toxic, drug-induced), hemolytic jaundice, liver cell dysfunction (cirrhosis, infection, metastases)

Urine Calcium

Normal: 6.2 mmol/dL (<250 mg/24 h)
Elevated in: primary hyperparathyroidism, hypervitaminosis D, bone metastases, multiple myeloma, increased calcium intake, steroids, prolonged immobilization, sarcoidosis, Paget's disease, idiopathic hypercalciuria, renal tubular acidosis
Decreased in: hypoparathyroidism, pseudohypoparathyroidism, vitamin D deficiency, vitamin D–resistant rickets, diet low in calcium, drug therapy (thiazide diuretics, oral contraceptives), familial hypocalciuric hypercalcemia, renal osteodystrophy, potassium citrate therapy

Urine cAMP

Elevated in: hypercalciuria, familial hypocalciuric hypercalcemia, primary hyperparathyroidism, pseudohypoparathyroidism, rickets
Decreased in: vitamin D intoxication, sarcoidosis

Urine Catecholamines

Normal:
  Norepinephrine: <100 mcg/24 h
  Epinephrine: <10 mcg/24 h
Elevated in: pheochromocytoma, neuroblastoma, severe stress

Urine Chloride

Normal range: 110 to 250 mEq/day
Elevated in: corticosteroids, Bartter's syndrome, diuretics, metabolic acidosis, severe hypokalemia
Decreased in: chloride depletion (vomiting), colonic villous adenoma, chronic renal failure, renal tubular acidosis

Urine Copper

Normal: <40 mcg/24 h

Urine Cortisol, Free

Normal range: 10 to 110 mcg/24 h
Elevated in: refer to Cortisol (Plasma)

Urine Creatinine (24-Hour)

Normal range:
  Male: 0.8 to 1.8 g/day
  Female: 0.6 to 1.6 g/day
Note: Useful test as an indicator of completeness of 24-hour urine collection.

Urine Crystals

Uric acid: acid urine, hyperuricosuria, uric acid nephropathy
Sulfur: antibiotics containing sulfa
Calcium oxalate: ethylene glycol poisoning, acid urine, hyperoxaluria
Calcium phosphate: alkaline urine
Cystine: cystinuria

Urine Eosinophils

Normal: absent
Present in: interstitial nephritis, acute tubular necrosis, urinary tract infection, kidney
transplant rejection, hepatorenal syndrome

Urine 5-Hydroxyindole-Acetic Acid (Urine 5-HIAA)

Normal range: 2 to 8 mg/24 h
Elevated in: carcinoid tumors, ingestion of certain foods (bananas, plums, tomatoes,
avocados, pineapples, eggplant, walnuts), drug therapy (monoamine oxidase inhibitors, phenacetin, methyldopa, glycerol guaiacolate, acetaminophen, salicylates,
phenothiazines, imipramine, methocarbamol, reserpine, methamphetamine)

Urine Glucose (Qualitative)

Normal: absent
Present in: diabetes mellitus, renal glycosuria (decreased renal threshold for glucose),
glucose intolerance

Urine Hemoglobin, Free

Normal: absent
Present in: hemolysis (with saturation of serum haptoglobin binding capacity and
renal threshold for tubular absorption of hemoglobin)

Urine Hemosiderin

Normal: absent
Present in: paroxysmal nocturnal hemoglobinuria, chronic hemolytic anemia, hemochromatosis, blood transfusion, thalassemias

Urine Indican

Normal: absent
Present in: malabsorption resulting from intestinal bacterial overgrowth

Urine Ketones (Semiquantitative)

Normal: absent
Present in: diabetic ketoacidosis, alcoholic ketoacidosis, starvation, isopropanol
ingestion

Urine Metanephrines

Normal range: 0 to 2 mg/24 h
Elevated in: pheochromocytoma, neuroblastoma, caffeine use, drug therapy (phenothiazines, monoamine oxidase inhibitors), stress

Urine Myoglobin

Normal: absent
Present in: severe trauma, hyperthermia, polymyositis or dermatomyositis, carbon monoxide poisoning, drug use (narcotic and amphetamine toxicity), hypothyroidism, muscle ischemia

Urine Nitrite

Normal: absent
Present in: urinary tract infections

Urine Occult Blood

Normal: negative
Positive in: trauma to urinary tract, renal disease (glomerulonephritis, pyelonephritis), renal or ureteral calculi, bladder lesions (carcinoma, cystitis), prostatitis, prostatic carcinoma, menstrual contamination, hematopoietic disorders (hemophilia, thrombocytopenia), anticoagulant therapy, acetylsalicylic acid use
Note: Hematuria without erythrocyte casts or significant albuminuria suggests the possibility of renal or bladder cancers.

Urine Osmolality

Normal range: 50 to 1200 mOsm/kg
Elevated in: syndrome of inappropriate antidiuretic hormone, dehydration, glycosuria, adrenal insufficiency, high-protein diet
Decreased in: diabetes insipidus, excessive water intake, intravenous hydration with 5% dextrose in water, acute renal insufficiency, glomerulonephritis

Urine pH

Normal range: 4.6 to 8 (average 6)
Elevated in: bacteriuria, vegetarian diet, renal failure with inability to form ammonia, drug therapy (antibiotics, sodium bicarbonate, acetazolamide)
Decreased in: acidosis (metabolic, respiratory), drug therapy (ammonium chloride, methenamine mandelate), diabetes mellitus, starvation, diarrhea

Urine Phosphate

Normal range: 0.8 to 2 g/24 h
Elevated in: acute tubular necrosis (diuretic phase), chronic renal disease, uncontrolled diabetes mellitus, hyperparathyroidism, hypomagnesemia, metabolic acidosis, metabolic alkalosis, neurofibromatosis, adult-onset vitamin D–resistant hypophosphatemic osteomalacia
Decreased in: acromegaly, acute renal failure, decreased dietary intake, hypoparathyroidism, respiratory acidosis

Urine Potassium

Normal range: 25 to 100 mEq/24 h
Elevated in: aldosteronism (primary, secondary), glucocorticoids therapy, alkalosis, renal tubular acidosis, excessive dietary potassium intake
Decreased in: acute renal failure, potassium-sparing diuretic use, diarrhea, hypokalemia

Urine Protein (Quantitative)

Normal: <150 mg/24 h
Elevated in: renal disease (glomerular, tubular, interstitial), congestive heart failure, hypertension, neoplasms of renal pelvis and bladder, multiple myeloma, Waldenström's macroglobulinemia

Urine Sodium (Quantitative)

Normal range: 40 to 220 mEq/day
Elevated in: diuretic administration; high sodium intake; salt-losing nephritis; acute tubular necrosis; vomiting; Addison's disease; syndrome of inappropriate antidiuretic hormone; hypothyroidism; congestive heart failure; hepatic failure; chronic renal failure; Bartter's syndrome; glucocorticoid deficiency; interstitial nephritis caused by analgesic abuse; mannitol, dextran, or glycerol therapy; milk-alkali syndrome; decreased renin secretion; postobstructive diuresis
Decreased in: increased aldosterone, glucocorticoid excess, hyponatremia, prerenal azotemia, decreased salt intake
Comments: Box 2.4 summarizes use of urine electrolytes in the differential diagnosis of hypokalemia.

BOX 2.4	Urine Electrolytes* in the Differential Diagnosis of Hypokalemia	
CONDITION	**URINE ELECTROLYTE**	
	NA+	**CL−**
Vomiting		
Recent	High[†]	Low[‡]
Remote	Low	Low
Diuretics		
Recent	High	High
Remote	Low	Low
Diarrhea or Laxative Abuse	Low	High
Bartter's or Gitelman's Syndrome	High	High

Cl−, Chloride; Na+, sodium.
*Do not use the urine electrolytes in this fashion during polyuric states.
[†]High = urine concentration >15 mmol/L.
[‡]Low = urine concentration <15 mmol/L.
(From Vincent JL, Abraham E, Moore FA, et al: *Textbook of critical care,* ed 6, Philadelphia, Saunders, 2011.)

Urine Specific Gravity

Normal range: 1.005 to 1.03
Elevated in: dehydration, excessive fluid losses (vomiting, diarrhea, fever), ingestions of radiograph contrast media, diabetes mellitus, congestive heart failure, syndrome of inappropriate antidiuretic hormone, adrenal insufficiency, decreased fluid intake
Decreased in: diabetes insipidus, renal disease (glomerulonephritis, pyelonephritis), excessive fluid intake or intravenous hydration

Urine Vanillylmandelic Acid (VMA)

Normal: <6.8 mg/24 h
Elevated in: pheochromocytoma; neuroblastoma; ganglioblastoma; drug therapy (isoproterenol, methocarbamol, levodopa, sulfonamides, chlorpromazine); severe stress; ingestion of bananas, chocolate, vanilla, tea, coffee
Decreased in: drug therapy (monoamine oxidase inhibitors, reserpine, guanethidine, methyldopa)

Varicella Zoster Virus (VZV) Serologic Testing

Test description: This test can be performed on whole blood, tissue, skin lesions, and cerebrospinal fluid.

Vasoactive Intestinal Peptide (VIP)

Normal: <50 pg/mL
Elevated in: pancreatic VIPomas, neuroblastoma, pancreatic islet call hyperplasia, liver disease, multiple endocrine neoplasia I, ganglioneuroma, ganglioneuro-blastoma

Venereal Disease Research Laboratories (VDRL)

Normal range: negative
Positive test: syphilis, other treponemal diseases (yaws, pinta, bejel)
Note: A false-positive test may be seen in patients with systemic lupus erythematosus and other autoimmune diseases, infectious mononucleosis, human immunodeficiency virus, atypical pneumonia, malaria, leprosy, typhus fever, rat-bite fever, or relapsing fever.

VIP See Vasoactive Intestinal Peptide

Viscosity (Serum)

Normal range: 1.4 to 1.8 relative to water (1.10–1.22 centipoise)
Elevated in: monoclonal gammopathies (Waldenström's macroglobulinemia, multiple myeloma), hyperfibrinogenemia, systemic lupus erythematosus, rheumatoid arthritis, polycythemia, leukemia

Vitamin B$_{12}$

Decreased in: pernicious anemia, dietary (strict lacto-ovo vegetarians, food faddists), malabsorption (achlorhydria, gastrectomy, ileal resection, Crohn's disease of terminal ileum, pancreatic insufficiency, drug therapy [omeprazole and other protein pump inhibitors, metformin, cholestyramine]), chronic alcoholism, *Helicobacter pylori* infection

Vitamin D, 1,25 Dihydroxy Calciferol, Vitamin D 25(OH) D (25-Hydroxyvitamin D)

Normal range: 16 to 65 pg/mL for 1,25(OH)D; 50 to 70 mg/mL for 25(OH)D
Elevated in: Excessive oral intake, tumor calcinosis, primary hyperparathyroidism, sarcoidosis, tuberculosis, idiopathic hypercalciuria

Decreased in: dietary deficiency, low sunlight exposure, postmenopausal osteoporosis, chronic renal failure, hypoparathyroidism, tumor-induced osteomalacia, rickets, tumor-induced osteomalacia, elevated blood lead levels

Vitamin K

Normal range: 0.10 to 2.20 ng/mL
Decreased in: primary biliary cirrhosis, anticoagulants, antibiotics, cholestyramine, gastrointestinal disease, pancreatic disease, cystic fibrosis, obstructive jaundice, hypoprothrombinemia, hemorrhagic disease of the newborn

Von Willebrand's Factor

Normal: Levels vary according to blood type:
Blood type O: 50 to 150 U/dL
Blood type non-O: 90 to 200 U/dL
Decreased in: von Willebrand's disease (however, in type II von Willebrand's disease the antigen may be normal but the function impaired)

WBCs See Complete Blood Cell (CBC) Count

Westergren See Erythrocyte Sedimentation Rate

White Blood Cell Count See Complete Blood Cell (CBC) Count

References

Auerbach PS, Constance BB, Freer L. Field guide to wilderness medicine. 4th ed. St. Louis: Mosby; 2013.

Ballinger A. Essentials of Kumar and Clark's clinical medicine. 5th ed. Edinburgh: Saunders; 2012.

Besser GM, Thorner MO. Comprehensive clinical endocrinology. 3rd ed. St. Louis: Mosby; 2002.

Bolognia JL, Jorizzo JL, Rapini RP. Dermatology. St. Louis: Mosby; 2003.

Cohen J, Powderly WG. Infectious diseases. 2nd ed. St. Louis.

Dhingra R, Gona P, Nam BH, et al. C-reactive protein, inflammatory conditions, and cardiovascular disease risk. Am J Med 2007;120(12):1054–62.

Ferri F. Ferri's best test. 2nd ed. St. Louis: Mosby; 2010.

Ferri FF. Practical guide to the care of the medical patient. 8th ed. St. Louis: Mosby; 2011.

Goldman L, Schafer AI. Goldman's Cecil medicine. 24th ed. Philadelphia: Saunders; 2012.

Hoffmann R, Benz EJ, Silberstein LE, et al. Hematology: basic principles and practice. 6th ed. Philadelphia: Saunders; 2013.

Henry JB, Davey FR, Herman CJ, et al., editors. Clinical diagnosis and management by laboratory methods. 23th ed. Philadelphia: Saunders; 2017.

Lehmann CA. Saunders manual of clinical laboratory science. Philadelphia: WB Saunders; 1998.

Hochberg MC, Silma AJ, Smolen JS, et al., editors. Rheumatology. 3rd ed. St. Louis: Mosby; 2003.

Young NS, Gerson SL, High KA. Clinical hematology. St. Louis: Mosby; 2006.

Jeremias A, Gibson CM. Narrative review: alternative cause for elevated cardiac troponin levels when acute coronary syndromes are excluded. Ann Intern Med 2005;142(9):786–91.

Jones JS, Klein E. Four no more: the 'PSA cutoff era' is over. Cleve Clin J Med 2008;75(1):30–2.

McKie PM, Burnett JC. B-type natriuretic peptide as a biomarker beyond heart failure: speculations and opportunities. Mayo Clin Proc 2005;80(8):1029–36.

Pagana KD, Pagana TJ. Mosby's diagnostic and laboratory test reference. 8th ed. St. Louis: Mosby; 2006.

Ravel R. Clinical laboratory medicine. 6th ed. St. Louis: Mosby; 1995.

Sarnak MJ, Katz R, Stehman-Breen CO, et al. Cystatin C concentration as a risk factor for heart failure in older adults. Ann Intern Med 2005;142(7):497–505.

Talley NJ, Martin CJ. Clinical gastroenterology. 2nd ed. Sidney: Churchill Livingstone; 2006.

Johns Hopkins Hospital, Arcara KM, Tschudy MM. The Harriet Lane handbook. 19th ed. Philadelphia: Mosby; 2012.

Vincent JL, Abraham E, Moore FA, et al. Textbook of critical care. 6th ed. Philadelphia: Saunders; 2011.

Wu AHB. Tietz clinical guide to laboratory tests. 4th ed. Philadelphia: WB Saunders; 2006.

Diseases and Disorders

This section includes the diagnostic modalities (imaging and laboratory tests) and algorithms useful to diagnose the following 145 diseases and disorders. It is assumed that the patient has had a detailed history and physical examination before any testing sequence is initiated.

These algorithms are designed to assist clinicians in the evaluation and treatment of patients. They may not apply to all patients with a particular disease or disorder, and they are not intended to replace a clinician's individual judgment. Please note that specific findings in the patient's history and physical examination may significantly alter any of the proposed testing sequences.

1. Abdominal abscess
2. Abdominal aortic aneurysm
3. Achalasia
4. Acid-base disorders
5. Acute kidney injury
6. Addison's disease (adrenal insufficiency)
7. Adrenal mass
8. Alkaline phosphatase elevation
9. ALT/AST elevation
10. Amenorrhea, primary
11. Amenorrhea, secondary
12. Anemia, macrocytic
13. Anemia, microcytic
14. Antinuclear antibody (ANA)–positive
15. Aortic dissection
16. Appendicitis
17. Ascites
18. Avascular necrosis
19. Back pain, acute, lumbosacral (LS) area
20. Bilirubin elevation
21. Bleeding disorder, congenital
22. Brain abscess
23. Breast mass
24. Carcinoid syndrome
25. Cardiomegaly on chest radiograph
26. Cholangitis
27. Cholecystitis
28. Cholelithiasis
29. Complex regional pain syndrome (reflex sympathetic dystrophy [RSD])
30. Constipation
31. Creatinine phosphokinase (CPK) elevation
32. Cushing's syndrome
33. Deep vein thrombosis (DVT)
34. Delayed puberty
35. Delirium
36. Diarrhea
37. Disseminated intravascular coagulation (DIC)
38. Diverticulitis
39. Dyspepsia
40. Dyspnea
41. Dysuria
42. Ectopic pregnancy
43. Edema, lower extremity
44. Endocarditis, infective
45. Endometriosis

46. Epiglottitis
47. Fatigue
48. Fever of undetermined origin (FUO)
49. Galactorrhea
50. Genital lesions/ulcers
51. Goiter
52. Granulomatosis with polyangiitis
53. Gynecomastia
54. Hearing loss
55. Hematochezia
56. Hematuria
57. Hemochromatosis
58. Hemoptysis
59. Hepatomegaly
60. Hirsutism
61. Hyperaldosteronism
62. Hypercalcemia
63. Hyperkalemia
64. Hypermagnesemia
65. Hypernatremia
66. Hyperphosphatemia
67. Hyperthyroidism
68. Hypocalcemia
69. Hypoglycemia
70. Hypogonadism
71. Hypokalemia
72. Hypomagnesemia
73. Hyponatremia
74. Hypophosphatemia
75. Hypothyroidism
76. Immunodeficiency
77. Infertility
78. Jaundice
79. Joint effusion
80. Liver function test elevations
81. Liver mass
82. Lymphadenopathy, generalized
83. Malabsorption, suspected
84. Meningioma
85. Mesenteric ischemia
86. Mesothelioma
87. Metabolic acidosis
88. Metabolic alkalosis
89. Microcytosis
90. Multiple myeloma
91. Multiple sclerosis
92. Myalgias
93. Muscle weakness
94. Neck mass
95. Neuropathy
96. Neutropenia
97. Osteomyelitis
98. Pancreatic mass
99. Pancreatitis, acute
100. Parapharyngeal abscess

101. Pelvic mass
102. Peripheral arterial disease (PAD)
103. Pheochromocytoma
104. Pituitary adenoma
105. Pleural effusion
106. Polyarteritis nodosa
107. Polycythemia
108. Portal vein thrombosis
109. Precocious puberty
110. Proteinuria
111. Pruritus, generalized
112. Pulmonary embolism
113. Pulmonary hypertension
114. Pulmonary nodule
115. Purpura
116. Renal artery stenosis
117. Renal mass
118. Rotator cuff tear
119. Sarcoidosis
120. Scrotal mass
121. Small-bowel obstruction
122. Spinal epidural abscess
123. Splenomegaly
124. Stroke
125. Subarachnoid hemorrhage
126. Subclavian steal syndrome
127. Subdural hematoma
128. Superior vena cava syndrome
129. Syncope
130. Testicular torsion
131. Thoracic outlet syndrome
132. Thrombocytopenia
133. Thrombocytosis
134. Thyroid nodule
135. Thyroiditis
136. Tinnitus
137. Transient ischemic attack (TIA)
138. Urethral discharge
139. Urolithiasis
140. Urticaria
141. Vaginal discharge
142. Vertigo
143. Viral hepatitis
144. Weight gain
145. Weight loss, involuntary

Acronyms

A-a: alveolar-arterial
AAA: abdominal aortic aneurysm
Ab: antibody
ABG: arterial blood gas
ABI: ankle brachial index
ACE: angiotensin-converting enzyme
ACTH: adrenocorticotropic hormone
ADA: adenosine deaminase

AFB: acid-fast bacteria
ALT: alanine aminotransferase
AMA: antimitochondrial antibody
AMP: adenosine monophosphate
ANA: antinuclear antibody
ASMA: anti–smooth muscle antibody
AST: aspartate aminotransferase
BNP: B-type natriuretic peptide
BUN: blood urea nitrogen
c-ANCA: cytoplasmic antineutrophil cytoplasmic antibodies
CBC: complete blood cell
CDI: *Clostridium difficile* infection
CHF: congestive heart failure
CMV: cytomegalovirus
CPK: creatinine phosphokinase
C&S: culture and sensitivity
CSF: cerebrospinal fluid
CT: computed tomography
DHEAS: dehydroepiandrosterone sulfate
DIC: disseminated intravascular coagulation
DNA: deoxyribonucleic acid
Ds: double strand
DS: dehydroepiandrosterone
DVT: deep vein thrombosis
EB: Epstein-Barr
ECG: electrocardiogram
ECM: erythema chronicum migrans
ELISA: enzyme-linked immunosorbent assay
EMG: electromyogram
EPS: electrophysiologic
ERCP: endoscopic retrograde cholangiopancreatography
ESR: erythrocyte sedimentation rate
EUS: endoscopy ultrasound
FBS: fasting blood sugar
FENa: fractional excretion of sodium
FNAB: fine-needle aspiration biopsy
FSH: follicle-stimulating hormone
FUO: fever of undetermined origin
GB: gallbladder
GFR: glomerular filtration rate
GGT: γ-glutamyl transferase
GGTP: γ-glutamyl transpeptidase
GnRH: gonadotropin-releasing hormone
HBsAg: hepatitis B surface antigen
hCG: human chorionic gonadotropin
HCV: hepatitis C virus
HIV: human immunodeficiency virus
HSV: herpes simplex virus
IEP: immunoelectrophoresis
Ig: immunoglobulin
IGF: insulin-like growth factor
INR: International Normalized Ratio
IV: intravenous
IVP: intravenous pyelogram
K: potassium
KOH: potassium hydroxide

LDH: lactate dehydrogenase
LGV: lymphogranuloma venereum
LH: luteinizing hormone
LKM: liver kidney microsomal
LS: lumbosacral
LP: lumbar puncture
Na: sodium
MIBG: metaiodobenzylguanidine
MRA: magnetic resonance angiogram
MRCP: magnetic resonance cholangiopancreatography
MRDTI: magnetic resonance direct thrombus imaging
MRI: magnetic resonance imaging
OHP: hydroxyprogesterone
O&P: orthotic and prosthetic
OR: operating room
PA: posteroanterior
PAC: plasma aldosterone concentration
p-ANCA: perinuclear antineutrophil cytoplasmic antibody
PCOS: polycystic ovary syndrome
PCR: polymerase chain reaction
PCreat: plasma creatinine
PE: pulmonary embolism
PET: positron emission tomography
PFA: platelet function analysis
PNa: plasma sodium
PPD: purified protein derivative
PRA: plasma renin activity
PSA: prostate-specific antigen
PT: prothrombin time
PTH: parathyroid hormone
PTT: partial thromboplastin time
RAIU: radioactive iodine uptake
RBC: red blood cell
RDW: red blood cell distribution width
RF: rheumatoid factor
RNP: ribonucleoprotein
r/o: rule out
RSD: reflex sympathetic dystrophy
SBE: subacute bacterial endocarditis
TB: tuberculosis
Tc: technetium
TEE: transesophageal echocardiogram
TIA: transient ischemic attack
TIBC: total iron-binding capacity
TRH: thyrotropin-releasing hormone
TSH: thyroid-stimulating hormone
TT: thrombin time
TTKG: transtubular potassium gradient
UGI: upper gastrointestinal
UCreat: urine creatinine
UNa: urine sodium
UOsmo: urine osmolarity
VDRL: venereal disease research laboratories
V/Q: ventilation-perfusion
WBC: white blood cell

1. Abdominal Abscess (Fig. 3.1)

Diagnostic Imaging

Best Test(s)
- CT of abdomen with contrast (Fig. 3.2, *A, B*)

Ancillary Tests
- Ultrasound of abdomen is useful in young women and children

Laboratory Evaluation

Best Test(s)
- Gram stain and C&S of abscess

Ancillary Tests
- CBC count with differential
- Blood culture × 2
- ALT, AST
- BUN, creatinine, glucose

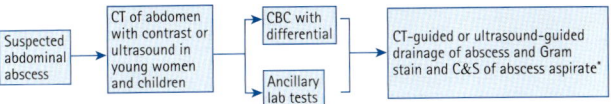

Fɪɢ. 3.1 Diagnostic algorithm. *Aspiration of hepatic amebic abscess is not indicated unless there is no response to treatment or a pyogenic cause is being considered.

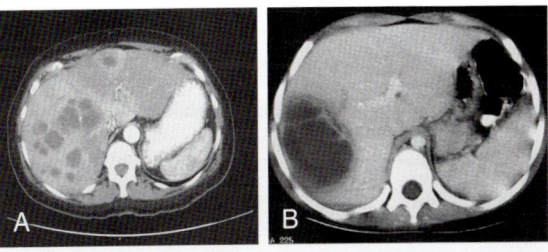

Fɪɢ. 3.2 A, CT showing multiple pyogenic abscesses in the liver. B, Computed tomography scan demonstrates a low-attenuation mass in the right lobe of the liver with a prominent halo. (A, Courtesy Dr. Mukesh Harisinghani, Boston; B, From Kuhn JP, Slovis TL, Haller JO: *Caffey's pediatric diagnostic imaging*, vol 2, ed 10, Philadelphia, Mosby, 2004, p. 1473.)

2. Abdominal Aortic Aneurysm (Fig. 3.3)

Diagnostic Imaging

Best Test(s)
- Ultrasound of abdominal aorta (Fig. 3.4) is best initial screening test; CT is more accurate test

Ancillary Tests
- CT of abdominal aorta with IV contrast for preoperative imaging and size estimation, and to diagnose perforation/tear
- Angiography for detailed arterial anatomy before surgery

Laboratory Evaluation

Best Test(s)
- None

Ancillary Tests
- Serum creatinine level before IV contrast

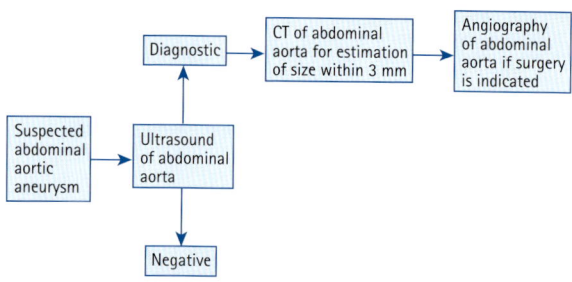

Fɪɢ. 3.3 Diagnostic algorithm.

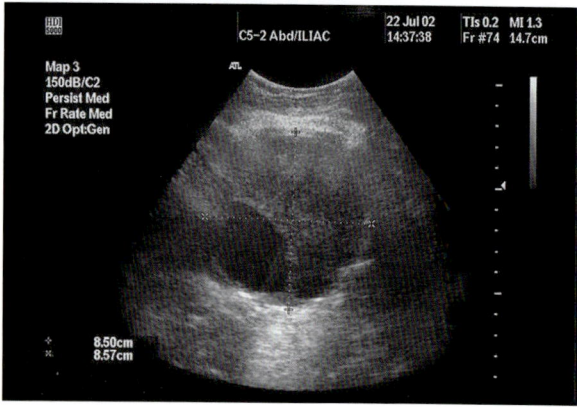

Fɪɢ. 3.4 Ultrasound appearance of an abdominal aortic aneurysm, seen in cross section. Sonography is highly accurate in diagnosing and measuring infrarenal aortic aneurysms. (Courtesy M. Ellis.)

3. Achalasia (Fig. 3.5)

Diagnostic Imaging

Best Test(s)
- Barium swallow with fluoroscopy (Figs. 3.6, 3.7)

Ancillary Tests
- Esophageal manometry to confirm diagnosis. It reveals esophageal aperistalsis and abnormal relaxation of lower esophageal sphincter.
- Upper endoscopy to exclude mechanical obstruction of the esophagus in the region of the lower sphincter

Laboratory Evaluation

Best Test(s)
- None

Ancillary Tests
- CBC
- Serum albumin for nutritional assessment

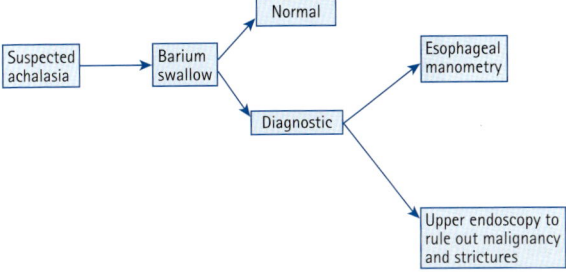

Fig. 3.5 Diagnostic algorithm.

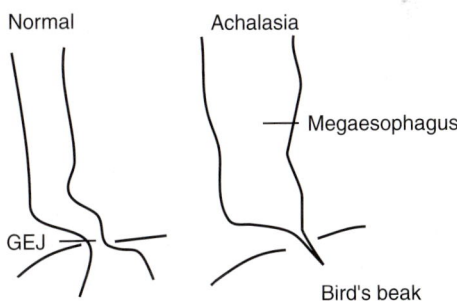

Fig. 3.6 Achalasia. *GEJ,* Gastroesophageal junction. (From Weissleder R, Wittenberg J, Harisinghani MG, Chen JW: *Primer of diagnostic imaging,* ed 5, St. Louis, Mosby, 2011.)

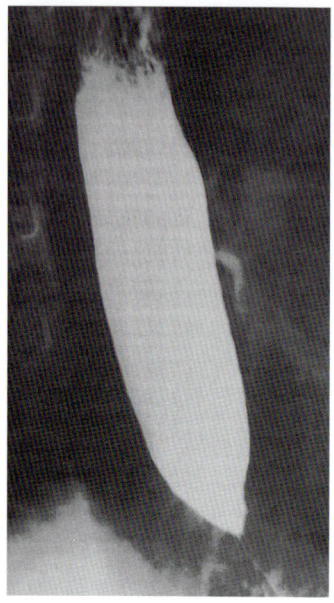

F<small>IG</small>. **3.7** Radiograph of esophageal achalasia showing the typical tapered ("bird beaked") appearance at the cardioesophageal junction and retention of food and fluid within a dilated and adynamic esophagus. (From Talley NJ, Martin CJ: *Clinical gastroenterology*, ed 2, Sidney, Churchill Livingstone, 2006.)

4. Acid-Base Disorders (Figs. 3.8 and 3.9)

Diagnostic Imaging

Best Test(s)
• None

Ancillary Tests
• Chest x-ray

Laboratory Evaluation

Best Test(s)
• ABGs

Ancillary Tests
• Serum electrolytes
• Urine electrolytes

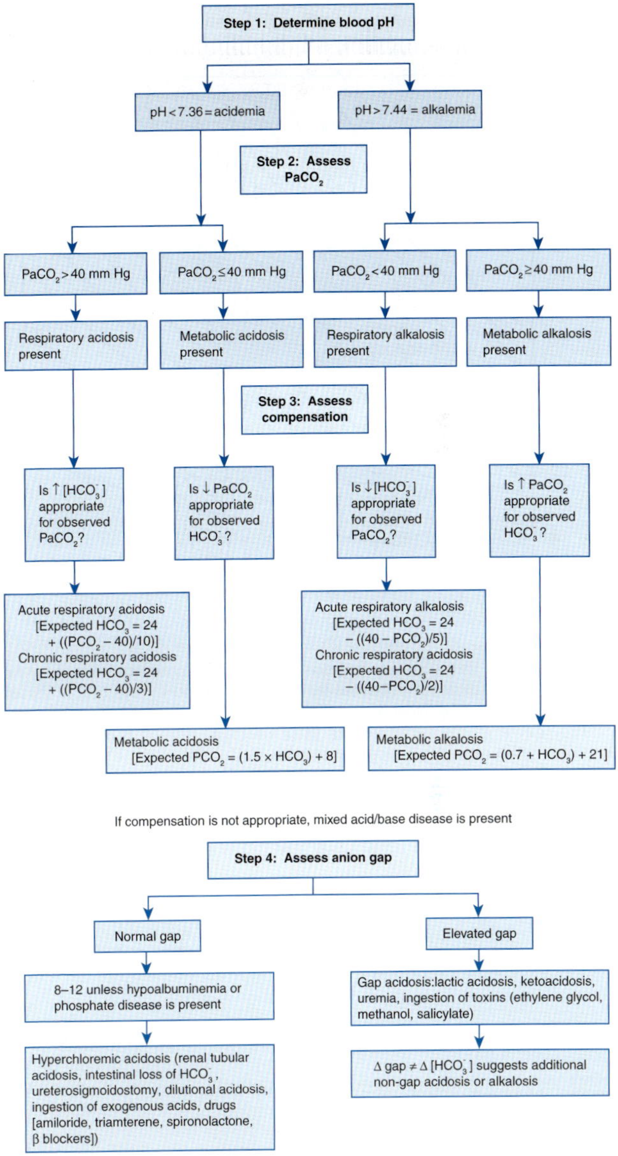

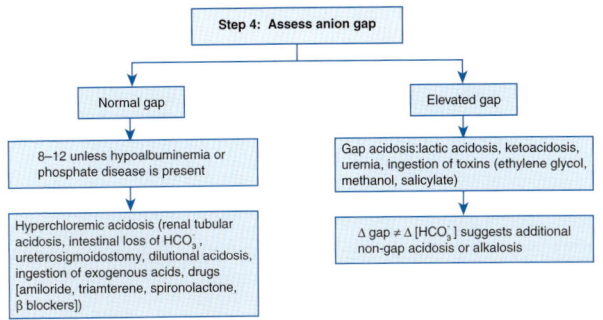

Step 1: Determine blood pH

pH < 7.36 = acidemia   pH > 7.44 = alkalemia

Step 2: Assess PaCO₂

PaCO₂ > 40 mm Hg → Respiratory acidosis present

PaCO₂ ≤ 40 mm Hg → Metabolic acidosis present

PaCO₂ < 40 mm Hg → Respiratory alkalosis present

PaCO₂ ≥ 40 mm Hg → Metabolic alkalosis present

Step 3: Assess compensation

Is ↑ [HCO₃] appropriate for observed PaCO₂?

Is ↓ PaCO₂ appropriate for observed HCO₃⁻?

Is ↓ [HCO₃] appropriate for observed PaCO₂?

Is ↑ PaCO₂ appropriate for observed HCO₃?

Acute respiratory acidosis [Expected HCO₃ = 24 + ((PCO₂ − 40)/10)]
Chronic respiratory acidosis [Expected HCO₃ = 24 + ((PCO₂ − 40)/3)]

Acute respiratory alkalosis [Expected HCO₃ = 24 − ((40 − PCO₂)/5)]
Chronic respiratory acidosis [Expected HCO₃ = 24 − ((40 − PCO₂)/2)]

Metabolic acidosis [Expected PCO₂ = (1.5 × HCO₃) + 8]

Metabolic alkalosis [Expected PCO₂ = (0.7 + HCO₃) + 21]

If compensation is not appropriate, mixed acid/base disease is present

Step 4: Assess anion gap

Normal gap

8–12 unless hypoalbuminemia or phosphate disease is present

Hyperchloremic acidosis (renal tubular acidosis, intestinal loss of HCO₃, ureterosigmoidostomy, dilutional acidosis, ingestion of exogenous acids, drugs [amiloride, triamterene, spironolactone, β blockers])

Elevated gap

Gap acidosis: lactic acidosis, ketoacidosis, uremia, ingestion of toxins (ethylene glycol, methanol, salicylate)

Δ gap ≠ Δ [HCO₃] suggests additional non-gap acidosis or alkalosis

FIG. 3.8 Determining acid-base status. *HCO3⁻*, Bicarbonate; *Paco₂*, partial pressure of arterial carbon dioxide. (From Cameron JL, Cameron AM: Acid-base disorders, *Current surgical therapy*, ed 12, Philadelphia, Saunders, 2017.)

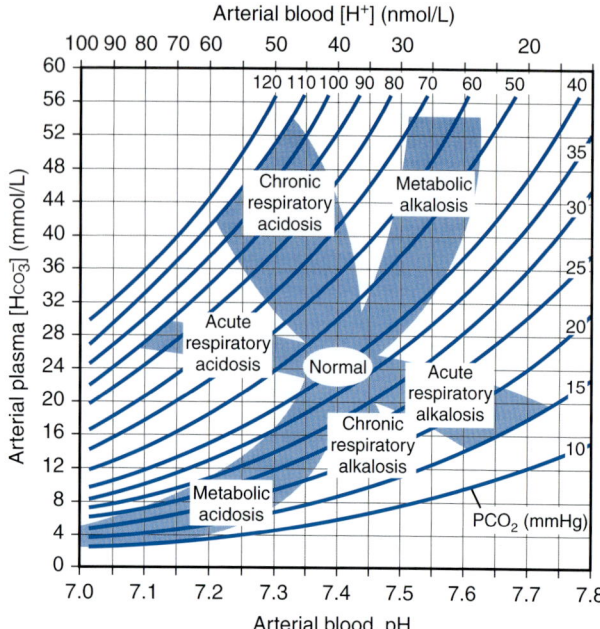

FIG. 3.9 Acid-base nomogram. Shaded areas represent 95% confidence limits of normal respiratory and metabolic compensations for primary disturbances. Points outside shaded areas represent a mixed disorder, assuming absence of laboratory error. *HCO3⁻*, Bicarbonate; *Pco₂*, carbon dioxide partial pressure. (From Vincent JL, Abraham E, Moore FA, et al: *Textbook of critical care*, ed 6, Philadelphia, Saunders, 2011.)

5. Acute Kidney Injury (Fig. 3.10)

Diagnostic Imaging

Best Test(s)
- Renal ultrasound

Ancillary Tests
- Chest radiograph (r/o pleural effusion, pulmonary/renal syndromes [e.g., Goodpasture, granulomatosis with polyangiitis])
- CT scan of kidneys without contrast (r/o suspected obstruction)

Laboratory Evaluation

Best Test(s)
- Urinalysis

Ancillary Tests
- Serum osmolality, urine osmolality, urine creatinine
- Serum electrolytes, calcium, phosphate, uric acid, magnesium
- Calculate FENa = [UNa/PNa/Ucreat/Pcreat × 100]; FEna <1 in prerenal, >2 in intrinsic renal failure

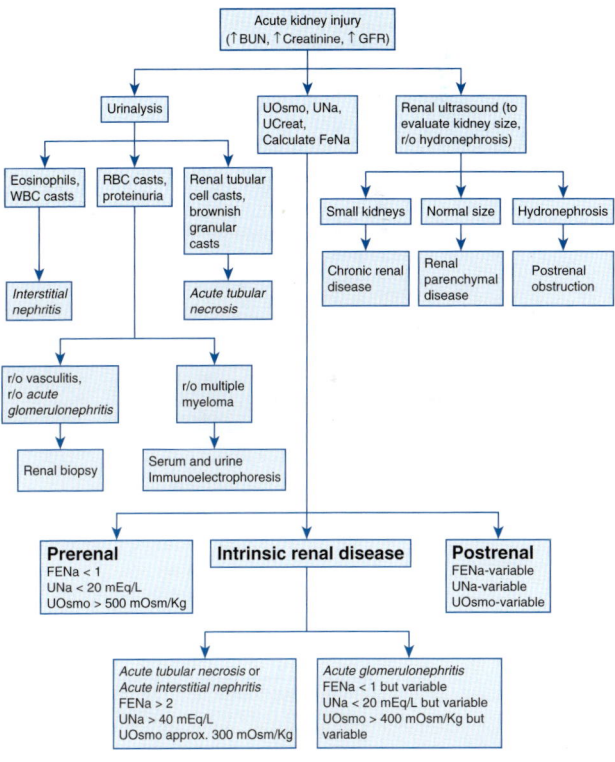

Fig. 3.10 Diagnostic algorithm.

6. Addison Disease (Adrenal Insufficiency) (Fig. 3.11)

Diagnostic Imaging

Best Test(s)
- None

Ancillary Tests
- CT or MRI of adrenals with contrast
- Chest x-ray

Laboratory Evaluation

Best Test(s)
- IV cosyntropin test, basal cortisol and serial measurement of cortisol, ACTH level

Ancillary Tests
- Serum electrolytes (hyponatremia, hyperkalemia)
- FBS, BUN, creatinine
- CBC (anemia)

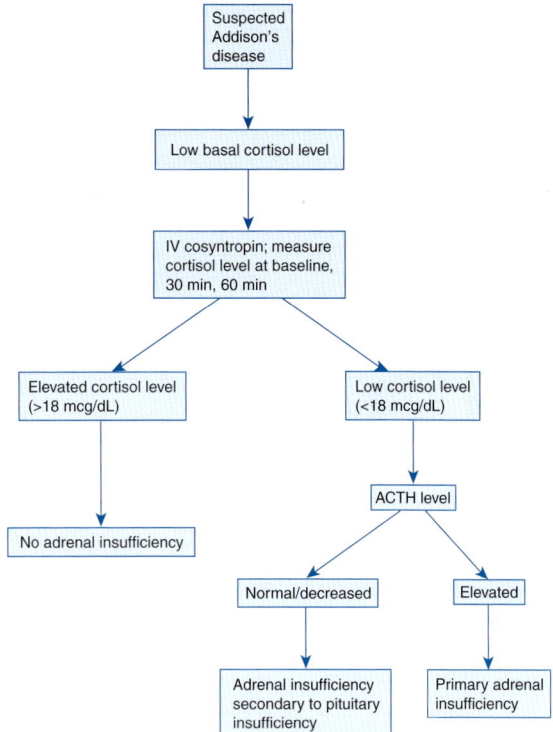

Fɪɢ. 3.11 Diagnostic algorithm.

7. Adrenal Mass (Fig. 3.12)

Diagnostic Imaging

Best Test(s)
- MRI of adrenal gland with contrast

Ancillary Tests
- CT of adrenal gland with and without contrast (Fig. 3.13) if MRI is contraindicated

Laboratory Evaluation

Best Test(s)
- Serum electrolytes

Ancillary Tests
- If symptoms of pheochromocytoma, obtain plasma-free metanephrine level, 24-hour urine collection for metanephrines
- If cushingoid appearance, obtain overnight dexamethasone suppression test
- If signs of virilization or feminization, order 24-hour urine for 17-ketosteroids and plasma DHEAS
- If hypertension is present with associated hypokalemia, evaluate for aldosteronism

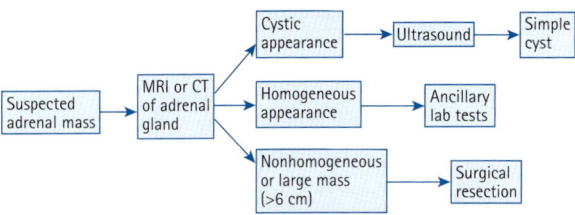

Fig. 3.12 Diagnostic algorithm.

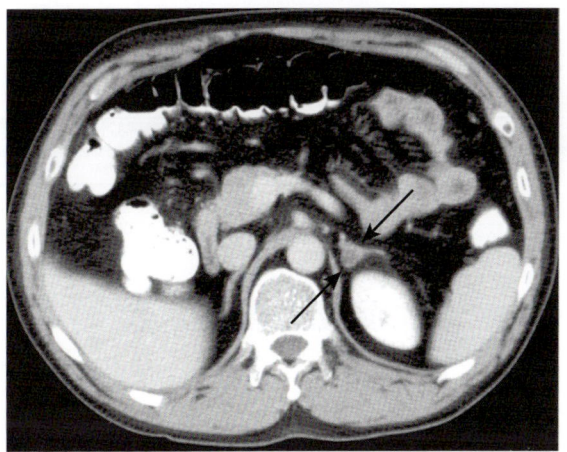

Fig. 3.13 Adrenal adenoma. On a computed tomography scan, incidental note was made of a small (approximately 1 cm) left adrenal mass (*arrows*). This small size combined with a relatively low-density (dark) center is consistent with a benign adrenal adenoma. Note that any adrenal mass larger than several centimeters should be suggestive of malignancy. (From Mettler FA: *Essentials of radiology*, ed 3, Philadelphia, Saunders, 2014.)

8. Alkaline Phosphatase Elevation (Fig. 3.14)

Diagnostic Imaging

Best Test(s)
- CT of liver

Ancillary Tests
- Ultrasound of liver
- Bone scintigraphy (nuclear bone scan) if Paget's disease of bone is suspected
- Plain x-rays of abnormal areas identified on bone scan to exclude metastatic disease and confirm Paget's disease

Laboratory Evaluation

Best Test(s)
- GGT

Ancillary Tests
- Serum calcium, phosphate
- ALT, AST

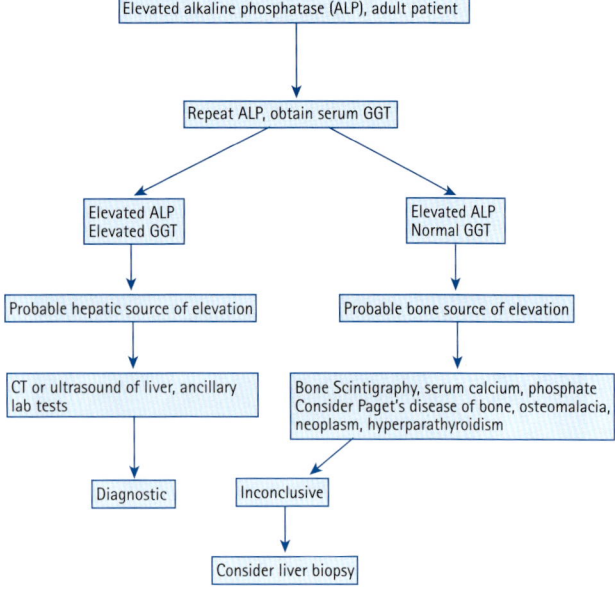

Fig. 3.14 Diagnostic algorithm.

9. ALT/AST Elevation (Fig. 3.15)

Diagnostic Imaging

Best Test(s)
• CT of liver

Ancillary Tests
• Ultrasound of liver

Laboratory Evaluation

Best Test(s)
• None

Ancillary Tests
• Ferritin/transferrin saturation
• Viral hepatitis serology
• GGT, alkaline phosphatase, bilirubin
• AMA, ASMA, ANA

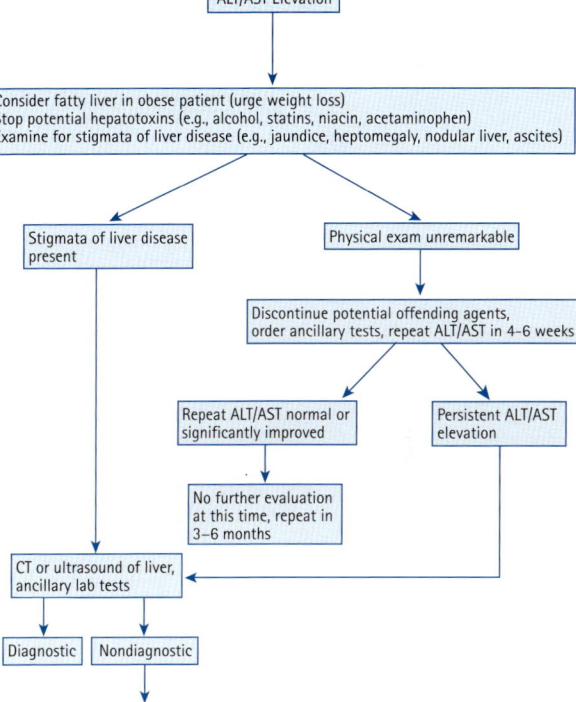

Fɪɢ. 3.15 Diagnostic algorithm.

10. Amenorrhea, Primary (Fig. 3.16)

Diagnostic Imaging

Best Test(s)
- MRI of pituitary/hypothalamus with gadolinium when hypothalamic/pituitary lesion is suspected

Ancillary Tests
- Pelvic ultrasound

Laboratory Evaluation

Best Test(s)
- Serum hCG
- FSH
- Prolactin
- TSH

Ancillary Tests
- Progesterone challenge test if initial labs are normal. Test will differentiate between estrogen-deficient state (no bleeding) and estrogen-sufficient state (withdrawal bleeding).
- Karyotype (r/o Turner syndrome) in patients with primary ovarian insufficiency

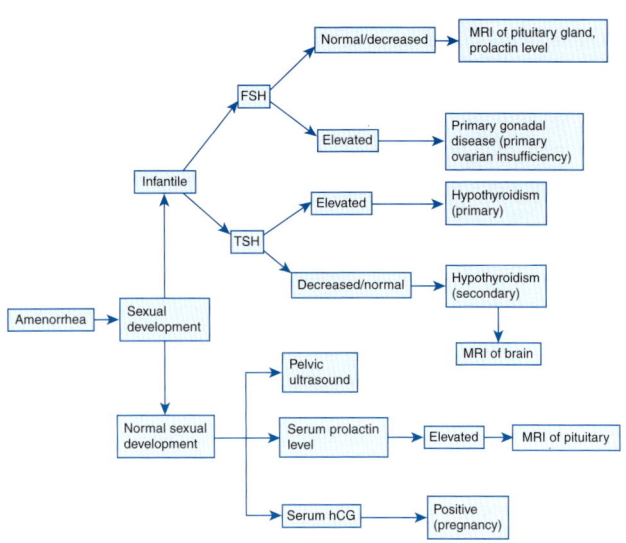

Fɪɢ. 3.16 Diagnostic algorithm.

11. Amenorrhea, Secondary (Fig. 3.17)

Diagnostic Imaging

Best Test(s)
- MRI of pituitary/hypothalamus with gadolinium when hypothalamic/pituitary lesion is suspected

Ancillary Tests
- Pelvic ultrasound
- CT or MRI of adrenals

Laboratory Evaluation

Best Test(s)
- Serum hCG
- Prolactin
- FSH
- Progesterone challenge test to differentiate between estrogen-deficient state (no bleeding) and estrogen-sufficient state (withdrawal bleeding)

Ancillary Tests
- LH
- Testosterone, DHEAS
- TSH

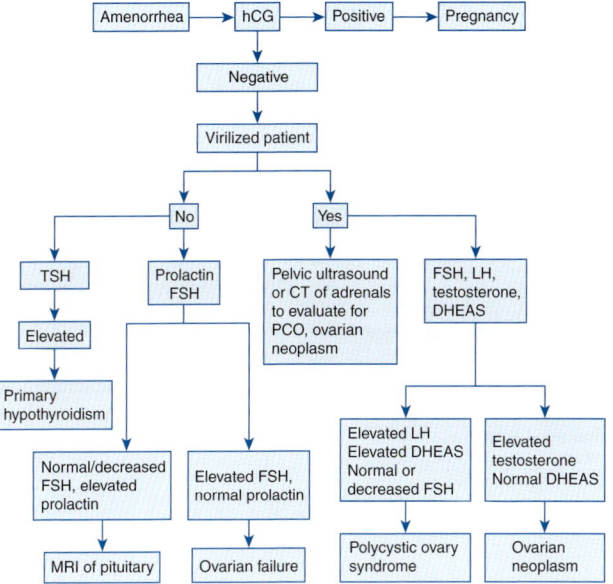

Fig. 3.17 Diagnostic algorithm.

12. Anemia, Macrocytic (Fig. 3.18)

Diagnostic Imaging

Best Test(s)
• None

Ancillary Tests
• None

Laboratory Evaluation

Best Test(s)
• Reticulocyte count

Ancillary Tests
• Serum B_{12} level, RBC folate level
• ALT, AST, GGTP
• TSH
• Direct and indirect antiglobulin test
• Stool for OB
• Bone marrow examination

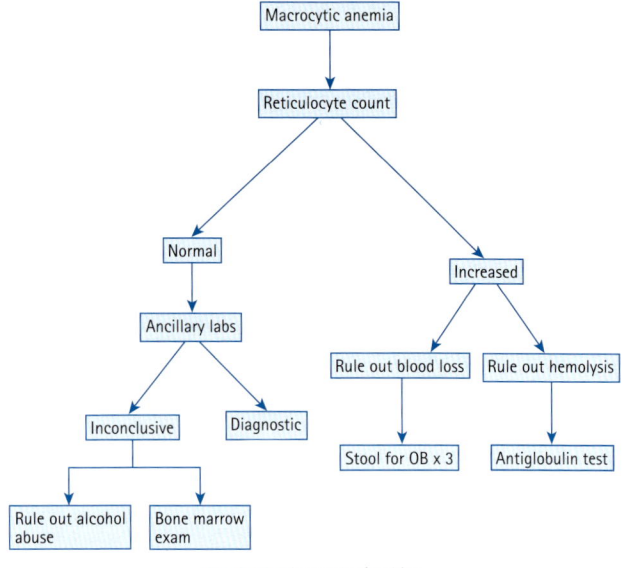

FIG. 3.18 Diagnostic algorithm.

13. Anemia, Microcytic (Fig. 3.19)

Diagnostic Imaging

Best Test(s)
• None

Ancillary Tests
• None

Laboratory Evaluation (Table 3.1)

Best Test(s)
• Reticulocyte count
• Stool for occult blood test × 3

Ancillary Tests
• Ferritin level
• TIBC, serum iron
• Hemoglobin electrophoresis
• Serum lead level

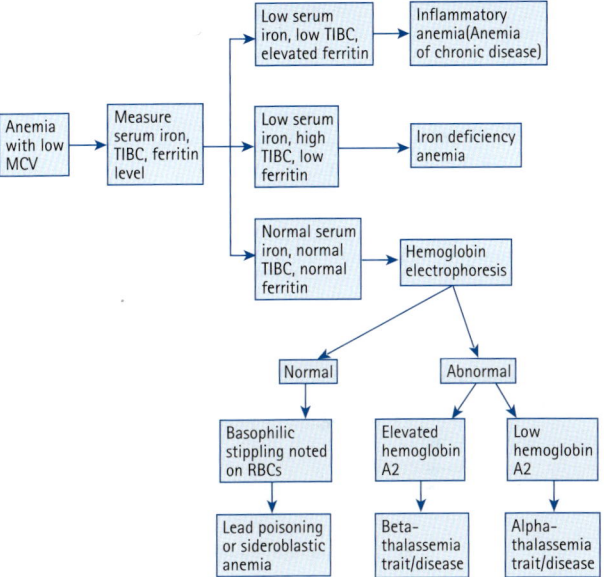

Fig. 3.19 Diagnostic algorithm.

TABLE 3.1	Laboratory Differentiation of Microcystic Anemias			
ABNORMALITY	**FERRITIN**	**SERUM IRON**	**TIBC**	**RDW**
Iron deficiency	↓	↓	↑	↑
Inflammatory anemia	N/↑	↓	↓	N
Sideroblastic anemia	N/↑	↑	N	N
Thalassemia	N/↑	N/↑	N/↓	N/↑

RDW, red cell distribution width; *TIBC,* total iron-binding capacity.

14. Antinuclear Antibody (ANA)–Positive (Fig. 3.20)

Diagnostic Imaging

Best Test(s)
- None

Ancillary Tests
- None

Laboratory Evaluation

Best Test(s)
- ANA pattern evaluation

Ancillary Tests
- Anti-Ds DNA Ab
- Anti-Smith Ab
- Anti-RNP Ab
- Anti-SS-A, Anti-SS-B
- ESR
- CBC

Box 3.1 summarizes major profiles of autoantibodies in systemic rheumatic diseases.

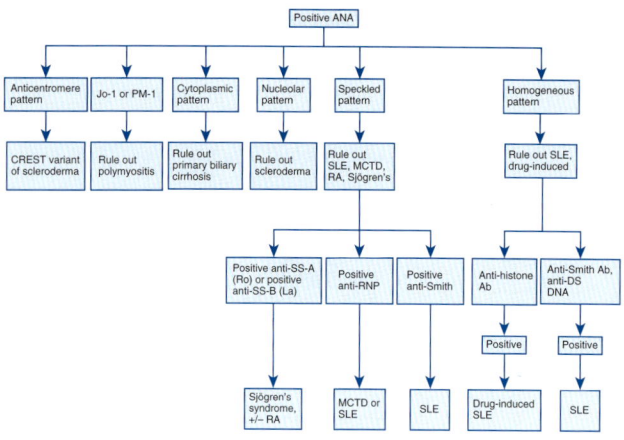

FIG. 3.20 Diagnostic algorithm.

BOX 3.1	Major Profiles of Autoantibodies in Systemic Rheumatic Diseases

1. **Systemic lupus erythematosus.** Multiple ANAs are frequently seen in systemic lupus erythematosus, often with high levels of anti-dsDNA and antinucleosome antibodies in active disease, together with low complement fractions and high novel biomarker levels, like SIGLEC-1. Distinctiveness of anti-Sm, anti-rRNP, anti-C1q, antinucleosomes, and anti-PCNA.
2. **Drug-induced lupus.** Restriction of ANAs to antihistone and/or antinucleosome antibodies.
3. **MCTD.** Restriction of ANA to U1nRNP (by disease definition).
4. **Sjögren's syndrome.** Characterized primarily by the presence of antibodies to SS-A/Ro and SS-B/La.
5. **Systemic sclerosis.** Profile consisting of antibodies to Scl-70/topo I, the centromere/kinetochore antigens, anti-RNA polymerases, and other nucleolar antigens such as fibrillarin and Th/To.
6. **Rheumatoid arthritis.** Frequent presence of rheumatoid factor and anti-citrullinated proteins (AKA, APF, anti-CCP).
7. **PM/DM.** Presence of Jo-1, Mi-2, and PM-Scl autoantibodies.
8. **Undifferentiated connective tissue diseases.** Autoantibodies, such as anticentromeres, anti-CCP, and anti-dsDNA, may antedate overt clinical disease by many years.

MCTD, mixed connective tissue disease; *PM/DM*, polymyositis/dermatomyositis.

(From McPherson RA, Pincus MR: *Henry's clinical diagnosis and management by laboratory methods*, ed 23, St. Louis, Elsevier, 2017.)

15. Aortic Dissection (Fig. 3.21)

Diagnostic Imaging

Best Test(s)
- CT (sensitivity 83%–100%); CT of aorta (Fig. 3.22) is generally readily available and performed as the initial diagnostic modality in suspected aortic dissection

Ancillary Tests
- MRI (sensitivity 90%–100%); difficult test for unstable, intubated patient
- TEE (sensitivity 97%–100%); can also detect aortic insufficiency and pericardial effusion
- Aortography (sensitivity 80%–90%); involves IV contrast; allows visualization of coronary arteries
- Chest radiograph (Fig. 3.23), ECG

Laboratory Evaluation

Best Test(s)
- None

Ancillary Tests
- CBC
- BUN, creatinine

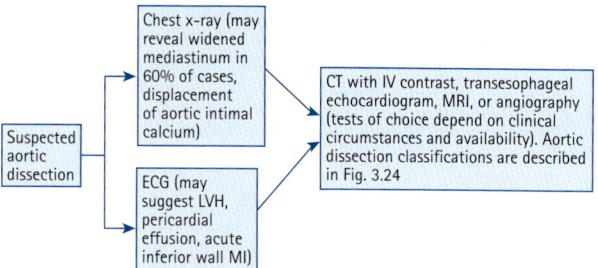

Fig. 3.21 Diagnostic algorithm.

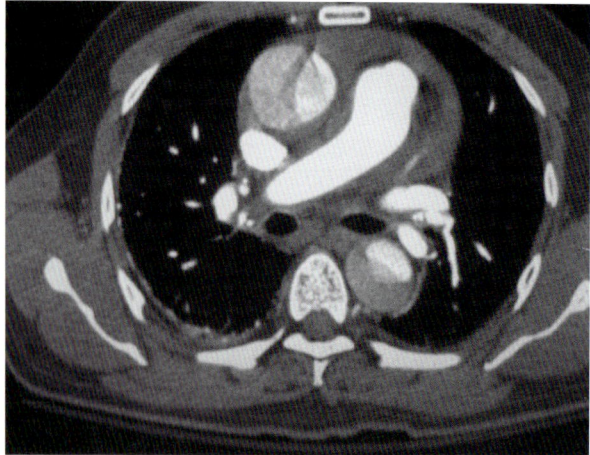

Fig. 3.22 Computed tomography scan demonstrating the true lumen and false lumen. (From Marx JA, Hockberger R, Walls R: *Rosen's emergency medicine*, ed 7, Philadelphia, Mosby, 2010.)

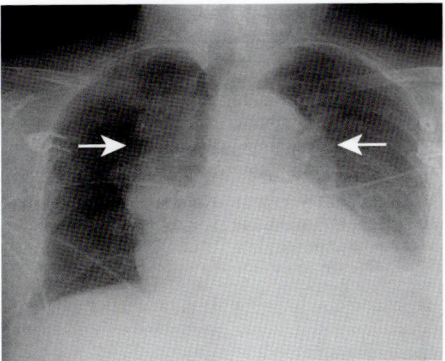

Fɪɢ. 3.23 Chest radiograph in acute type A aortic dissection demonstrating a widened mediastinum and enlargement of the ascending and descending aortic shadows (*arrows*). (From Bonow RO, Mann DL, Zipes DP, et al: *Braunwald's heart disease*, ed 9, Philadelphia, Saunders, 2012.)

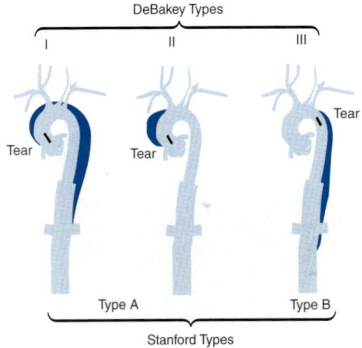

Fɪɢ. 3.24 Aortic dissection classification.

16. Appendicitis (Fig. 3.25)

Diagnostic Imaging

Best Test(s)
- CT of appendix (Fig. 3.26, *B*) with oral and IV contrast

Ancillary Tests
- Ultrasound of pelvis (see Fig. 3.26, *A*) may be used instead of CT in children and women of reproductive age when CT is unavailable or contraindicated

Laboratory Evaluation

Best Test(s)
- CBC with differential
- Urinalysis

Ancillary Tests
- Serum pregnancy test in women of reproductive age

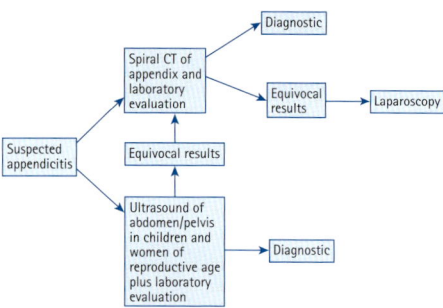

Fig. 3.25 Diagnostic algorithm.

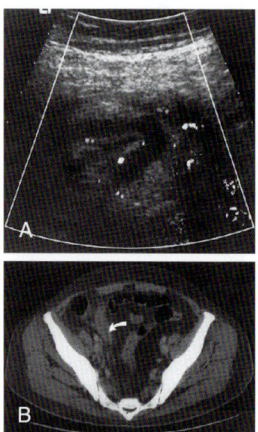

Fig. 3.26 Appendicitis. A, Ultrasound shows a thickened hypoechoic tubular blind-ended structure in the right iliac fossa. The surrounding fat is hyperechoic. B, Computed tomography shows the thickened, inflamed appendix (*arrow*). (Courtesy Dr. A McLean, St. Bartholomew's Hospital, London. From Grainger RG, Allison DJ, Adam A, Dixon AK [eds]: *Grainger and Allison's diagnostic radiology*, ed 4, Philadelphia, Churchill Livingstone, 2001.)

17. Ascites (Fig. 3.27)

Diagnostic Imaging

Best Test(s)
- Ultrasound of abdomen/pelvis (Fig. 3.28) is best initial diagnostic test

Ancillary Tests
- CT of abdomen/pelvis

Laboratory Evaluation

Best Test(s)
- Calculation of serum-ascites albumin gradient (SAAG) (Table 3.2); analysis of paracentesis fluid for LDH, glucose, albumin, total protein

Ancillary Tests
- Paracentesis fluid analysis for cell count and differential, Gram stain, AFB stain, bacterial and fungal cultures, amylase (Table 3.3)
- CBC, ALT, AST, BUN, creatinine

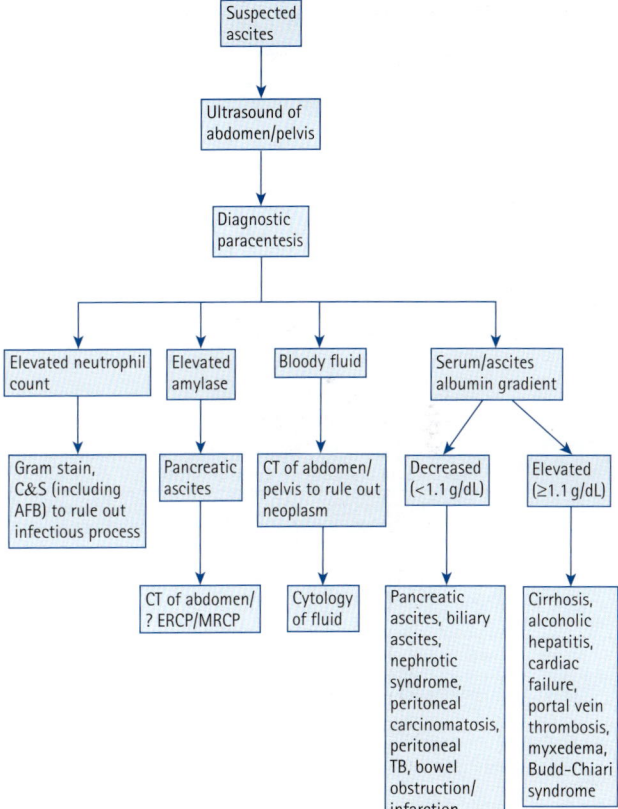

Fɪɢ. **3.27** Diagnostic algorithm.

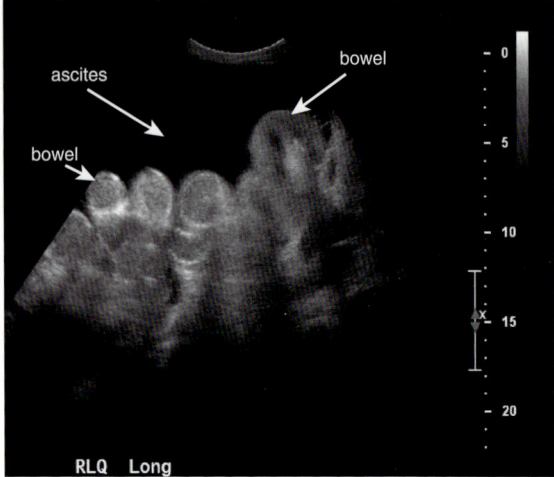

FIG. 3.28 Ascites, ultrasound. Ultrasound is useful for detection of ascites. Simple fluids such as ascites are excellent sound transmission media, reflecting almost no sound waves. As a consequence, they appear quite hypoechoic (black) on ultrasound. This view of the right lower quadrant shows loops of bowel surrounded by fluid. During the ultrasound, the bowel loops would be seen to undergo peristalsis and drift back and forth in the ascitic fluid with patient movement. Ultrasound cannot distinguish the composition of the fluid—ascites, liquid blood, liquid bile, urine, and infectious fluids have a similar appearance, with a few exceptions. Blood may coagulate and form septations within the fluid collection. Infectious fluids also frequently form loculated fluid collections that may be recognized on ultrasound, although the exact composition cannot be determined. (From Broder JS: *Diagnostic imaging for the emergency physician*, Philadelphia, Saunders, 2011.)

TABLE 3.2	Using the Serum-Ascites Albumin Gradient and the Ascites Total Protein Level to Diagnose the Cause of Ascites	
CONDITION	**SERUM-ASCITES ALBUMIN GRADIENT***	**ASCITES TOTAL PROTEIN LEVEL†**
Cirrhosis	High	Low
Malignant ascites	Low	High
Cardiac ascites	High	High

*High is greater than 1.1 g/dL; low is less than 1.1 g/dL.
†High is greater than 2.5 g/dL; low is less than 2.5 g/dL.
From Goldman L, Schafer AI: *Goldman's Cecil medicine*, ed 24, Philadelphia, Saunders, Elsevier, 2012.

TABLE 3.3	Characteristics of Ascitic Fluid in Various Conditions							
BIOLOGY	APPEARANCE	TOTAL PROTEIN (G/DL)	LDH (IU)	SPECIFIC GRAVITY	GLUCOSE (MG/DL)	WBCS/MM³	RBCS/MM³	AMYLASE
Neoplasm	Bloody Clear Chylous	>2.0	>200	Variable	<60	↑	↑↑	
Cirrhosis	Straw colored	<2.5	<200	<1.016	<60	↓	↓	
Nephrosis	Straw colored	<2.5	<200	<1.016	>60	↓	↓	
CHF	Straw colored	<2.5	<200	<1.016	>60	↓	↓	
Pyogenic	Turbid	>2.5	>200	>1.016	>60	↑↑ PMNs		
Pancreatic	Hemorrhagic Turbid Chylous	>2.5	>200	Variable	>60	Variable	Variable	↑↑

CHF, congestive heart failure; *LDH*, lactate dehydrogenase; *PMN*, polymorphonuclear neutrophil; *RBC*, red blood cell; *WBC*, white blood cell.
From Ferri F: *Practical guide to the care of the medical patient*, ed 8, St. Louis, Mosby, Elsevier, 2010.

18. Avascular Necrosis (Fig. 3.29)

Diagnostic Imaging

Best Test(s)
- MRI of affected joint (Fig. 3.30, *A*)

Ancillary Tests
- Bone scan if MRI is contraindicated or not readily available
- Plain films of affected joint usually insensitive in early course but more evident as disease progresses (Fig. 3.30, *B*)

Laboratory Evaluation

Best Test(s)
- None

Ancillary Tests
- CBC with differential
- ESR (nonspecific)
- ANA

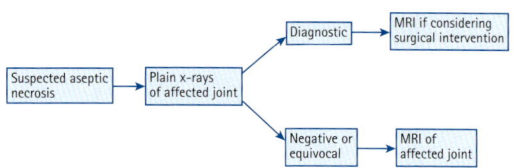

Fɪɢ. 3.29 Diagnostic algorithm.

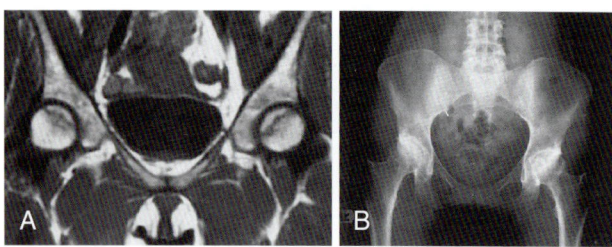

Fɪɢ. 3.30 Aseptic necrosis of the hip in a renal transplant recipient. A, Early changes consisting of low-intensity oblique lines are noted by magnetic resonance imaging. B, Late changes of avascular necrosis by radiograph show narrowing of the hip joint space, sclerosis of the femoral head, and flattening of the left femoral head. (From Johnson RJ, Feehally J: *Comprehensive clinical nephrology*, ed 2, St. Louis, Mosby, 2000.)

19. Back Pain, Acute, Lumbosacral (LS) Area (Fig. 3.31)

Diagnostic Imaging

Best Test(s)
• MRI of LS spine

Ancillary Tests
• Plain radiographs of spine
• CT scan of LS spine if MRI is contraindicated
• Ultrasound or CT of abdominal aorta if AAA is suspected

Laboratory Evaluation

Best Test(s)
• None

Ancillary Tests
• CBC
• ESR
• Urinalysis

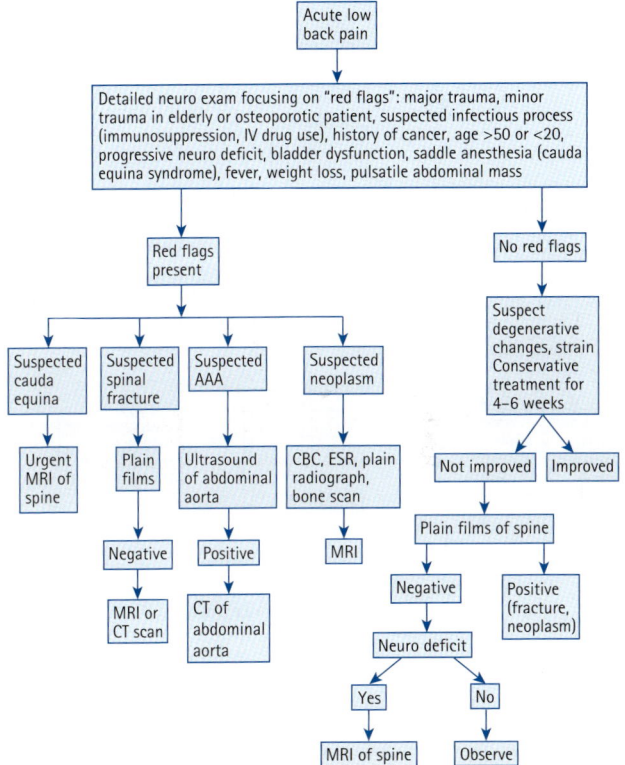

Fɪɢ. 3.31 Diagnostic algorithm.

20. Bilirubin Elevation (Fig. 3.32)

Diagnostic Imaging

Best Test(s)
• None

Ancillary Tests
• Ultrasound of abdomen
• CT of abdomen
• ERCP
• MRCP

Laboratory Evaluation

Best Test(s)
• Bilirubin fractionation

Ancillary Tests
• Alkaline phosphatase
• ALT, AST, PT (INR)
• CBC
• Antiglobulin test, haptoglobin
• Viral hepatitis panel
• ANA
• LDH

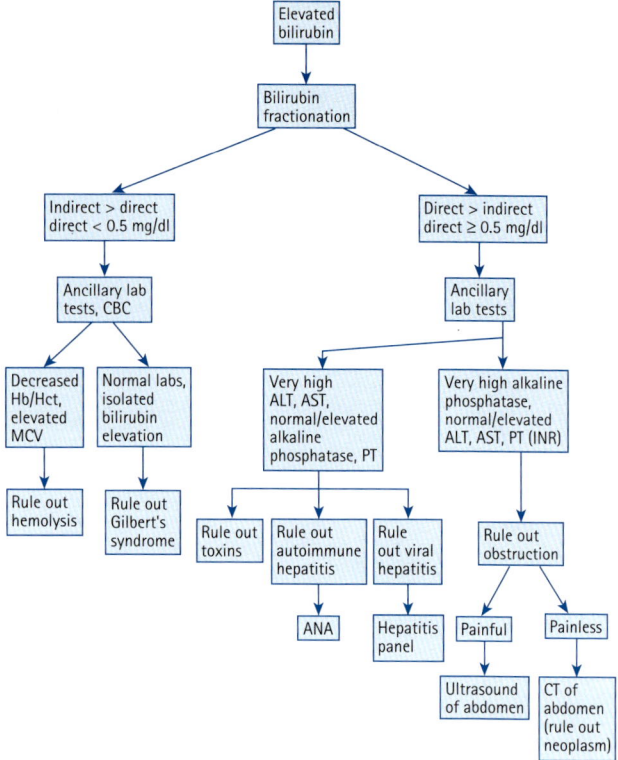

Fig. 3.32 Diagnostic algorithm.

21. Bleeding Disorder, Congenital (Fig. 3.33)

Diagnostic Imaging

Best Test(s)
- None

Ancillary Tests
- None

Laboratory Evaluation

Best Test(s)
- PT (INR), PTT

Ancillary Tests
- Platelet count
- Clot stability test
- TT
- Factor VIII, IX assay
- PFA 100 assay
- Fibrinogen level
- Factor II, V, X, XIII

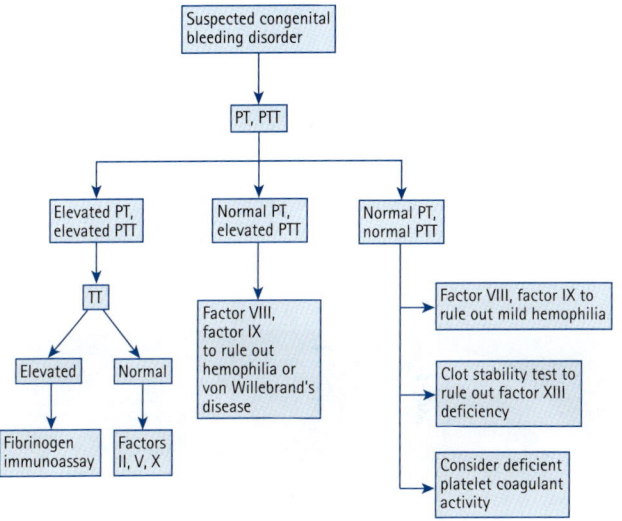

Fig. 3.33 Diagnostic algorithm.

22. Brain Abscess (Fig. 3.34)

Diagnostic Imaging

Best Test(s)
- MRI of brain with contrast

Ancillary Tests
- CT scan with IV contrast (Fig. 3.35) if MRI is contraindicated
- Echocardiography if bacterial endocarditis is suspected source of septic emboli to brain

Laboratory Evaluation

Best Test(s)
- CBC with differential

Ancillary Tests
- Blood cultures (10% positive)
- ESR

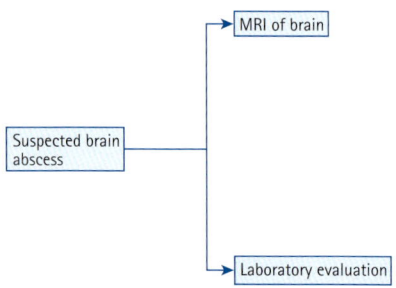

Fɪɢ. **3.34** Diagnostic algorithm.

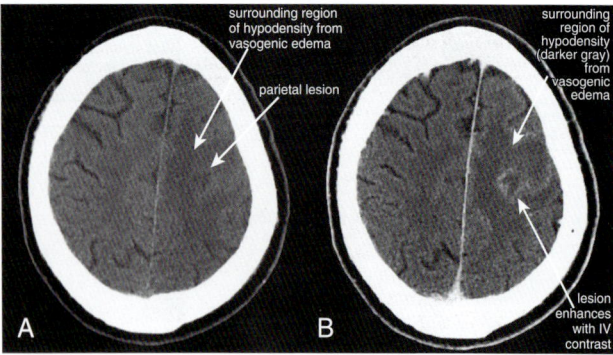

Fɪɢ. **3.35** Computed tomography (CT) with contrast. A, Note the large, wall-enhancing abscess in the left frontal lobe causing a shift of the brain to the right. B, The patient had no neurologic signs until just before the CT scan because the abscess is located in the frontal lobe, a "silent" area of the brain. (Kliegman RM, Stanton B, St. Geme J, et al: *Nelson textbook of pediatrics*, ed 19, Philadelphia, Saunders, 2011.)

23. Breast Mass (Fig. 3.36)

Diagnostic Imaging

Best Test(s)
- Diagnostic mammogram (Fig. 3.38, A)
- Breast ultrasound (Fig. 3.38, B)

Ancillary Tests
- MRI of breast in selected cases (e.g., prior breast cancer, breast implant)

Laboratory Evaluation

Best Test(s)
- Breast biopsy of suspicious lesion (Fig. 3.37)

Ancillary Tests
- Aspiration and cytologic examination of breast cyst

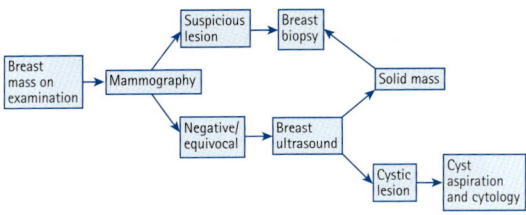

Fig. 3.36 Diagnostic algorithm.

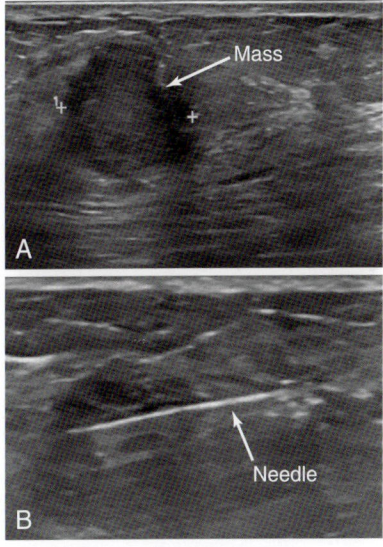

Fig. 3.37 An ultrasound image shows a nonpalpable superficial breast mass (A). A second image shows the linear echogenic biopsy needle penetrating the mass (B). (From Mettler FA: *Essentials of radiology*, ed 3, Philadelphia, Saunders, 2014.)

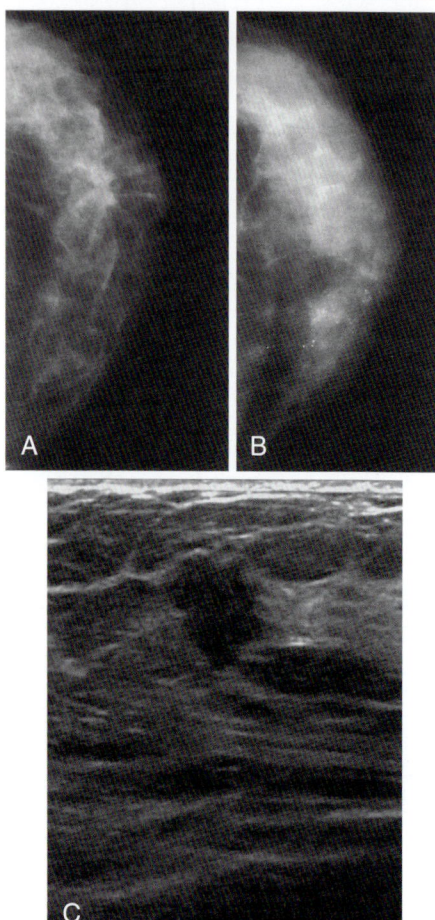

Fɪɢ. **3.38** Mammogram and ultrasound findings of breast disease. A, A stellate mass in the breast. The combination of a density with spiculated borders and distortion of surrounding breast architecture suggests a malignancy. B, Clustered microcalcifications. Fine, pleomorphic, and linear calcifications that cluster together suggest the diagnosis of ductal carcinoma in situ. C, An ultrasound image of breast cancer. The mass is solid, containing internal echoes and displaying an irregular border. Most malignant lesions are taller than they are wide. (From Townsend CM, Beauchamp RD, Evers BM, Mattox KL [eds]: *Sabiston textbook of surgery*, ed 17, Philadelphia, Saunders, 2004.)

24. Carcinoid Syndrome (Fig. 3.39)

Diagnostic Imaging

Best Test(s)
- Radionuclide iodine-113–labeled somatostatin scan (123-ISS) (octreotide scan) (Fig. 3.40)
- CT scan of abdomen, pelvis, and chest with oral and IV contrast to localize tumor and detect metastases

Ancillary Tests
- Echocardiogram or cardiac MRI in suspected cardiac carcinoid

Laboratory Evaluation

Best Test(s)
- 24-hour urine for HIAA

Ancillary Tests
- ALT, AST
- Serum electrolytes, BUN, creatinine
- Alkaline phosphatase
- Endoscopic biopsy of gastric carcinoids

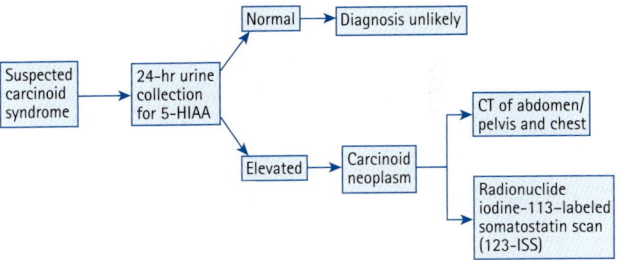

Fig. 3.39 Diagnostic algorithm.

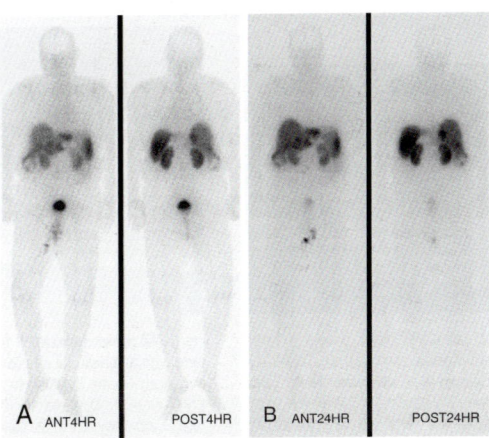

Fig. 3.40 Octreoscan illustrating uptake pattern in liver metastases from a small-bowel carcinoid. (From Cameron JL, Cameron AM: Small-bowel carcinoid/neuroendocrine tumors, *Current surgical therapy*, ed 10, Philadelphia, Saunders, 2011.)

25. Cardiomegaly on Chest Radiograph (Fig. 3.41)

Diagnostic Imaging

Best Test(s)
- Echocardiogram (Fig. 3.42)

Ancillary Tests
- Cardiac MRI if pericardial thickening or mass
- ECG

Laboratory Evaluation

Best Test(s)
- None

Ancillary Tests
- TSH
- Creatinine, ALT
- ESR
- Urinalysis

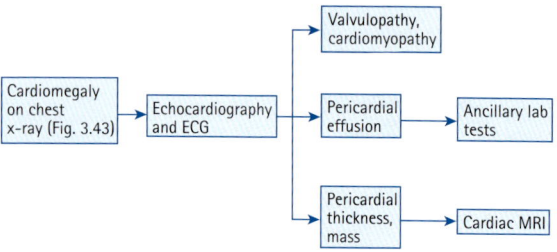

Fig. 3.41 Diagnostic algorithm.

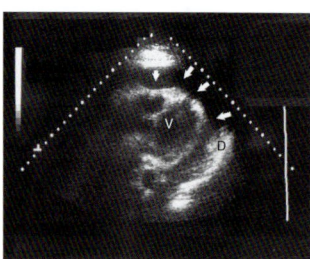

Fig. 3.42 Echocardiogram shows a large pericardial effusion (*arrows*). *D,* Diaphragmatic pericardium; *V,* ventricle. (From Sellke FW, del Nido PJ, Swanson SJ: *Sabiston & Spencer surgery of the chest,* ed 9, Philadelphia, Elsevier, 2016.)

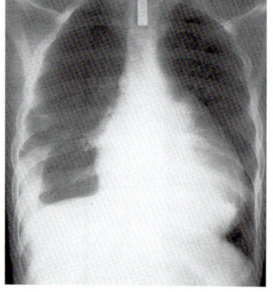

Fig. 3.43 Chest radiograph from a patient with a pericardial effusion showing a typical "water bottle" heart. (From Crawford MH, DiMarco JP, Paulus WJ [eds]: *Cardiology,* ed 2, St. Louis, Mosby, 2004.)

26. Cholangitis (Fig. 3.44)

Diagnostic Imaging

Best Test(s)
- Ultrasound of abdomen is preferred initial screening test; allows visualization of gallbladder and bile ducts to differentiate extrahepatic obstruction from intrahepatic cholestasis; insensitive but specific for visualization of common duct stones

Ancillary Tests
- CT of abdomen (Fig. 3.45) is less accurate for gallstones but more sensitive than ultrasound for visualization of bilirubin distal part of common bile duct; also allows better definition of neoplasm
- MRCP or ERCP indicated if CT or ultrasound is inconclusive; MRCP can be used to visualize common bile duct and level of obstruction; ERCP can also be used to confirm obstruction and its level; it also allows collection of specimen for culture and cytologic examination and provides relief of obstruction

Laboratory Evaluation

Best Test(s)
- CBC with differential
- Blood culture × 2

Ancillary Tests
- Serum amylase, lipase
- ALT, AST, alkaline phosphatase

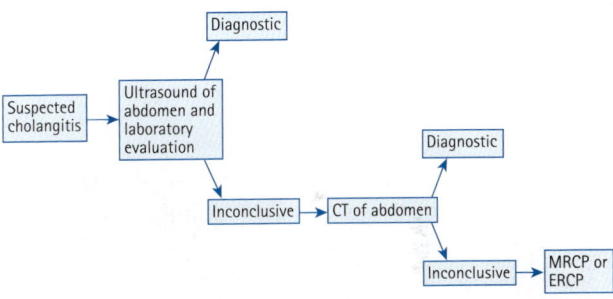

FIG. 3.44 Diagnostic algorithm.

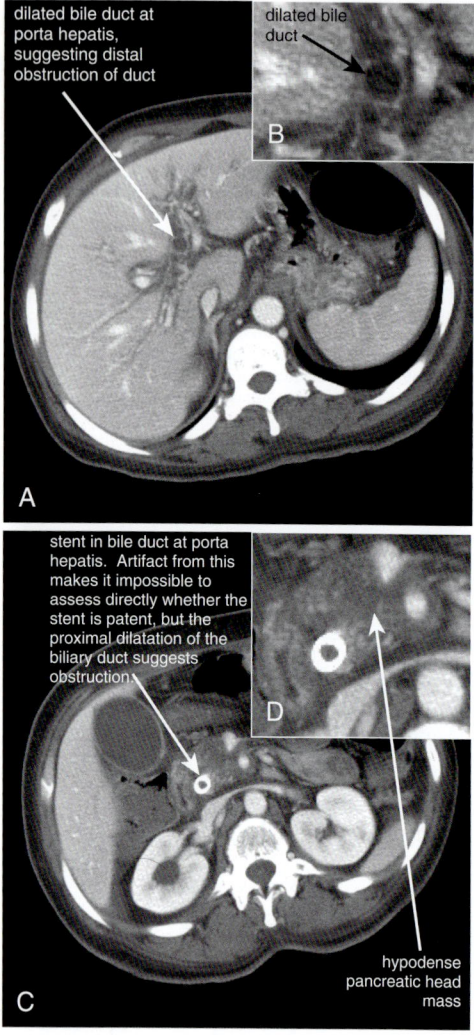

Fig. 3.45 Ascending cholangitis, dilated bile ducts, computed tomography with intravenous contrast. Obstruction of the biliary ducts can lead to devastating ascending infection—cholangitis. This is largely a clinical diagnosis, because classic imaging findings such as pneumobilia can lag behind clinical deterioration. Fever, jaundice, abdominal pain, and altered mental status should prompt immediate antibiotic therapy and surgical consultation, regardless of imaging findings. A, C, Axial images. B, D, Close-ups from A and C, respectively. This patient with a known pancreatic mass and a previously placed biliary stent presented with right upper quadrant abdominal pain. She became febrile during her emergency department stay. Her scan shows biliary ductal dilatation, suggesting obstruction of her stent. In addition to antibiotic therapy, a procedure to relieve obstruction is needed. Her blood cultures subsequently grew *Escherichia coli*, consistent with ascending cholangitis. (From Broder JS: *Diagnostic imaging for the emergency physician*, Philadelphia, Saunders, 2011.)

Labels within figure:
- dilated bile duct at porta hepatis, suggesting distal obstruction of duct
- dilated bile duct (B)
- stent in bile duct at porta hepatis. Artifact from this makes it impossible to assess directly whether the stent is patent, but the proximal dilatation of the biliary duct suggests obstruction.
- hypodense pancreatic head mass

27. Cholecystitis (Fig. 3.46)

Diagnostic Imaging

Best Test(s)
- Ultrasound of gallbladder (Fig. 3.47)

Ancillary Tests
- Nuclear imaging (HIDA) scan (Fig. 3.48) is useful for suspected acalculous cholecystitis or when ultrasound is inconclusive
- CT of abdomen useful in suspected abscess, neoplasm, or pancreatitis

Laboratory Evaluation

Best Test(s)
- CBC with differential

Ancillary Tests
- Alkaline phosphatase
- ALT, AST, bilirubin
- Serum amylase

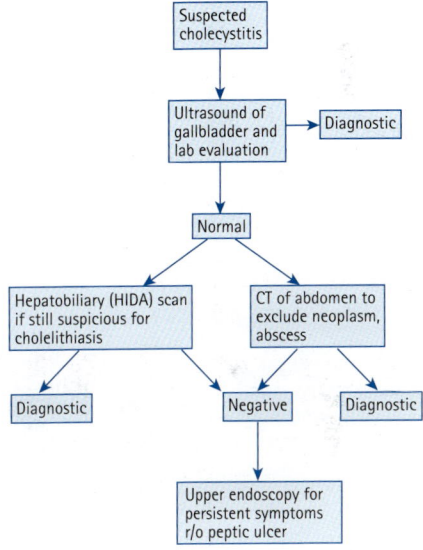

Fɪɢ. 3.46 Diagnostic algorithm.

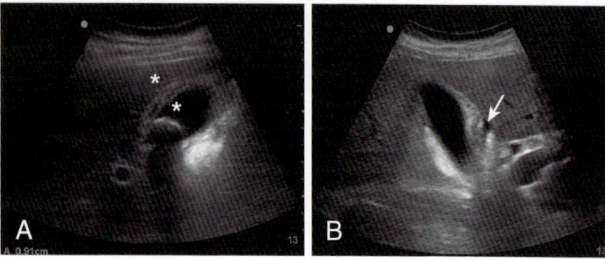

Fig. 3.47 Acute calculous cholecystitis. A, Gallstone with posterior shadowing and anterior wall thickening (measured to be 9.1 mm with the calipers; stars indicate the ends of the calipers measuring the gallbladder wall). B, Acute cholecystitis with pericholecystic fluid (*arrow*). A gallstone was impacted in the gallbladder neck. (From Adams JG et al: *Emergency medicine: clinical essentials*, ed 2, Philadelphia, Elsevier, 2013.)

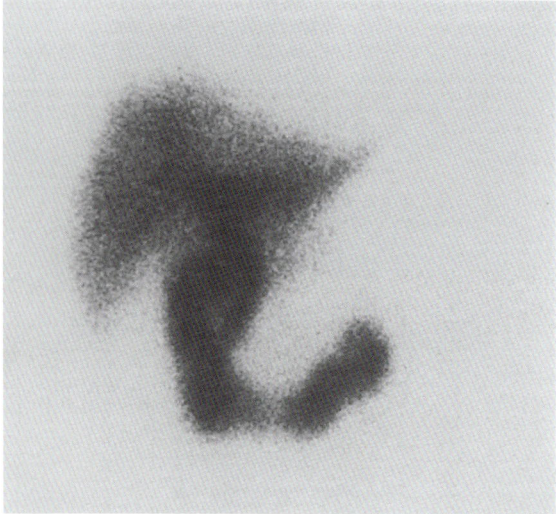

Fig. 3.48 Acute cholecystitis. Following the intravenous administration of 200 MBq (5 mCi) of Tc-99m–HIDA and a stimulus of cholecystokinin, the region of the liver and gallbladder is imaged. Intrahepatic bile ducts are visualized, as is excretion through the common duct into the small bowel. The gallbladder is not seen. This patient had gallstones demonstrated by ultrasound and confirmed at surgery. The pathologic diagnosis was acute cholecystitis. (From Grainger RG, Allison DJ, Adam A, Dixon AK [eds]: *Grainger and Allison's diagnostic radiology*, ed 4, Philadelphia, Churchill Livingstone, 2001.)

28. Cholelithiasis (Fig. 3.49)

Diagnostic Imaging

Best Test(s)
- Ultrasound of gallbladder (Fig. 3.50) will detect stones and biliary sludge (sensitivity 95%, specificity 90%)

Ancillary Tests
- CT of abdomen useful in patients with inconclusive ultrasound to r/o neoplasm or abscess mimicking cholelithiasis; however, it is less sensitive than ultrasound for cholelithiasis

Laboratory Evaluation

Best Test(s)
- None

Ancillary Tests
- Lipid panel
- ALT, alkaline phosphatase, bilirubin
- CBC, amylase

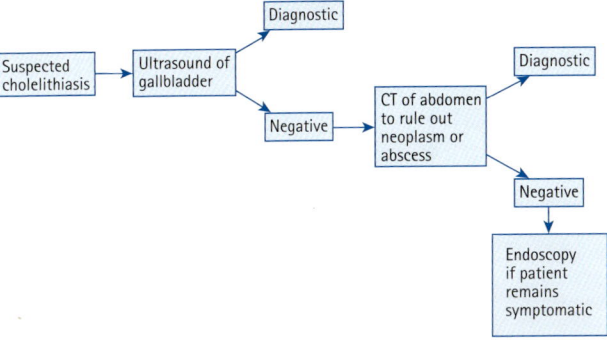

FIG. 3.49 Diagnostic algorithm.

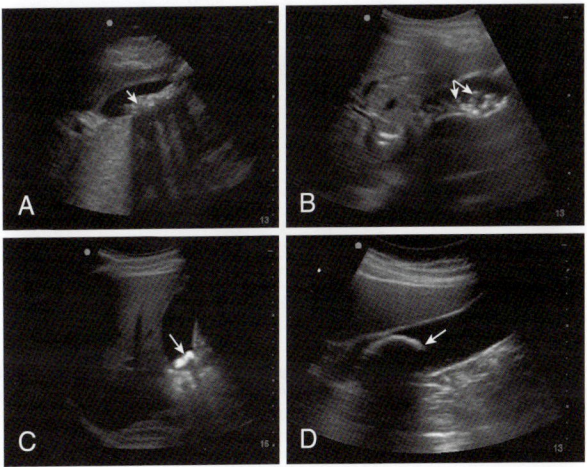

FIG. 3.50 Examples of cholelithiasis (A to D) depicting the different appearances of gallstones (*arrows*). (From Adams JG et al: *Emergency medicine: clinical essentials*, ed 2, Philadelphia, Elsevier, 2013.)

29. Complex Regional Pain Syndrome (Reflex Sympathetic Dystrophy [RSD]) (Fig. 3.51)

Diagnostic Imaging

Best Test(s)
- None

Ancillary Tests
- Three-phase radionuclide bone scan (Fig. 3.52)
- Radiograph of affected limb (r/o other cause of patient's symptoms)

Laboratory Evaluation

Best Test(s)
- None

Ancillary Tests
- CBC, ESR
- ANA
- FBS

Fɪɢ. 3.51 Diagnostic algorithm.

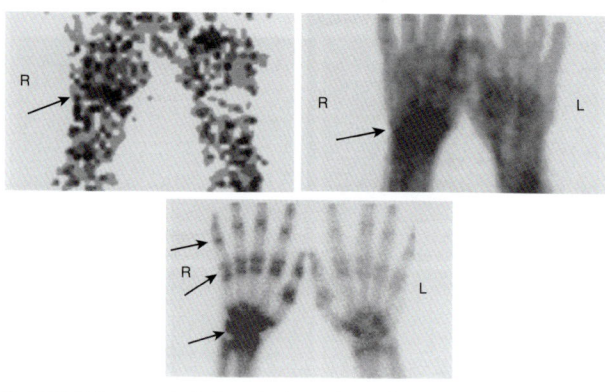

Fɪɢ. 3.52 Reflex sympathetic dystrophy in a patient with a history of right wrist fracture. Three-phase bone scintography shows increased tracer delivery to the right distal upper extremity diffusely *(arrow)* on flow *(top)* and blood pool *(middle)* images. Delayed images *(bottom)* reveal diffuse abnormal uptake in the wrist *(lower arrow)* and increased activity in a juxtaarticular distribution *(upper arrows)*. (From DeLee D, Drez D: *DeLee and Drez's orthopaedic sports medicine,* 2 ed. Philadelphia, 2003, Saunders.)

30. Constipation (Fig. 3.53)

Diagnostic Imaging

Best Test(s)
- CT colonography if patient refuses colonoscopy

Ancillary Tests
- Consider barium enema (if patient refuses colonoscopy or CT colonography)
- Pelvic ultrasound
- GI motility study

Laboratory Evaluation

Best Test(s)
- None

Ancillary Tests
- CBC
- TSH
- Serum calcium, electrolytes, BUN, creatinine, ALT

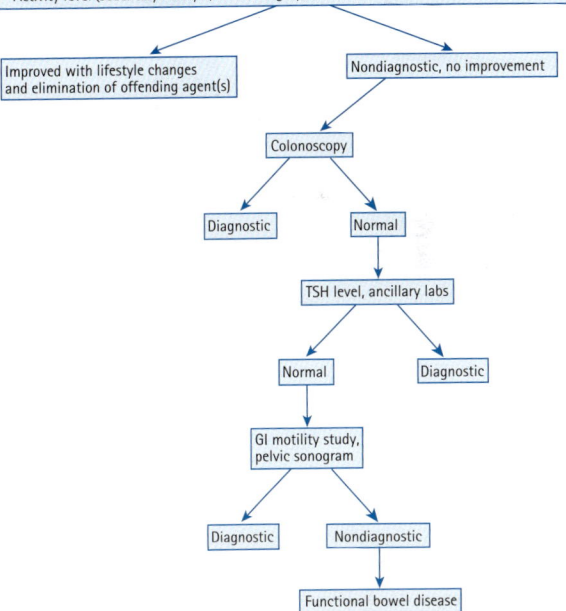

FIG. 3.53 Diagnostic algorithm.

31. Creatinine Phosphokinase (CPK) Elevation (Fig. 3.54)

Diagnostic Imaging

Best Test(s)
• None

Ancillary Tests
• None

Laboratory Evaluation

Best Test(s)
• CPK fractionation

Ancillary Tests
• Serum troponin levels
• CBC, electrolytes, BUN, ALT, creatinine
• TSH, urinalysis
• ANA
• ESR, CRP

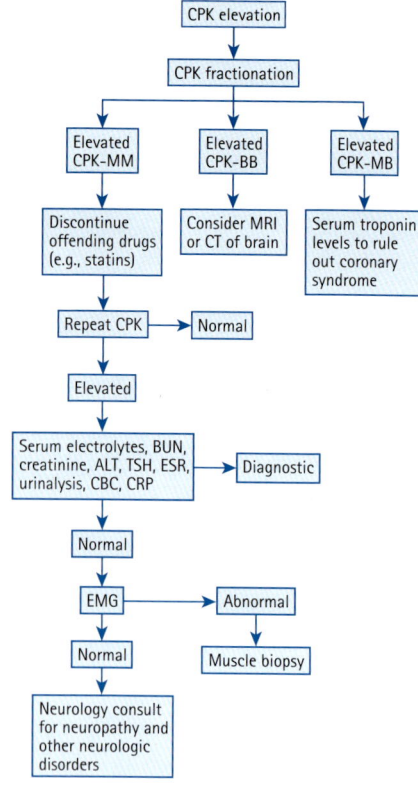

FIG. 3.54 Diagnostic algorithm.

32. Cushing's Syndrome (Fig. 3.56, Table 3.4)

Diagnostic Imaging*

Best Test(s)
- MRI or CT of adrenals with IV contrast in suspected adrenal Cushing's syndrome
- MRI of brain (Fig. 3.55) with gadolinium in suspected pituitary Cushing's syndrome

Ancillary Tests
- CT of chest in patients with ectopic ACTH production to r/o neoplasm of lung, kidney, or pancreas

Laboratory Evaluation

Best Test(s)
- 24-h urine-free cortisol (gold standard for diagnosis, a 3-fold to 4-fold increase above normal confirms diagnosis)
- Overnight dexamethasone suppression test
- Plasma cortisol
- Late-night salivary cortisol measurement

Ancillary Tests
- Electrolytes, creatinine, glucose
- CRH plus desmopressin stimulation test

*Imaging studies are indicated only after biochemical documentation of hypercortisolism.

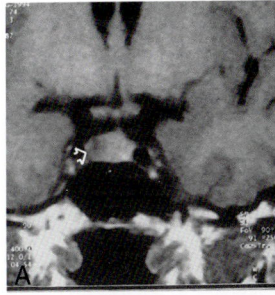 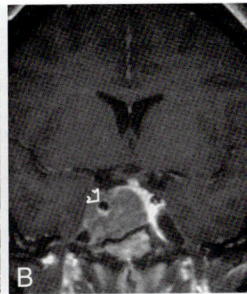

FIG. 3.55 A, Magnetic resonance imaging (MRI) scan of pituitary demonstrates the typical appearance of a pituitary microadenoma. A hypodense lesion is seen in the right side of the gland (*arrow*), with deviation of the pituitary stalk away from the lesion. After a biochemical diagnosis of Cushing's disease, this patient was cured by transsphenoidal hypophysectomy. B, MRI scan of the pituitary gland demonstrates a large macroadenoma (*arrow*) in a patient with Cushing's disease. In contrast to smaller tumors, large macroadenomas are invariably invasive and recur after surgery. (From Melmed S, Polonsky K, Larsen P, Kronenberg H: *Williams textbook of endocrinology*, ed 12, Philadelphia, Saunders, 2011.)

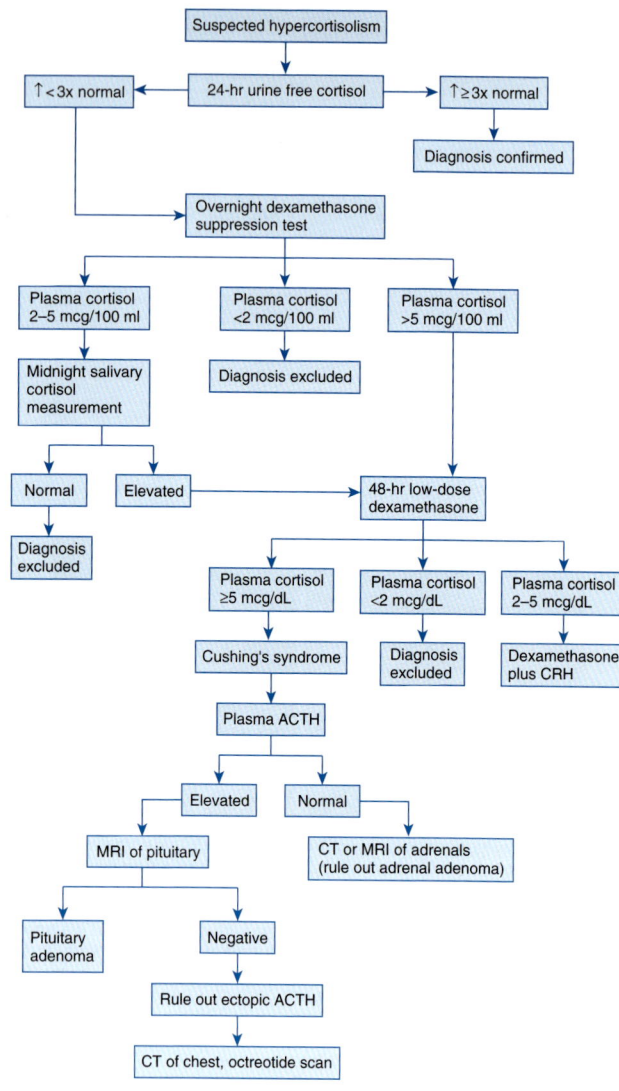

Fig. 3.56 Diagnostic algorithm.

TABLE 3.4   Tests Used in the Differential Diagnosis of Cushing's Syndrome*

ETIOLOGY	OVERNIGHT DEXAMETHASONE SUPPRESSION TEST	PLASMA ACTH	LOW-DOSE DEXAMETHASONE	HIGH-DOSE DEXAMETHASONE	CORTICOTROPIN-RELEASING HORMONE STIMULATION OF ACTH	PETROSAL-TO-PERIPHERAL ACTH RATIO
Normal	Suppression	Normal	Suppression		Normal	
Pituitary	No suppression	Normal or high	No suppression	Suppression	Normal or increased	>2
Ectopic	No suppression	High or normal	No suppression	No suppression	No response	<1.5
Adrenal	No suppression	Low	No suppression	No suppression	No response	

ACTH, adrenocorticotropic hormone.

*Classic responses are indicated. Certain cases of ectopic ACTH production are suppressed by high-dose dexamethasone or are stimulated by corticotropin-releasing hormone. In these cases, petrosal sinus sampling is the most reliable method for distinguishing pituitary and ectopic sources of ACTH.

From Goldman L, Schafer AI: *Goldman's Cecil medicine,* ed 24, Philadelphia, Saunders, Elsevier, 2012.

33. Deep Vein Thrombosis (DVT) (Fig. 3.57, Table 3.5)

Diagnostic Imaging

Best Test(s)
- Compressive duplex ultrasonography of affected extremity (Fig. 3.58)

Ancillary Tests
- Contrast venography; "gold standard" but invasive and painful
- MRDTI; very accurate and noninvasive but limited by cost and availability

Laboratory Evaluation

Best Test(s)
- D-dimer assay by ELISA

Ancillary Tests
- PT, PTT, platelet count
- Coagulopathy workup (e.g., protein C, protein S, antithrombin III, factor V Leiden, lupus anticoagulant) in patients with suspected coagulopathy

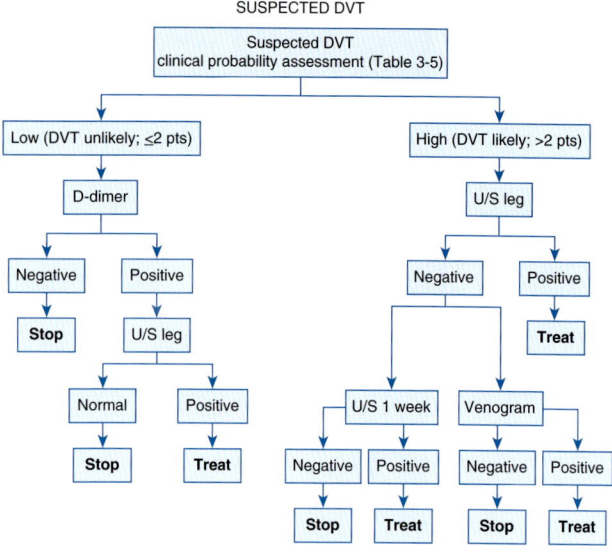

Fig. 3.57 Integrated strategy for diagnosis of deep venous thrombosis (DVT) using clinical probability assessment, measurement of D-dimer, and ultrasonography of the legs as primary diagnostic tests. If the clinical probability is low (i.e., DVT is unlikely and D-dimer is negative), no further investigations are required. If D-dimer is positive, proceed to ultrasonography of the legs; then either treat the DVT or stop the investigations. If the clinical probability is high (i.e., DVT is likely), D-dimer measurement need not be carried out; proceed directly to ultrasonography of legs. If negative, options are to repeat ultrasound in 1 week or, in some cases with high suspicion for DVT, to perform an ascending venogram. (From Vincent JL, Abraham E, Moore FA, et al: *Textbook of critical care*, ed 6, Philadelphia, Saunders, 2011.)

TABLE 3.5	Wells' Clinical Assessment Model for the Pretest Probability of Lower-Extremity Deep Vein Thrombosis	
		SCORE
Active cancer (treatment ongoing or within previous 6 months or palliative)		1
Paralysis, paresis, or recent plaster immobilization of the lower extremities		1
Recently bedridden >3 days or major surgery within 4 weeks		1
Localized tenderness along the distribution of the deep venous system		1
Entire leg swollen		1
Calf swelling >3 cm asymptomatic side (measured 10 cm below tibial tuberosity)		1
Pitting edema confined to the symptomatic leg		1
Collateral superficial veins (nonvaricose)		1
Alternative diagnosis as likely or greater than likelihood of DVT		−2

DVT, deep vein thrombosis.
In patients with symptoms in both legs, the more symptomatic leg is used. Pretest probability is calculated as the total score: high >3; moderate 1 or 2; low <0.
From Crawford MH, DiMarco JP, Paulus WJ [eds]: *Cardiology*, ed 2, St. Louis, Mosby, 2004.

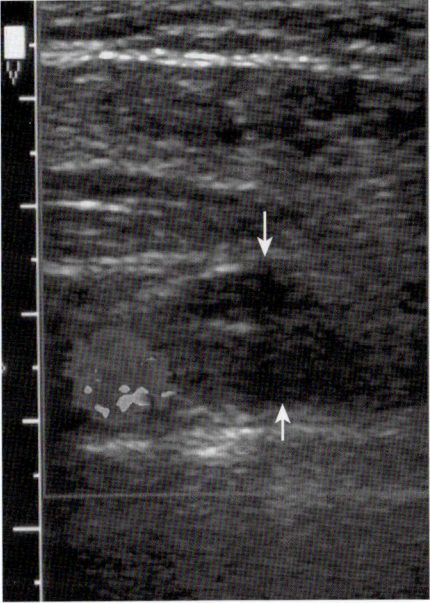

FIG. 3.58 Doppler ultrasound appearance of deep vein thrombosis. The superficial femoral vein is filled with echogenic material representing thrombus, and no flow can be identified in the vein on Doppler evaluation. Flow can be identified in the adjacent artery on color Doppler evaluation (*arrows*). (From Crawford MH, DiMarco JP, Paulus WJ [eds]: *Cardiology*, ed 2, St. Louis, Mosby, 2004.)

34. Delayed Puberty (Fig. 3.59)

Diagnostic Imaging

Best Test(s)
• None

Ancillary Tests
• MRI of pituitary with gadolinium
• Pelvic ultrasound (females)
• Bone age (hand and wrist film)

Laboratory Evaluation

Best Test(s)
• FSH, LH

Ancillary Tests
• Prolactin
• Serum free testosterone
• GnRH
• Chromosomal karyotyping
• IGF-1

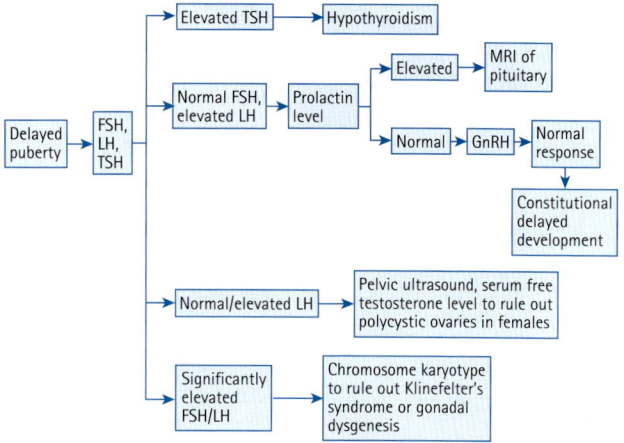

FIG. 3.59 Diagnostic algorithm.

35. Delirium (Fig. 3.60)

Diagnostic Imaging

Best Test(s)
- CT of head without contrast to r/o subdural hematoma, hemorrhage in patients with focal neurologic deficits on examination

Ancillary Tests
- Chest x-ray (rule out pneumonia)

Laboratory Evaluation

Best Test(s)
- Variable with clinical suspicion and physical examination (e.g., toxicologic screen in suspected drug abuse, CSF examination in suspected encephalitis or meningitis, CBC with differential, urinalysis, urine culture in suspected infectious process)

Ancillary Tests
- Blood culture × 2
- ALT, AST
- TSH, B_{12} level
- BUN, creatinine, urinalysis
- ABGs or pulse oximetry
- Serum electrolytes, glucose, calcium, phosphate, magnesium

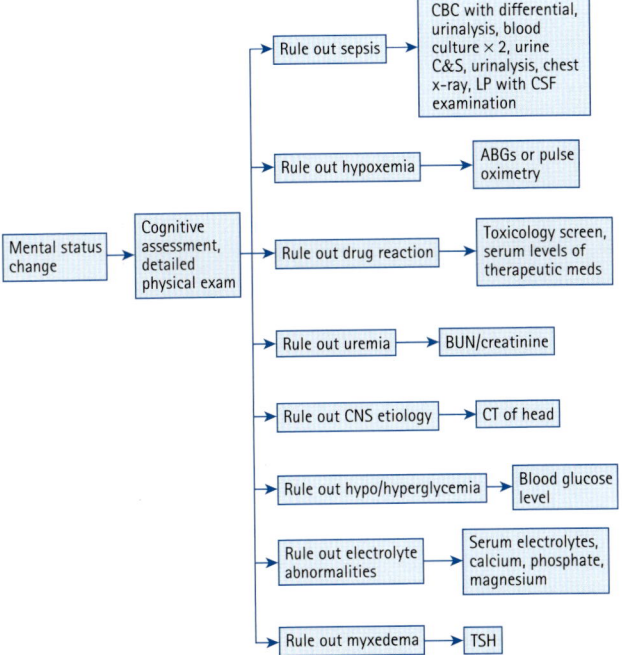

Fig. 3.60 Diagnostic algorithm.

36. Diarrhea (Fig. 3.61)

Diagnostic Imaging

Best Test(s)
- None

Ancillary Tests
- Small-bowel series if malabsorption is suspected
- Capsule endoscopy

Laboratory Evaluation

Best Test(s)
- Serum electrolytes, BUN, creatinine
- Stool Na^+, K^+
- CBC with differential
- Stool for *Clostridium difficile* toxin assay

Ancillary Tests
- Stool for O&P
- Stool for occult blood × 3
- Stool Sudan stain (in malabsorption, mucosal disease, pancreatic insufficiency, bile salt insufficiency)
- Stool osmolality (stool [Na plus K]) × 2
- ALT, AST
- TSH, free T_4
- IgA anti-tTG antibody to screen for celiac disease
- Albumin, total protein, glucose
- Stool cultures for *Escherichia coli, Shigella, Salmonella, Yersinia, Campylobacter, Entamoeba histolytica*
- Antigliadin IgA antibody, endomysial IgA antibody
- Colon biopsy

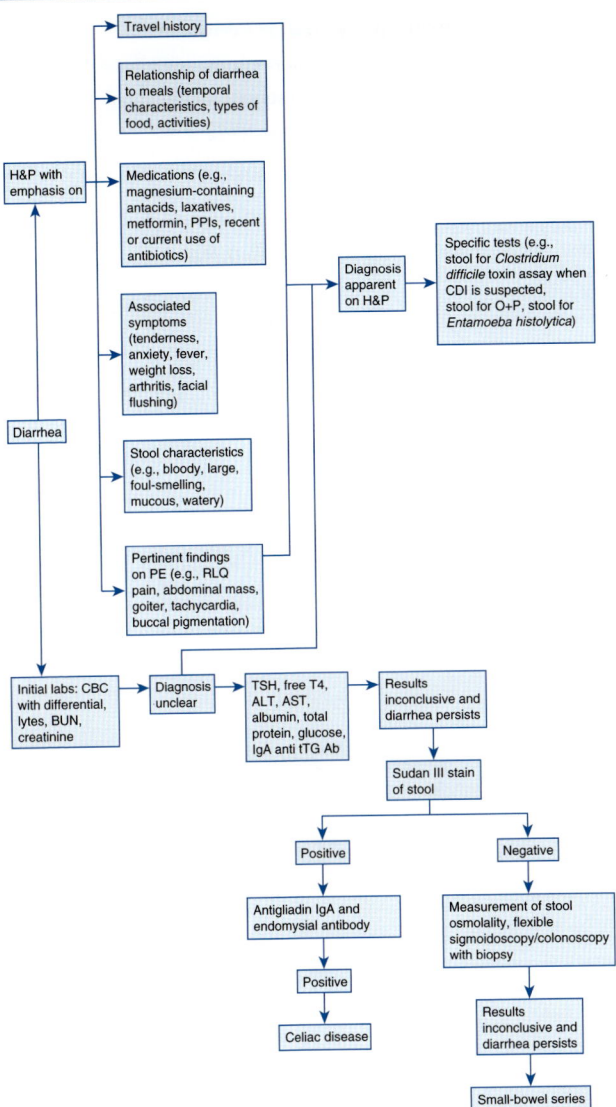

FIG. 3.61 Diagnostic algorithm.

37. Disseminated Intravascular Coagulation (DIC) (Fig. 3.62)

Diagnostic Imaging

Best Test(s)
- None

Ancillary Tests
- Chest radiograph to exclude infectious process in patients presenting with pulmonary symptoms

Laboratory Evaluation

Best Test(s)
- PT (INR), PTT, fibrin degradation products (FDPs), D-dimer, TT
- Fibrinogen level, platelet count
- CBC, peripheral blood smear

Ancillary Tests
- ALT, AST to r/o liver disease
- Factor V, VIII

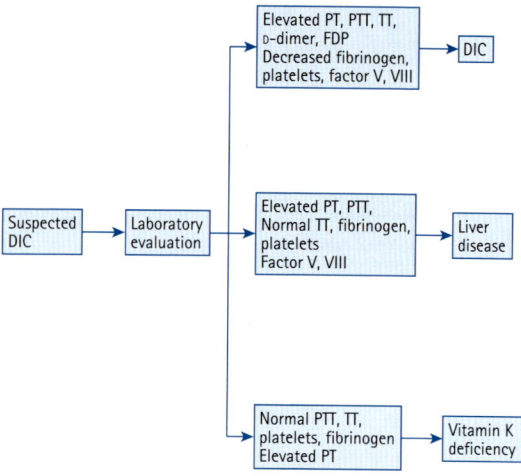

Fig. 3.62 Diagnostic algorithm.

38. Diverticulitis (Fig. 3.63)

Diagnostic Imaging

Best Test(s)
- CT scan of abdomen and pelvis with oral and IV contrast (Fig. 3.64); typical findings are thickening of bowel wall, pericolonic fat stranding, abscess formation

Ancillary Tests
- None

Laboratory Evaluation

Best Test(s)
- CBC with differential
- Blood culture × 2

Ancillary Tests
- Urinalysis, urine culture
- ALT, creatinine, BUN, electrolytes, amylase
- Serum pregnancy test in reproductive-age women

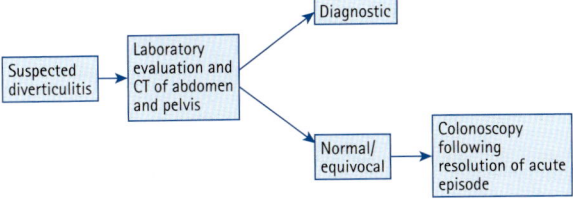

Fᴵɢ. 3.63 Diagnostic algorithm.

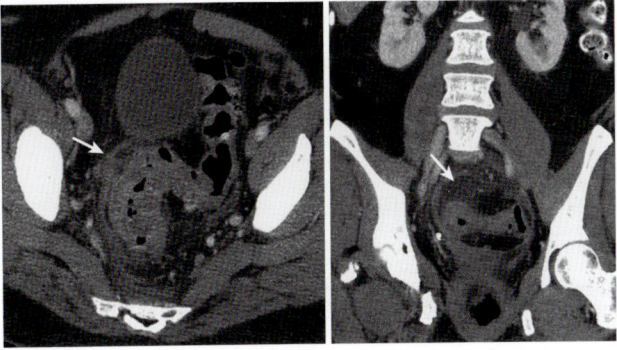

Fɪɢ. 3.64 Axial and coronal postcontrast images show a redundant sigmoid looping to the right with thick wall and pericolonic fat stranding (*arrow*) suggestive of diverticulitis. (From Fielding JR, Brown DL, Thurmond AS: *Gynecologic imaging*, Philadelphia, Saunders, 2011.)

39. Dyspepsia (Fig. 3.65)

Diagnostic Imaging

Best Test(s)
- Endoscopy

Ancillary Tests
- UGI series if patient refuses endoscopy
- Chest radiograph
- Ultrasound of neck

Laboratory Evaluation

Best Test(s)
- *H. pylori* stool antigen or breath test

Ancillary Tests
- CBC

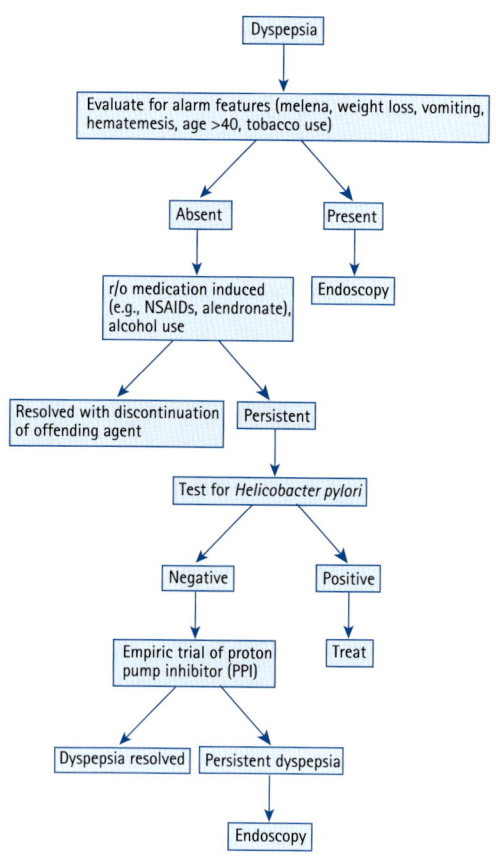

FIG. 3.65 Diagnostic algorithm.

40. Dyspnea (Fig. 3.66)

Diagnostic Imaging

Best Test(s)
- Chest radiograph

Ancillary Tests
- ECG
- Echocardiogram
- PFTs
- CT of chest to rule out pulmonary embolism

Laboratory Evaluation

Best Test(s)
- CBC (r/o anemia, infection)
- D-dimer

Ancillary Tests
- ABGs or pulse oximetry (r/o PE)
- TSH
- Serum electrolytes, BUN, creatinine
- BNP

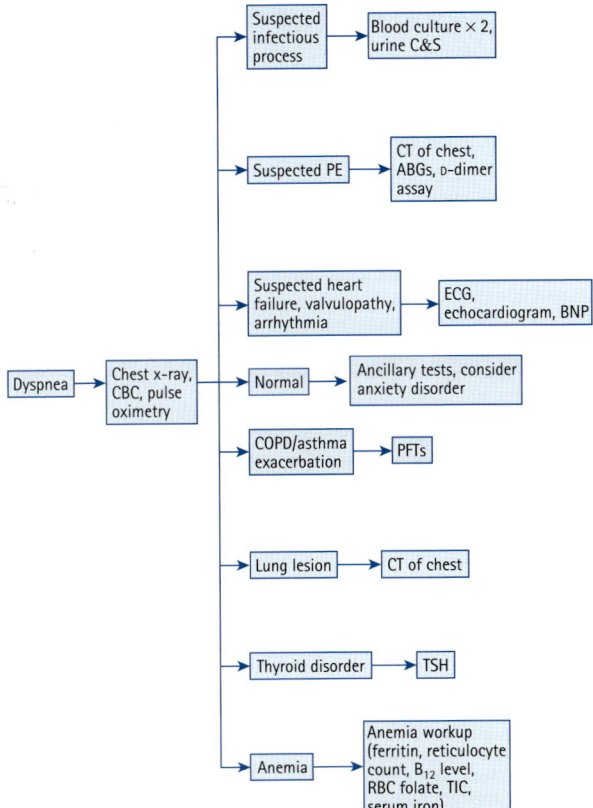

Fig. 3.66 Diagnostic algorithm.

41. Dysuria (Fig. 3.67)

Diagnostic Imaging

Best Test(s)
- None

Ancillary Tests
- Pelvic ultrasound

Laboratory Evaluation

Best Test(s)
- Urinalysis, urine C&S (Box 3.2)
- PCR assay for *Chlamydiae* or *Neisseria gonorrhoeae*

Ancillary Tests
- summarizes patients in which urine culture is indicated
- VDRL, HIV
- Gram stain and C&S of urethral discharge

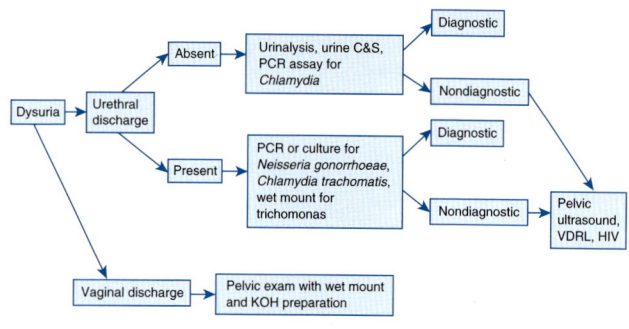

Fig. 3.67 Diagnostic algorithm.

BOX 3.2   Patient Groups in Which Urine Culture Is Indicated

- Children
- Adult men
- Immunocompromised patients
- Patients with "treatment failure" (i.e., with persistent urinary symptoms despite recently completed course of antibiotics)
- Patients with duration of symptoms more than 4 to 6 days
- Elderly patients at risk for bacteremia
- Ill-appearing patients with signs and symptoms suggestive of pyelonephritis or bacteremia
- Pregnant women
- Patients with known chronic or recurrent renal infection
- Patients with known anatomic urologic abnormalities
- Patients in whom urinary tract obstruction is suspected (e.g., stones, benign prostatic hypertrophy)
- Patients with serious medical diseases, including diabetes mellitus, sickle cell anemia, cancer, and other debilitating diseases
- Patients with alcoholism or drug dependence
- Recently hospitalized patients
- Patients taking antibiotics
- Patients who recently have undergone urinary tract instrumentation (e.g., cystoscopy, catheterization)

(From Marx JA, Hockberger R, Walls R: *Rosen's emergency medicine,* ed 8, Philadelphia, Saunders, 2014.)

42. Ectopic Pregnancy (Fig. 3.68)

Diagnostic Imaging

Best Test(s)
- Transvaginal ultrasound (Fig. 3.69)

Ancillary Tests
- CT of abdomen and pelvis with IV contrast (Fig. 3.70) in ruptured ectopic pregnancy with confirmed fetal loss

Laboratory Evaluation

Best Test(s)
- Serum hCG (quantitative)

Ancillary Tests
- Serum progesterone (decreased production in ectopic pregnancy)
- CBC
- Urinalysis

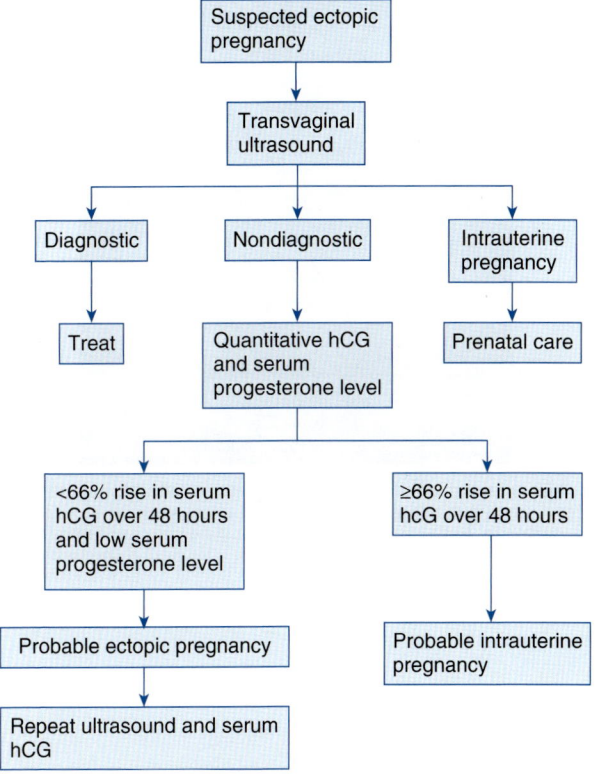

Fɪɢ. 3.68 Diagnostic algorithm.

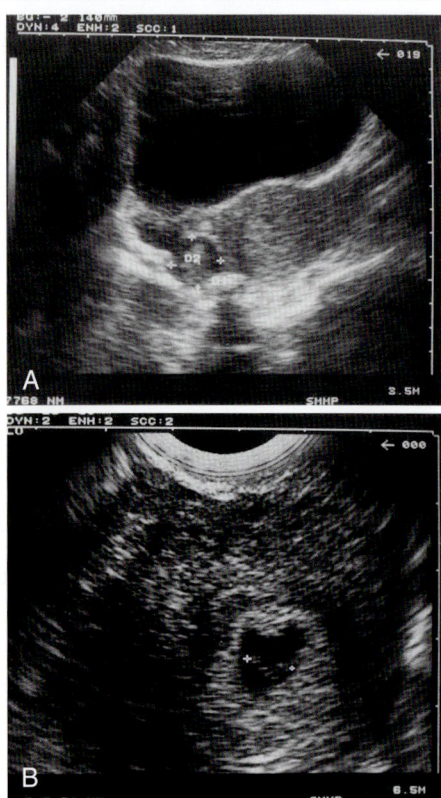

Fɪɢ. 3.69 Ultrasound scan showing ectopic pregnancy. A, Transabdominal scan show-ing empty uterus with a complex mass in the right adnexa measuring 21 × 22 mm. B, Transvaginal scan showing absence of gestational sac in the uterus and decidual reaction with marked endometrial thickening. There is free fluid in the pouch of Douglas (blood will be found there in ruptured ectopic pregnancy). (From Greer IA, Cameron IT, Kitchener HC, Prentice A: *Mosby's color atlas and text of obstetrics and gynecology*, London, Harcourt, 2000.)

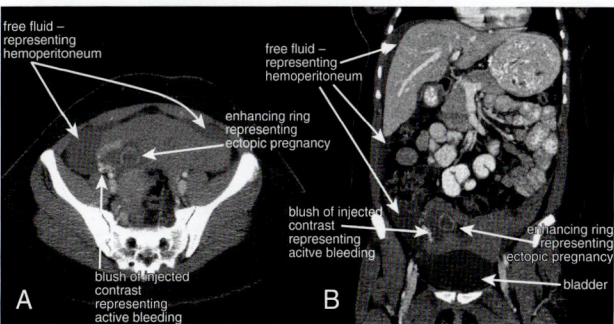

Fig. 3.70 Ruptured ectopic pregnancy: computed tomography (CT) with intravenous and oral contrast viewed on soft-tissue windows. This 20-year-old woman underwent abdominal CT for suspected appendicitis, given a history of right lower abdominal pain. A urine pregnancy test was not obtained before CT scan. A and B, Free fluid representing hemoperitoneum is seen in the pelvis and extending to a perihepatic subdiaphragmatic location (compare this dark fluid density with the urine-filled bladder). A ringlike enhancing structure is seen in the right pelvis. A bright blush of contrast is seen that is not contained within vessels or bowel—this represents active bleeding from a ruptured ectopic pregnancy. It is bright because of active extravasation of contrast, whereas blood that accumulated before contrast injection is dark. At laparotomy, 2 L of hemoperitoneum were evacuated, and active bleeding from the right fallopian tube was noted. A 4-cm structure that looked like a gestational sac with a possible tiny embryo within it was also found free in the abdomen. (From Broder JS: *Diagnostic imaging for the emergency physician*, Philadelphia, Saunders, 2011.)

43. Edema, Lower Extremity (Fig. 3.71)

Diagnostic Imaging

Best Test(s)
- Doppler ultrasound in suspected DVT

Ancillary Tests
- Plain x-ray films of extremities in patients with history of musculoskeletal trauma (r/o fracture)
- Echocardiogram (r/o CHF, valvulopathy)
- CT and/or ultrasound of pelvis

Laboratory Evaluation

Best Test(s)
- None

Ancillary Tests
- D-dimer by ELISA when DVT is suspected
- Creatinine, ALT, albumin
- TSH
- Urinalysis
- 24-hour urine protein to r/o nephrotic syndrome if urinalysis reveals proteinuria
- BNP
- Box 3.3 summarizes causes of lower-extremity edema

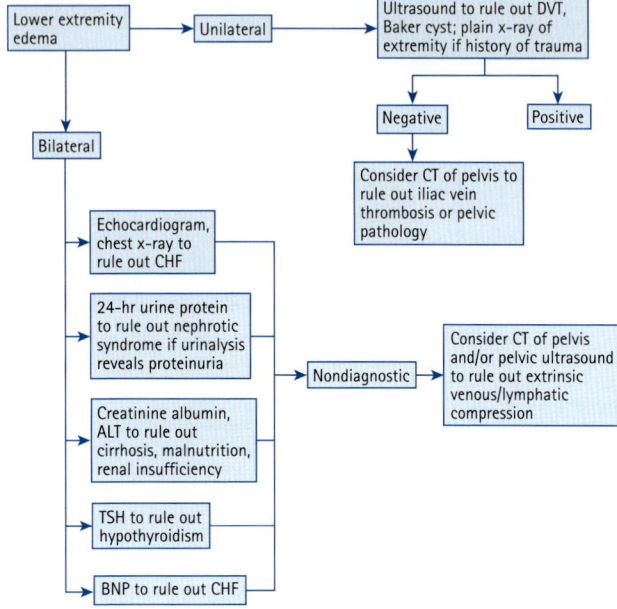

FIG. 3.71 Diagnostic algorithm.

BOX 3.3   Causes of Peripheral Edema

Heart failure
Hypoproteinemia
Liver cirrhosis
Nephrotic syndrome
Lymphedema
Malnutrition
Gravitational edema

(From Vincent JL, Abraham E, Moore FA, et al: *Textbook of critical care*, ed 6, Philadelphia, Saunders, 2011.)

44. Endocarditis, Infective (Fig. 3.72, Table 3.6)

Diagnostic Imaging

Best Test(s)
- Transesophageal echocardiogram (TEE), (Fig. 3.73)

Ancillary Tests
- Transthoracic echocardiography if TEE is not readily available or patient is uncooperative

Laboratory Evaluation

Best Test(s)
- Blood culture × 3

Ancillary Tests
- CBC with differential
- ESR (nonspecific)
- Urinalysis

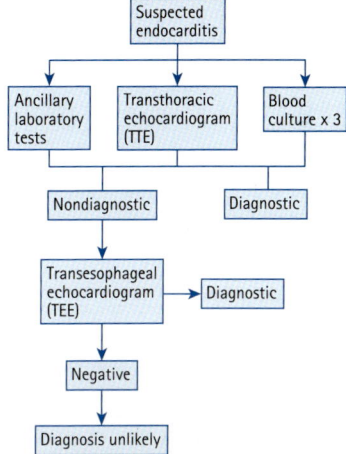

Fig. 3.72 Diagnostic algorithm.

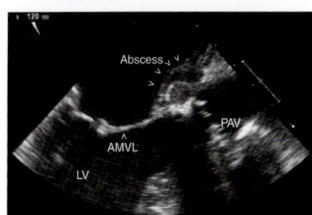

Fig. 3.73 A transesophageal 120-degree view in a patient with prosthetic aortic valve endocarditis and aortic root abscess. The posterior aortic root wall is thickened with lucent areas of liquefaction (abscess formation) from infection within the wall (*arrowheads*). The infection can extend down the intervalvular fibrosa to involve the anterior mitral valve leaflet (*AMVL*). Multiple small echo densities are attached to the prosthetic aortic valve (*PAV*) sewing ring, which are representative of prosthetic valve vegetations. *LV*, Left ventricle. (From Boxt LM, Abbara S: *The requisites: cardiac imaging*, ed 4, Philadelphia, Elsevier, 2016.)

| **TABLE 3.6** | Modified Duke Criteria for the Diagnosis of Infective Endocarditis |

MAJOR CRITERIA

Positive blood cultures for infective endocarditis

Typical microorganism for infective endocarditis from two separate blood cultures in the absence of a primary focus: *Streptococcus viridans, Streptococcus bovis*

HACEK group: *Haemophilus* species, *Actinobacillus actinomycetemcomitans, Cardiobacterium hominis, Eikenella corrodens,* and *Kingella kingae*

Community-acquired *Staphylococcus aureus* or enterococci

Persistently positive blood cultures, defined as recovery of a microorganism consistent with infective endocarditis from blood cultures drawn more than 12 hours apart, or all of three, or the majority of four or more separate blood cultures, with the first and last drawn at least 1 hour apart

Single positive blood culture for *Coxiella burnetii* or antiphase IgG antibody titer >1:800

Evidence for endocardial involvement

TTE (TEE in prosthetic valve) showing oscillating intracardiac mass on a valve or supporting structures, in the path of regurgitant jet or on implanted material, in the absence of an alternative anatomic explanation, *or*

Abscess, *or*

New partial dehiscence of a prosthetic valve

MINOR CRITERIA

Predisposition, e.g., prosthetic valve, intravenous drug use

Fever –38°C

Vascular phenomena

Immunologic phenomena

Microbiologic evidence—positive blood culture but not meeting major criteria

Ig, immunoglobulin; *TEE,* transesophageal echocardiogram; *TTE,* transthoracic echocardiogram.
From Ballinger A: *Kumar & Clark's essentials of clinical medicine,* ed 5, Edinburgh, Saunders, 2012.

45. Endometriosis (Fig. 3.74)

Diagnostic Imaging

Best Test(s)
- None

Ancillary Tests
- Laparoscopy
- Pelvic ultrasound for adnexal mass
- CT (Fig. 3.75) or MRI of abdomen/pelvis
- Colonoscopy

Laboratory Evaluation

Best Test(s)
- None

Ancillary Tests
- CA 125 limited overall value in the diagnosis of endometriosis (if result >35 U/mL, positive predictive value of 0.58 and negative predictive value of 0.96 for presence of endometriosis)

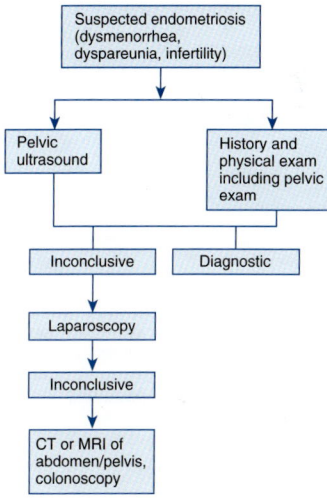

Fig. 3.74 Diagnostic algorithm.

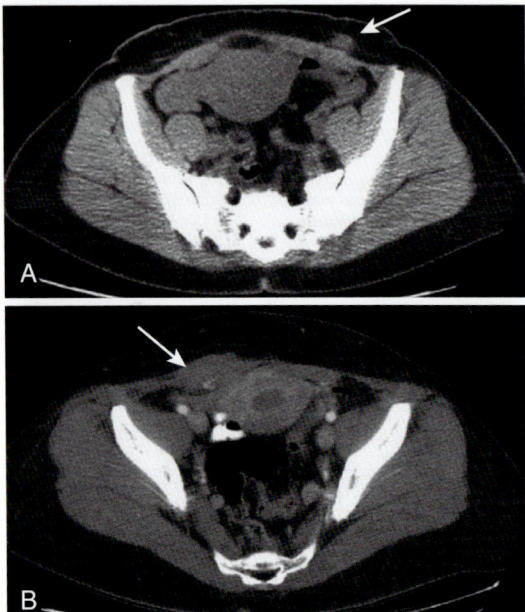

Fig. 3.75 A, Computed tomographic (CT) scan shows a subcutaneous lesion (*arrow*) in the anterior abdominal wall, which pathologic examination found to be endometriosis. B, CT scan shows a nonspecific mass involving the right rectus muscle (*arrow*) that was subsequently proven to be endometriosis. (From Fielding JR, Brown DL, Thurmond AS: Endometriosis, *Gynecologic imaging*, Philadelphia, Saunders, 2011.)

46. Epiglottitis (Fig. 3.76)

Diagnostic Imaging

Best Test(s)
- Lateral soft tissue radiograph of neck (Fig. 3.77)

Ancillary Tests
- Chest radiograph (evidence of pneumonia is present in >20% of epiglottitis cases)

Laboratory Evaluation

Best Test(s)
- None

Ancillary Tests
- CBC with differential
- Blood culture × 2
- Culture of epiglottitis (should be obtained only in controlled environment [e.g., OR])

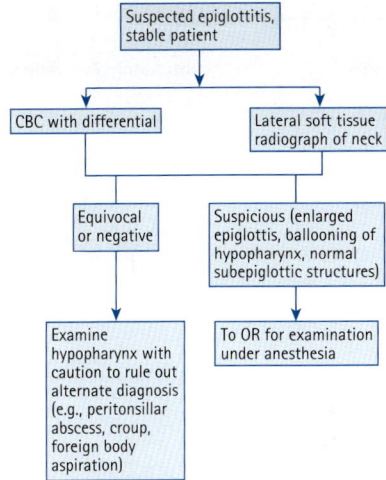

FIG. 3.76 Diagnostic algorithm.

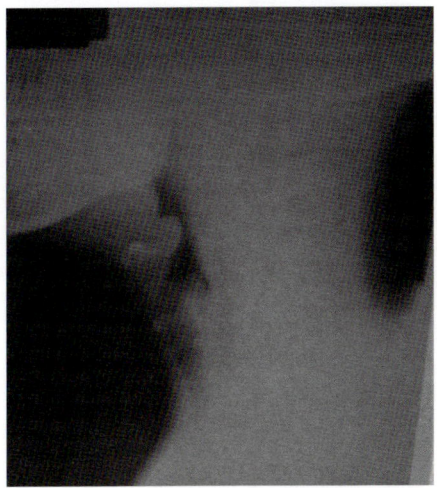

FIG. 3.77 Classic radiograph of epiglottitis with the thumb sign indicating a swollen epiglottis. (From Adams JG et al: *Emergency medicine: clinical essentials*, ed 2, Philadelphia, Elsevier, 2013.)

47. Fatigue (Fig. 3.78)

Diagnostic Imaging

Best Test(s)
- None

Ancillary Tests
- Polysomnography (sleep study)
- Chest radiograph
- ECG

Laboratory Evaluation

Best Test(s)
- CBC with differential

Ancillary Tests
- TSH, B_{12} level
- Electrolytes, BUN, creatinine
- ALT, FBS
- Monospot EB, CMV viral titers (rarely helpful)
- ESR (nonspecific)

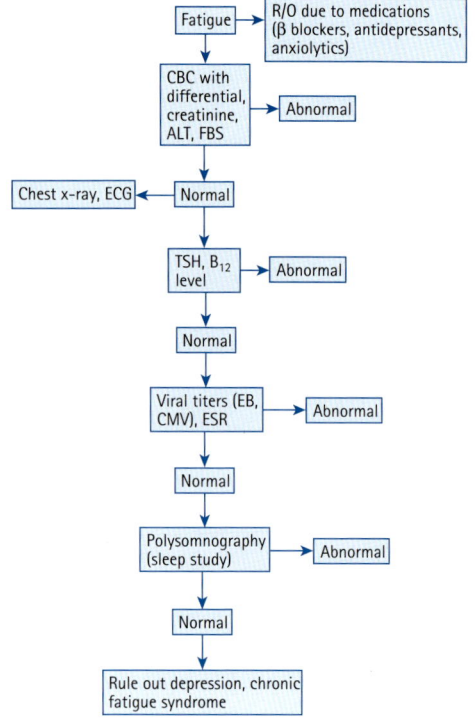

Fig. 3.78 Diagnostic algorithm.

48. Fever of Undetermined Origin (FUO) (Fig. 3.79)

Diagnostic Imaging

Best Test(s)
- None

Ancillary Tests
- Chest radiograph
- Echocardiogram (if blood cultures are positive or SBE suspected clinically)
- CT of chest/abdomen/pelvis

Laboratory Evaluation

Best Test(s)
- None

Ancillary Tests
- CBC with differential
- Blood culture × 2
- Urinalysis, urine C&S
- ESR, ANA
- ALT, AST, creatinine
- HIV, PPD
- c-ANCA, p-ANCA

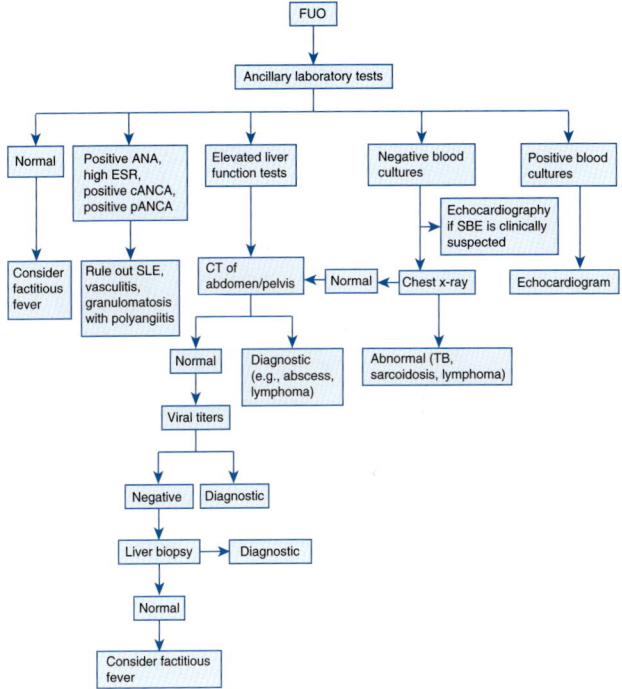

FIG. 3.79 Diagnostic algorithm.

49. Galactorrhea (Fig. 3.80)

Diagnostic Imaging

Best Test(s)
- MRI of pituitary with gadolinium (Fig. 3.81)

Ancillary Tests
- CT of pituitary with contrast if MRI is contraindicated

Laboratory Evaluation

Best Test(s)
- Serum prolactin level

Ancillary Tests
- TSH, hCG, serum creatinine
- TRH stimulation (useful in equivocal cases); normal response is increase in serum prolactin level by 100% within 1 hour of TRH infusion; failure to increase is suggestive of pituitary lesion
- Table 3.7 summarizes signs and symptoms of prolactinomas

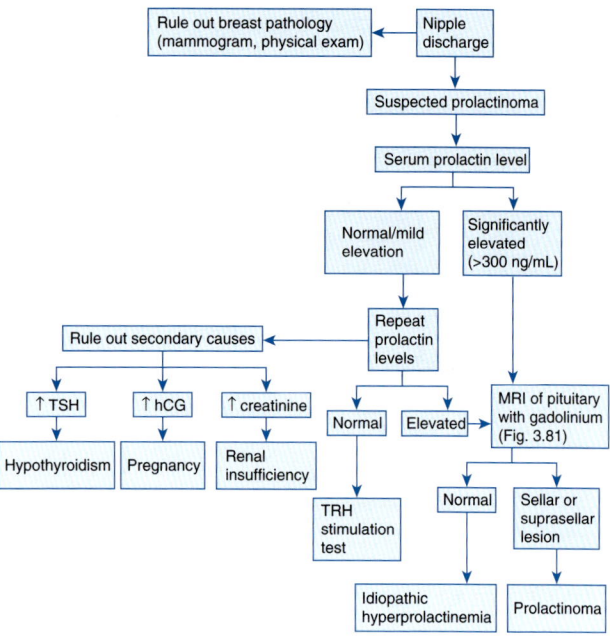

Fig. 3.80 Diagnostic algorithm.

TABLE 3.7   Signs and Symptoms of Prolactinomas

SIGNS AND SYMPTOMS ASSOCIATED WITH TUMOR MASS	SIGNS AND SYMPTOMS ASSOCIATED WITH HYPERPROLACTINEMIA
Visual field abnormalities	Amenorrhea, oligomenorrhea, infertility
Blurred vision or decreased visual acuity	Decreased libido, impotence, premature ejaculation, erectile dysfunction, oligospermia
Symptoms of hypopituitarism	Galactorrhea
Headaches	Osteoporosis
Cranial nerve palsies	
Pituitary apoplexy	
Seizures (temporal lobe)	
Hydrocephalus (rare)	
Unilateral exophthalmos (rare)	

From Melmed S, Polonsky K, Larsen P, Kronenberg H: *Williams textbook of endocrinology*, ed 12, Philadelphia, Saunders, 2011.

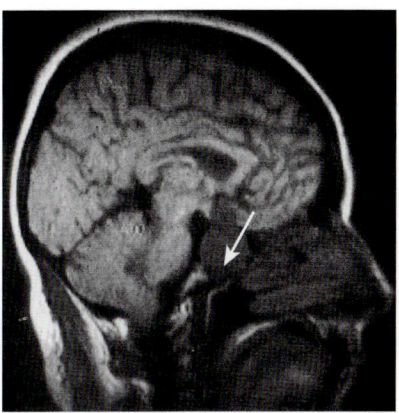

Fig. 3.81 T1 sagittal magnetic resonance imaging scan depicting a large pituitary prolactinoma in a 13-year-old girl presenting with headaches and galactorrhea. (From Grainger RG, Allison DJ, Adam A, Dixon AK [eds]: *Grainger and Allison's diagnostic radiology*, ed 4, Philadelphia, Churchill Livingstone, 2001.)

50. Genital Lesions/Ulcers (Fig. 3.82)

Diagnostic Imaging

Best Test(s)
• None

Ancillary Tests
• None

Laboratory Evaluation

Best Test(s)
• Variable with appearance of lesion

Ancillary Tests
• VDRL
• Biopsy of lesion
• LGV complement fixation
• *Chlamydia trachomatis* immunofluorescence
• Serology or viral cultures for HSV
• HIV

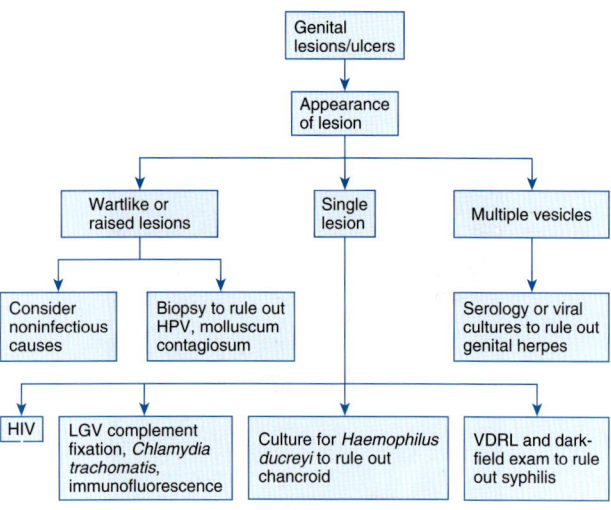

FIG. 3.82 Diagnostic algorithm.

51. Granulomatosis With Polyangiitis (Wegener Granulomatosis) (Fig. 3.83)

Diagnostic Imaging

Best Test(s)
- Chest radiograph (Fig. 3.84, *A*): May reveal bilateral multiple nodules, cavitated mass lesions, pleural effusion

Ancillary Tests
- Ultrasound of kidneys
- CT of sinuses and chest (see Fig. 3.84, *B*)

Laboratory Evaluation

Best Test(s)
- Biopsy of affected organ (e.g., lung, nasopharynx)

Ancillary Tests
- Serum c-ANCA level
- Urinalysis (hematuria, RBC casts, proteinuria)
- CBC (anemia, leukocytosis)
- BUN, creatinine (increased)
- ESR (increased)

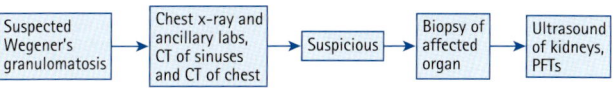

FIG. 3.83 Diagnostic algorithm.

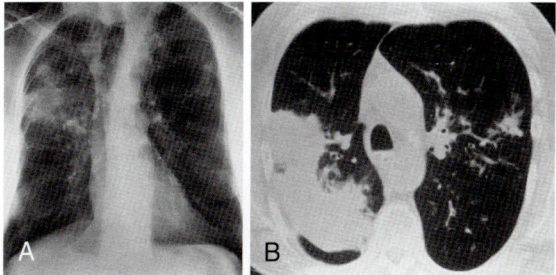

FIG. 3.84 Granulomatosis with polyangiitis. A, Chest radiograph showing bilateral patch air-space opacities; there is no evidence of cavitation. B, Computed tomography scan through the upper zones with multifocal regions of dense parenchymal opacification. (From Grainger RG, Allison DJ, Adam A, Dixon AK [eds]: *Grainger and Allison's diagnostic radiology*, ed 4, Philadelphia, Churchill Livingstone, 2001.)

52. Goiter (Fig. 3.85)

Diagnostic Imaging

Best Test(s)
• Ultrasound of thyroid

Ancillary Tests
• RAIU scan of thyroid

Laboratory Evaluation

Best Test(s)
• TSH, free T_4

Ancillary Tests
• CBC with differential
• Antimicrosomal Ab
• Thyroglobulin level

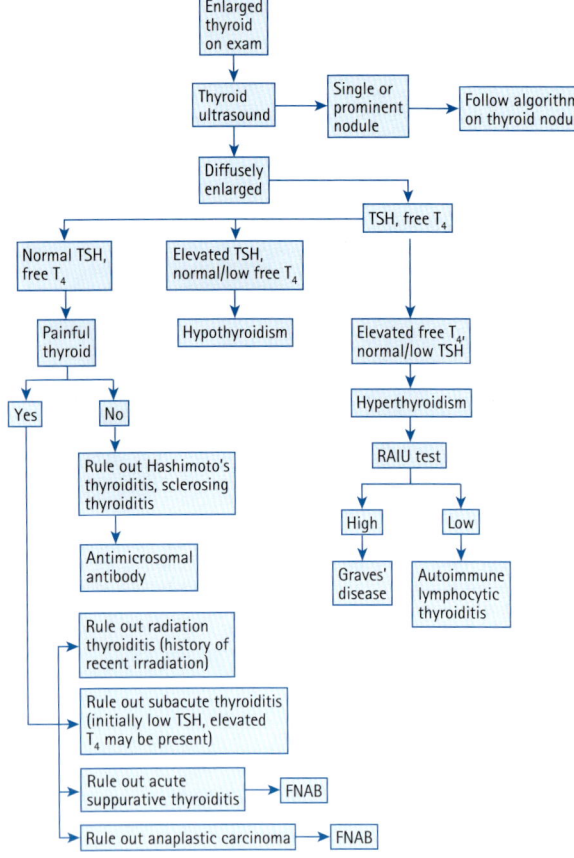

FIG. 3.85 Diagnostic algorithm.

53. Gynecomastia (Fig. 3.86)

Diagnostic Imaging

Best Test(s)
- None

Ancillary Tests
- Mammography/sonography if unilateral gynecomastia
- MRI of pituitary with contrast if elevated prolactin
- Ultrasound of testicles when testicular lesion is suspected

Laboratory Evaluation

Best Test(s)
- Serum testosterone, LH
- Prolactin

Ancillary Tests
- ALT, AST, creatinine, urinalysis
- Serum estradiol, hCG

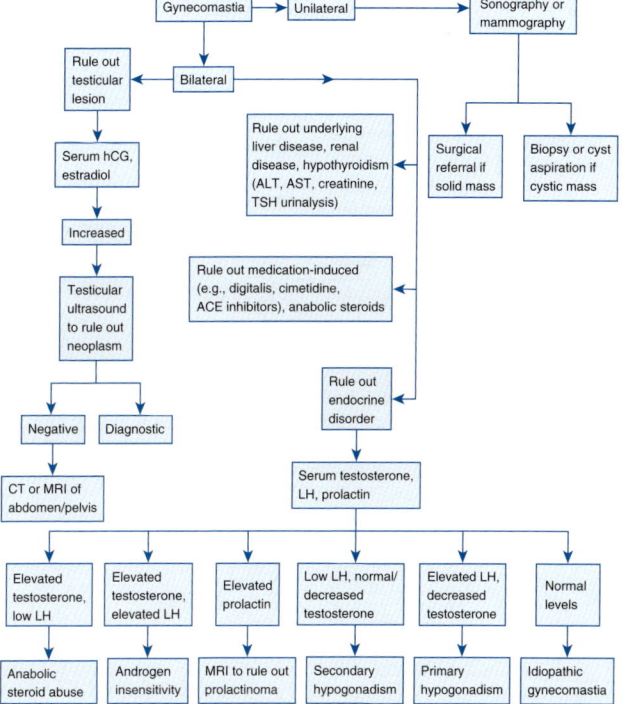

Fig. 3.86 Diagnostic algorithm.

54. Hearing Loss (Fig. 3.87)

Diagnostic Imaging

Best Test(s)
- Audiography

Ancillary Tests
- CT of head with contrast or MRI with contrast
- CT of temporal bone without contrast

Laboratory Evaluation

Best Test(s)
- None

Ancillary Tests
- CBC
- ALT, AST
- ANA, VDRL
- TSH

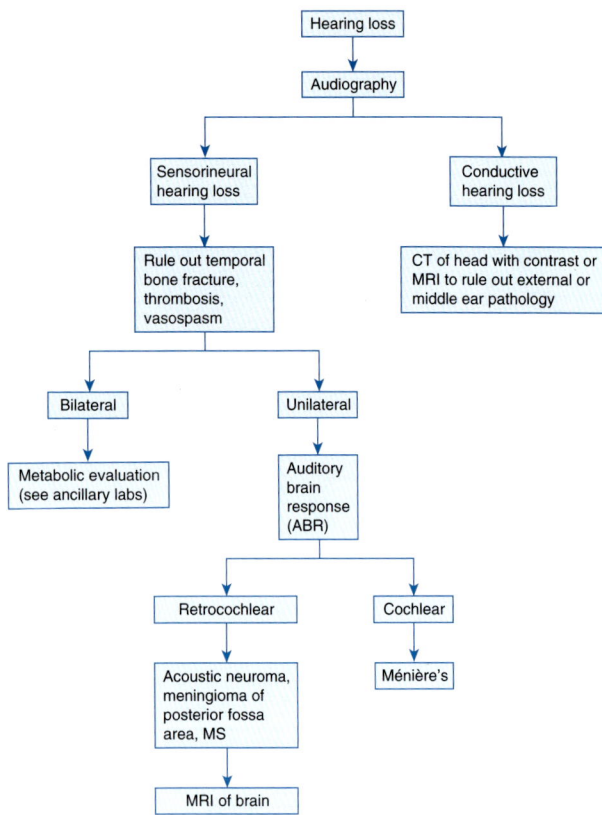

Fig. 3.87 Diagnostic algorithm.

55. Hematochezia (Fig. 3.88)

Diagnostic Imaging

Best Test(s)
- None

Ancillary Tests
- Endoscopic evaluation (anoscopy, flexible sigmoidoscopy)
- Capsule endoscopy
- RBC scintigraphy or angiography

Laboratory Evaluation

Best Test(s)
- CBC, platelet count
- Type and cross match
- Coagulation tests (INR, PTT)

Ancillary Test(s)
- Chemistry panel
- BUN, creatinine
- Liver biochemical tests

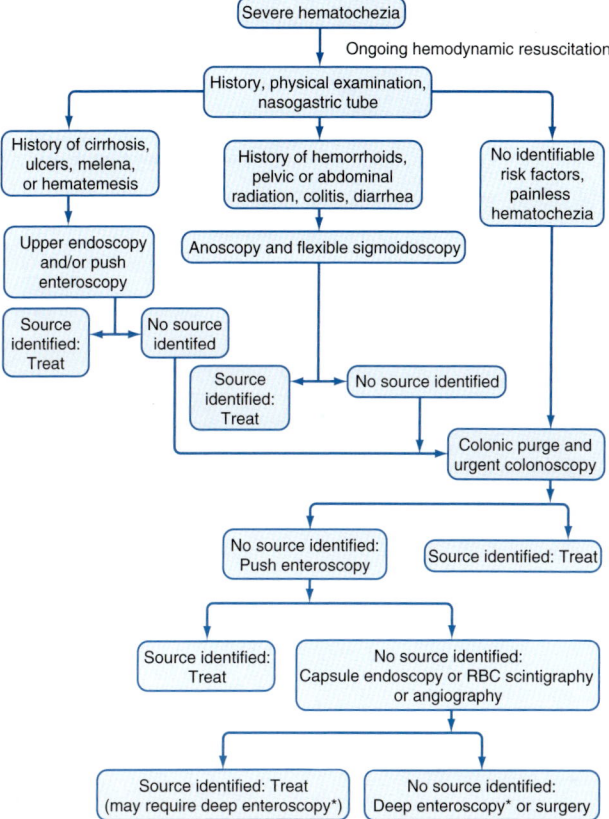

Fɪɢ. **3.88** Algorithm for the management of severe hematochezia modified according to patient's history. *Deep enteroscopy includes double-balloon enteroscopy, single-balloon enteroscopy, and spiral enteroscopy. *RBC,* red blood cell. (From Feldman M, Friedman LS, Brandt LJ: *Sleisenger and Fordtran's gastrointestinal and liver disease*, ed 10, Philadelphia, Saunders, 2016.)

56. Hematuria (Fig. 3.89)

Diagnostic Imaging

Best Test(s)
- Ultrasound of kidneys and urinary bladder

Ancillary Tests
- CT of abdomen and pelvis without contrast (Fig. 3.90); useful to exclude renal mass and calculi
- IVP (will detect most calculi and papillary necrosis)

Laboratory Evaluation

Best Test(s)
- Urine microscopy
- Urine C&S

Ancillary Tests
- Urine cytology × 3
- BUN, creatinine
- CBC, platelet count, PT, PTT

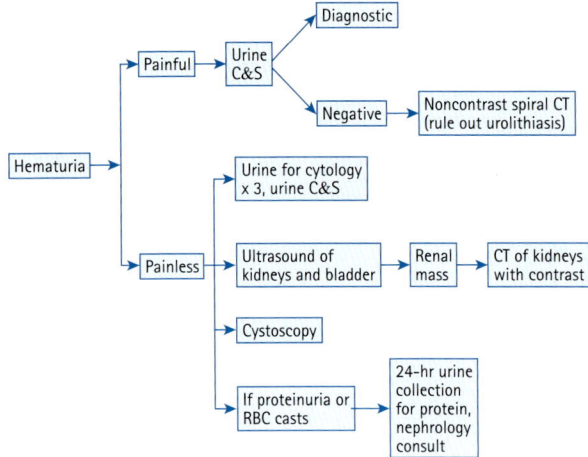

Fig. 3.89 Diagnostic algorithm.

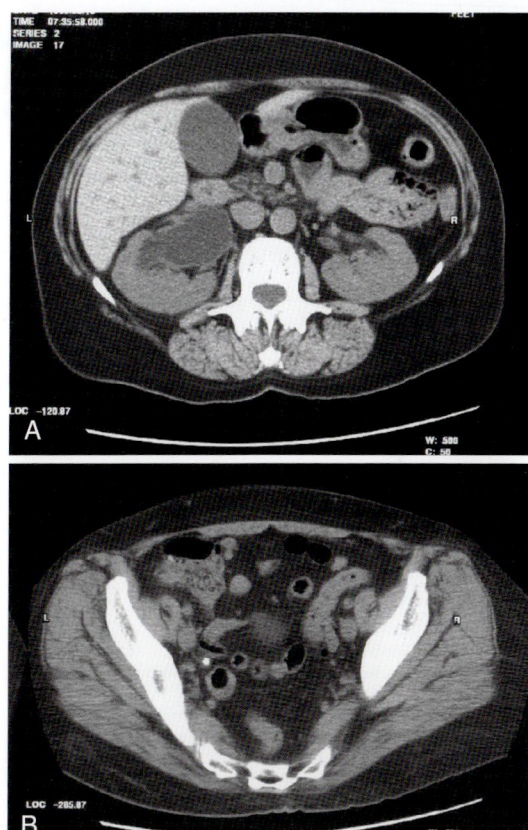

Fɪɢ. 3.90 Computed tomography images obtained in a patient with renal colic. A, Right-sided hydronephrosis. B, Right ureteral calculi. (From Marx JA, Hockberger R, Walls R: *Rosen's emergency medicine*, ed 7, Philadelphia, Mosby, 2010.)

57. Hemochromatosis (Fig. 3.91)

Diagnostic Imaging

Best Test(s)
- None

Ancillary Tests
- Noncontrast CT or MRI of liver; useful for excluding other causes of elevated liver enzymes; imaging of liver may reveal increased density of liver tissue and is also useful in screening for hepatocellular carcinoma (increased risk in patients with cirrhosis)

Laboratory Evaluation

Best Test(s)
- Plasma transferrin saturation is best screening test
- Plasma ferritin; good indicator of total body iron stores but may be elevated in many other conditions (e.g., inflammation, malignancy)
- Measurement of hepatic iron index (hepatic iron concentration/age) in liver biopsy specimen to confirm diagnosis

Ancillary Test(s)
- ALT, AST, alkaline phosphatase
- Genetic testing (HFE phenotyping for *C282Y* and *H63D* mutations)

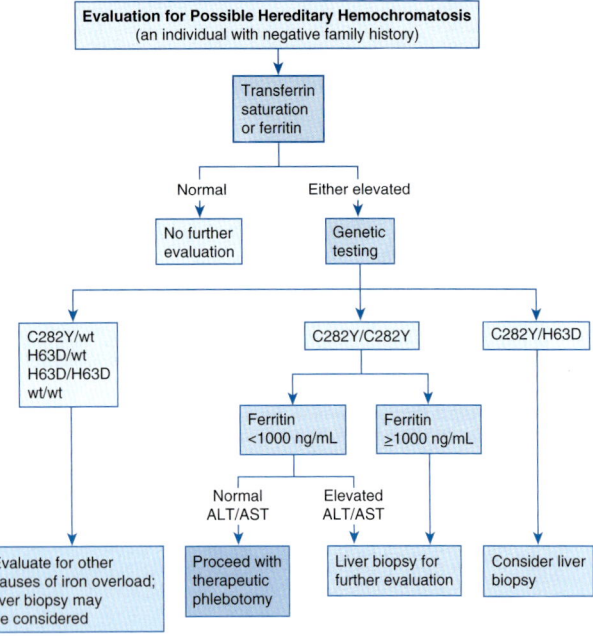

Fig. 3.91 Algorithm for evaluation of possible hereditary hemochromatosis in a person with a negative family history. *ALT*, Alanine transaminase; *AST*, aspartate transaminase. (From Goldman L, Schafer AI: *Goldman's Cecil medicine*, ed 25, Philadelphia, Saunders, 2015.)

58. Hemoptysis (Fig. 3.92)

Diagnostic Imaging

Best Test(s)
- Chest radiograph (Fig. 3.93) is best initial test

Ancillary Tests
- CT of chest
- Bronchoscopy
- Echocardiogram

Laboratory Evaluation

Best Test(s)
- None

Ancillary Tests
- Coagulation workup (PT [INR], PTT, platelets)
- Sputum C&S and Gram stain, AFB stain
- CBC
- ABGs or pulse oximetry

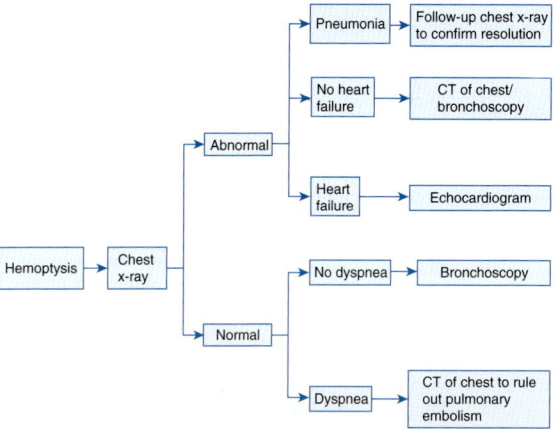

Fɪɢ. 3.92 Diagnostic algorithm.

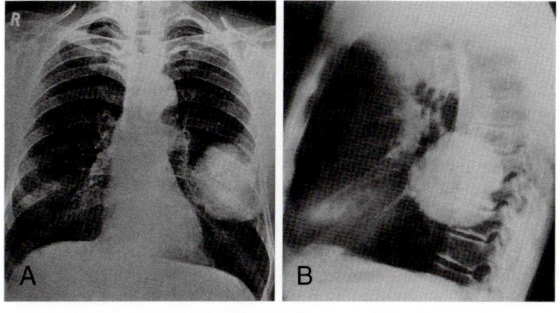

Fɪɢ. 3.93 Bronchial carcinoma in left lower lobe showing typical rounded, slightly lobular configuration. The mass shows a notch posteriorly. A, posterior anterior view. B, lateral view. (From Grainger RG, Allison DJ, Adam A, Dixon AK [eds]: *Grainger and Allison's diagnostic radiology*, ed 4, Philadelphia, Churchill Livingstone, 2001.)

59. Hepatomegaly (Fig. 3.94)

Diagnostic Imaging

Best Test(s)
• CT scan of abdomen

Ancillary Tests
• Ultrasound of abdomen if CT is contraindicated
• Echocardiography if congestive hepatomegaly is suspected

Laboratory Evaluation

Best Test(s)
• Liver biopsy

Ancillary Tests
• ALT, AST, alkaline phosphatase, GGTP
• Bilirubin, viral hepatitis panel, CBC
• FBS, serum albumin, INR
• ANA, AMA, transferrin saturation, ceruloplasmin level, serum copper, alpha-1 antitrypsin level

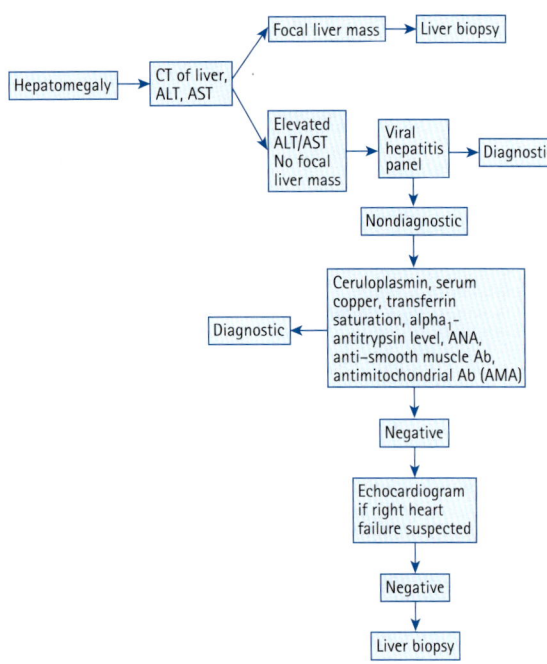

Fɪɢ. 3.94 Diagnostic algorithm.

60. Hirsutism (Fig. 3.95)

Diagnostic Imaging

Best Test(s)
- None

Ancillary Tests
- CT or MRI of adrenal gland with contrast (r/o adrenal tumors)
- Pelvic ultrasound (r/o ovarian tumor, PCOS)
- MRI of pituitary (r/o pituitary tumor)

Laboratory Evaluation

Best Test(s)
- Serum testosterone (total and free), DS, 17-OHP

Ancillary Tests
- Dexamethasone suppression test
- LH, FSH, FBS
- Prolactin
- TSH

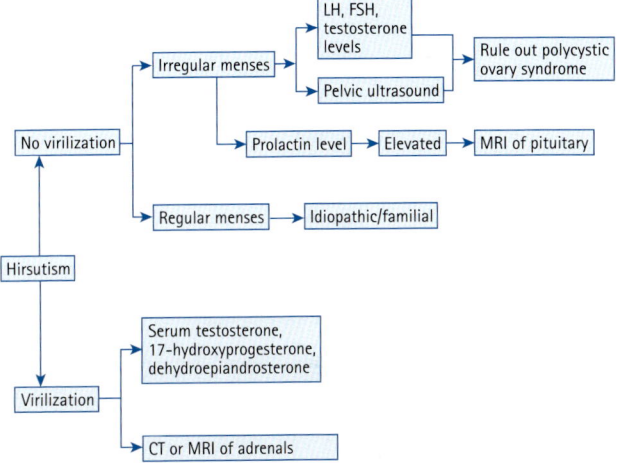

Fɪɢ. 3.95 Diagnostic algorithm.

61. Hyperaldosteronism (Fig. 3.96)

Diagnostic Imaging

Best Test(s)
- None

Ancillary Tests*
- CT scan of adrenals with contrast to localize neoplasm
- Adrenal scan with iodocholesterol (NP-59) or 6-beta-iodomethyl-19-norcholesterol

*Indicated only after biochemical confirmation of aldosteronism.

Laboratory Evaluation

Best Test(s)
- PAC
- PRA

Ancillary Tests
- Serum electrolytes
- Aldosterone suppression test

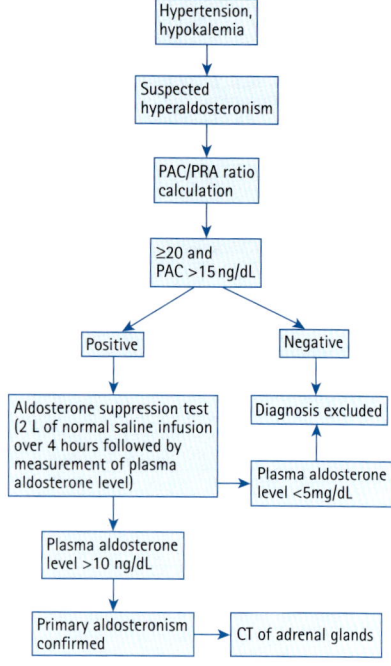

Fɪɢ. 3.96 Diagnostic algorithm.

62. Hypercalcemia (Fig. 3.97)

Diagnostic Imaging

Best Test(s)
- None

Ancillary Tests
- Radiograph of painful bones (r/o Paget disease, bone neoplasm, multiple myeloma)
- Tc-99m parathyroid scan (r/o parathyroid adenoma)
- Ultrasound of parathyroid glands
- Ultrasound of kidneys (r/o renal cell carcinoma)

Laboratory Evaluation

Best Test(s)
- Serum calcium level
- PTH level

Ancillary Tests
- Serum phosphate, magnesium, alkaline phosphatase, albumin
- Electrolytes, BUN, creatinine
- 24-hour urine collection for calcium
- Urinary cyclic AMP
- PSA (if prostate carcinoma is suspected)
- Serum and urine protein immunoelectrophoresis (if multiple myeloma suspected)

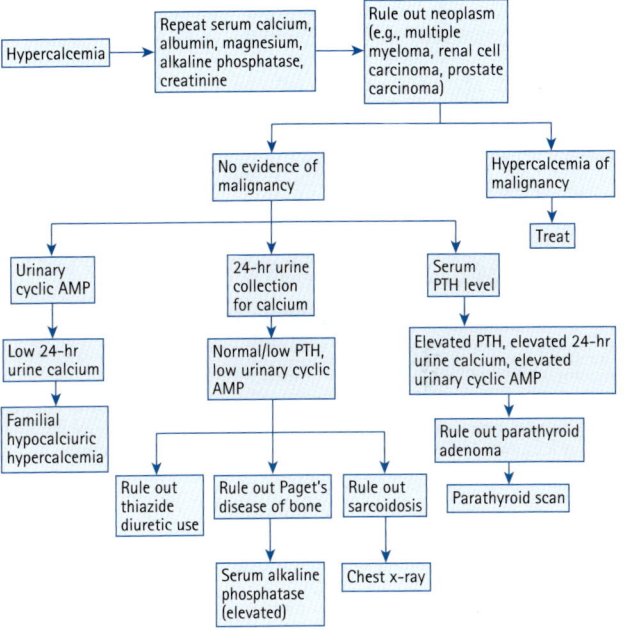

Fɪɢ. 3.97 Diagnostic algorithm.

63. Hyperkalemia (Fig. 3.98)

Diagnostic Imaging	Laboratory Evaluation
Best Test(s)	*Best Test(s)*
• None	• Heparinized potassium level
Ancillary Tests	*Ancillary Tests*
• None	• Serum electrolytes, BUN, plasma osmolality, urine osmolality creatinine, urine potassium, calculation of TTKG (see Fig. 3.98)
	• Glucose, CBC
	• CPK (when rhabdomyolysis is suspected)

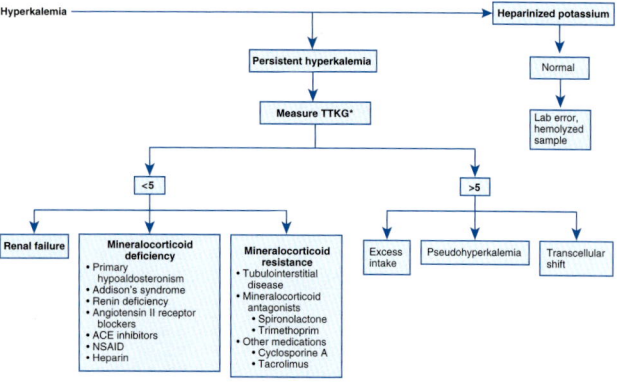

FIG. 3.98 Algorithm for treatment of hyperkalemia.

*Transtubular potassium gradient (TTKG) $= \dfrac{\text{UrineK}^+}{\text{PlasmaK}^+} \div \dfrac{\text{UrineOSM}}{\text{PlasmaOSM}}$

The transtubular potassium gradient (TTKG) is typically less than 5 when a renal cause of hyperkalemia is present. Note that this formula is valid only when UOSM >300 and UNa >25 (From Cameron JL, Cameron AM: Electrolyte disorders, *Current surgical therapy*, ed 10, Philadelphia, Saunders, 2011.)

64. Hypermagnesemia (Fig. 3.99)

Diagnostic Imaging

Best Test(s)
- None

Ancillary Tests
- None

Laboratory Evaluation

Best Test(s)
- 24-hour urine magnesium level

Ancillary Tests
- Serum electrolytes, calcium
- BUN, creatinine, glucose
- TSH
- CPK

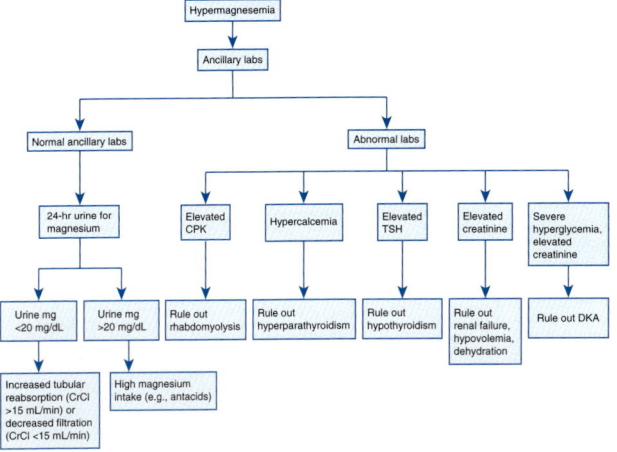

Fɪɢ. 3.99 Diagnostic algorithm.

65. Hypernatremia (Fig. 3.100)

Diagnostic Imaging

Best Test(s)
• None

Ancillary Tests
• None

Laboratory Evaluation

Best Test(s)
• Serum electrolytes, urine sodium

Ancillary Tests
• Urine osmolality
• BUN, creatinine

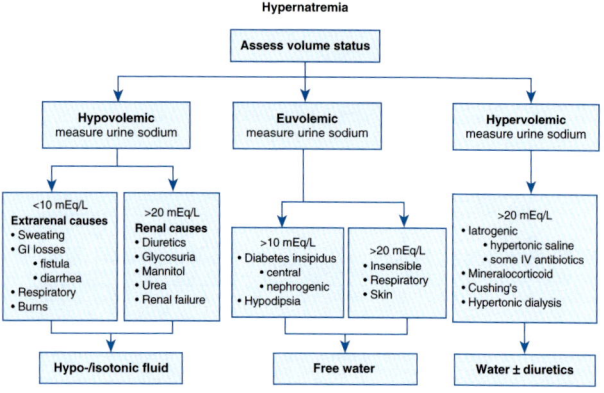

Fɪɢ. **3.100** Algorithm for treatment of hypernatremia. (From Cameron JL, Cameron AM: Electrolyte disorders, *Current surgical therapy*, ed 12, Philadelphia, Saunders, 2017.)

66. Hyperphosphatemia (Fig. 3.101)

Diagnostic Imaging

Best Test(s)
- None

Ancillary Tests
- None

Laboratory Evaluation

Best Test(s)
- 24-hour urine phosphate collection

Ancillary Tests
- Serum electrolytes, BUN, creatinine, magnesium, calcium, glucose
- Urinalysis, CPK

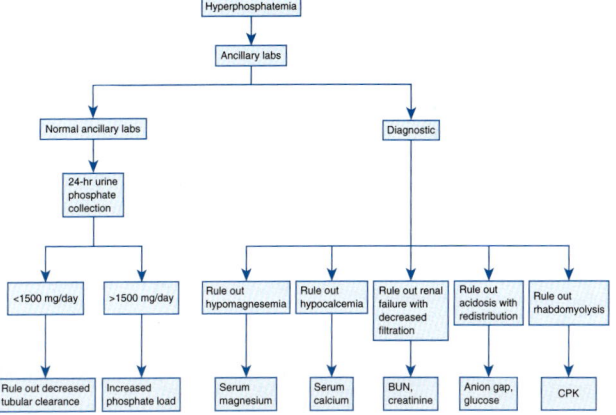

Fɪɢ. 3.101 Diagnostic algorithm.

67. Hyperthyroidism (Fig. 3.102)

Diagnostic Imaging

Best Test(s)
- None

Ancillary Tests
- Thyroid ultrasound
- RAIU thyroid scan

Laboratory Evaluation

Best Test(s)
- Free T_4
- TSH

Ancillary Tests
- Free T_3
- Serum thyroglobulin level
- Serum antimicrosomal antibodies

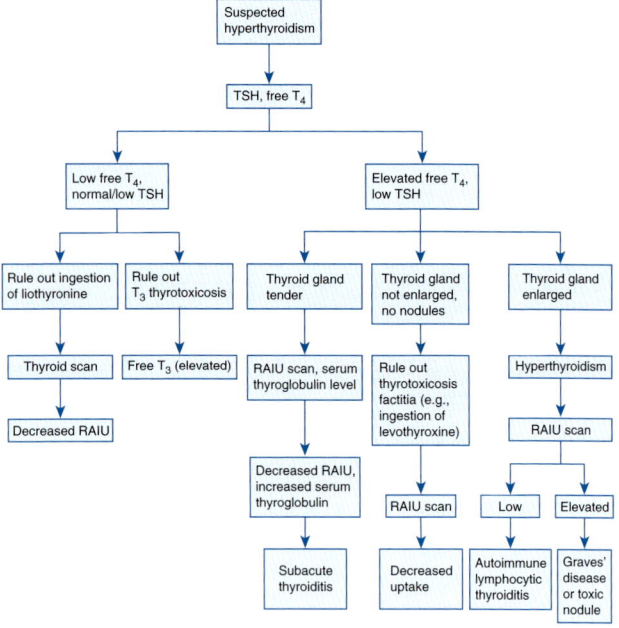

Fig. 3.102 Diagnostic algorithm.

68. Hypocalcemia (Fig. 3.103)

Diagnostic Imaging

Best Test(s)
• None

Ancillary Tests
• None

Laboratory Evaluation

Best Test(s)
• PTH

Ancillary Tests
• Serum albumin level
• Serum phosphate, magnesium level

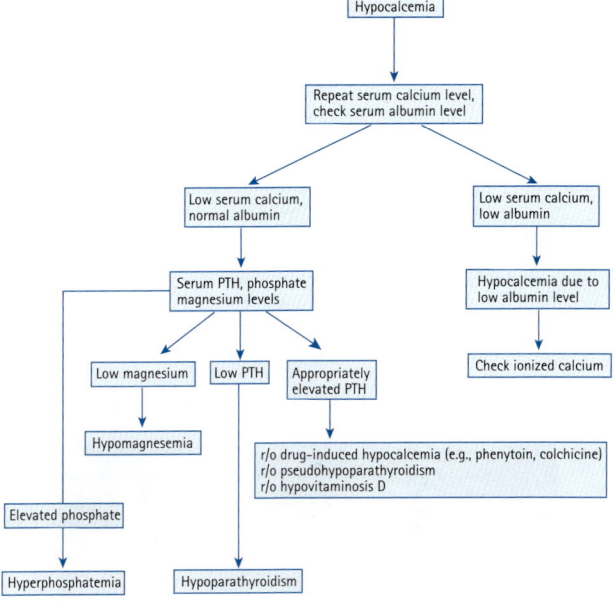

FIG. 3.103 Diagnostic algorithm.

69. Hypoglycemia (Fig. 3.104)

Diagnostic Imaging*

Best Test(s)
- Abdominal CT or MRI with contrast (Fig. 3.105)

Ancillary Tests
- Octreotide scan (somatostatin receptor scintigraphy) (Fig. 3.106)

Laboratory Evaluation

Best Test(s)
- FBS
- Insulin level
- C peptide
- Plasma sulfonylurea assay

Ancillary Tests
- None

*Indicated only after biochemical confirmation of insulinoma.

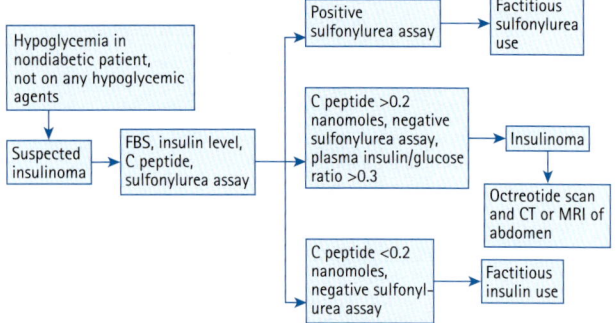

FIG. 3.104 Diagnostic algorithm.

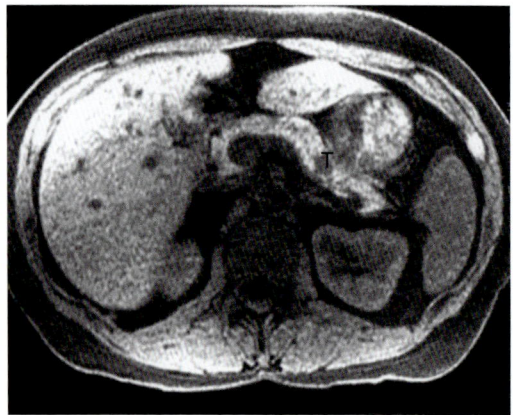

FIG. 3.105 Magnetic resonance imaging scan shows an insulinoma (*T*) in the tail of the pancreas. (From Cameron JL, Cameron AM: Management of pancreatic islet cell tumors excluding gastrinoma, *Current surgical therapy*, ed 10, Philadelphia, Saunders, 2011.)

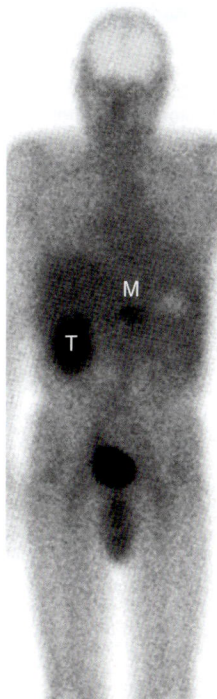

Fɪɢ. 3.106 Somatostatin receptor scintigraphy demonstrates primary tumor (*T*) and liver metastasis (*M*). (From Cameron JL, Cameron AM: Management of pancreatic islet cell tumors excluding gastrinoma, *Current surgical therapy*, ed 10, Philadelphia, Saunders, 2011.)

70. Hypogonadism (Fig. 3.107)

Diagnostic Imaging

Best Test(s)
- None

Ancillary Tests
- MRI of brain (pituitary) only after biochemical confirmation of secondary hypogonadism

Laboratory Evaluation

Best Test(s)
- Serum testosterone (total and free)
- LH, FSH

Ancillary Tests
- Serum prolactin
- hCG stimulation
- Chromosome karyotype
- Ferritin level, transferrin saturation
- Semen analysis
- Seminal fluid fructose
- Testicular biopsy

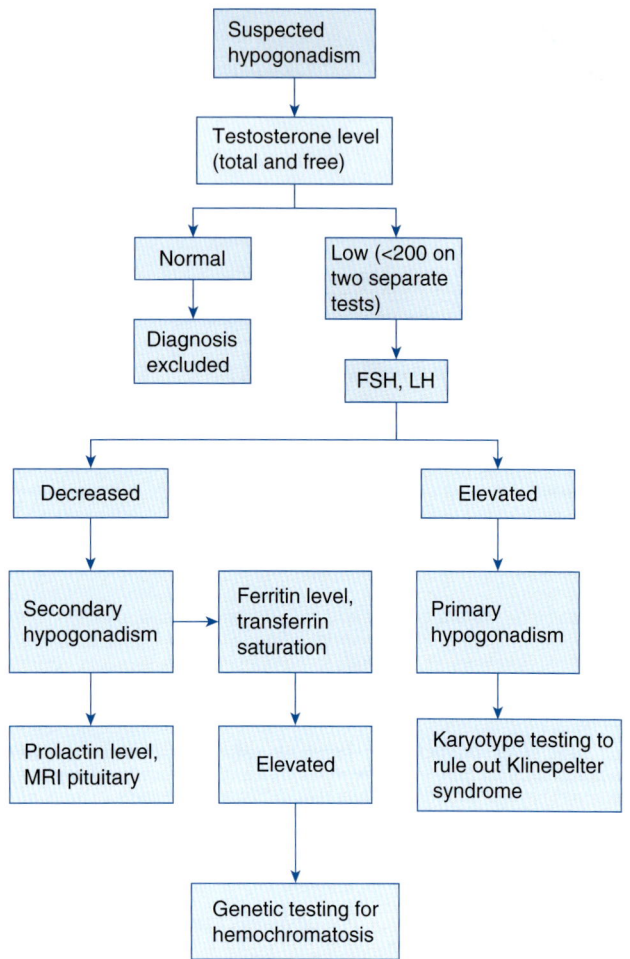

Fɪɢ. 3.107 Diagnostic algorithm.

71. Hypokalemia (Fig. 3.108)

Diagnostic Imaging

Best Test(s)
• None

Ancillary Tests
• ECG

Laboratory Evaluation

Best Test(s)
• 24-hour urine potassium excretion

Ancillary Tests
• Serum electrolyte, BUN, creatinine
• Urine chloride
• Plasma renin, aldosterone

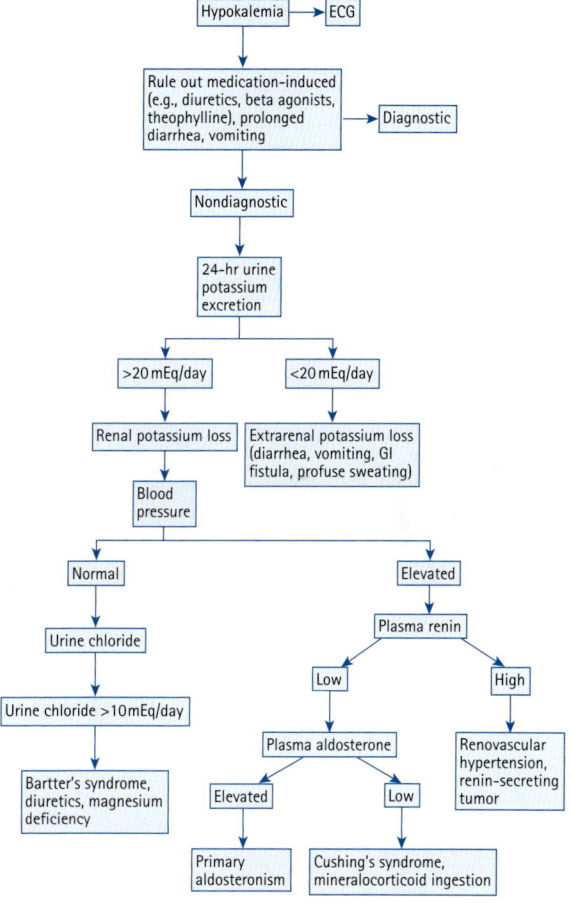

FIG. 3.108 Diagnostic algorithm.

72. Hypomagnesemia (Fig. 3.109)

Diagnostic Imaging

Best Test(s)
- None

Ancillary Tests
- None

Laboratory Evaluation

Best Test(s)
- 24-hour urine magnesium level

Ancillary Tests
- Serum electrolytes, calcium, phosphate, glucose, albumin
- BUN, creatinine
- TSH
- Urine sodium

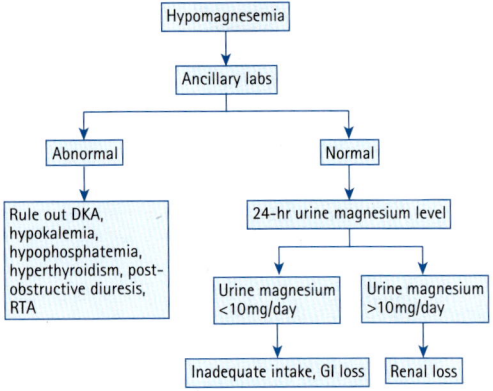

Fig. 3.109 Diagnostic algorithm.

73. Hyponatremia (Fig. 3.110)

Diagnostic Imaging

Best Test(s)
- None

Ancillary Tests
- None

Laboratory Evaluation

Best Test(s)
- Serum electrolytes
- Serum osmolality, urine osmolality, urine sodium

Ancillary Tests
- Serum glucose, BUN, creatinine
- Uric acid
- TSH
- Serum cortisol

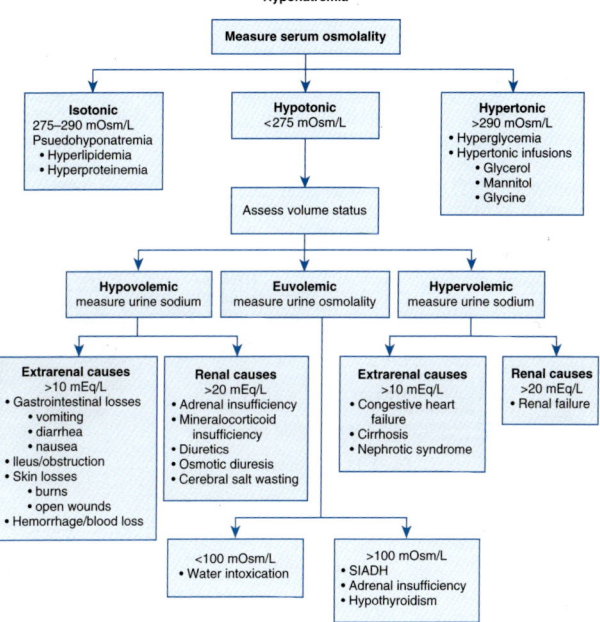

Fig. **3.110** Algorithm for treatment of hyponatremia. *SIADH,* syndrome of inappropriate antidiuretic hormone. (From Cameron JL, Cameron AM: Electrolyte disorders, *Current surgical therapy,* ed 12, Philadelphia, Saunders, 2017.)

74. Hypophosphatemia (Fig. 3.111)

Diagnostic Imaging

Best Test(s)
• None

Ancillary Tests
• None

Laboratory Evaluation

Best Test(s)
• 24-hour urine phosphate excretion

Ancillary Tests
• Serum calcium, electrolytes, creatinine, glucose
• ABGs

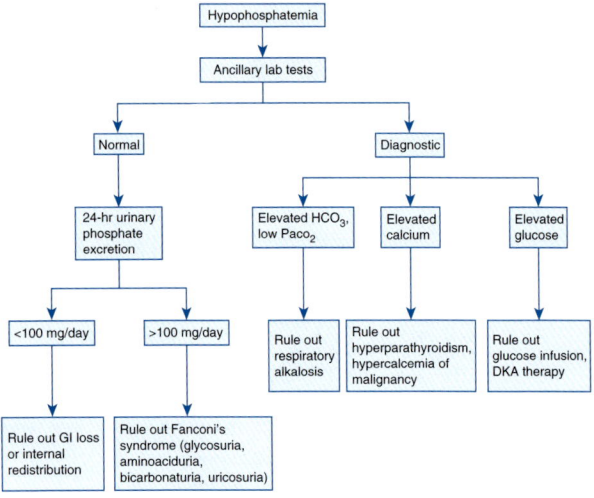

Fɪɢ. 3.111 Diagnostic algorithm.

75. Hypothyroidism (Fig. 3.112)

Diagnostic Imaging

Best Test(s)
- None

Ancillary Tests
- MRI of pituitary with contrast if secondary hypothyroidism is suspected

Laboratory Evaluation

Best Test(s)
- TSH

Ancillary Tests
- Antimicrosomal Ab if Hashimoto's thyroiditis is suspected
- Serum thyroglobulin level when subacute thyroiditis is suspected
- Free T_4

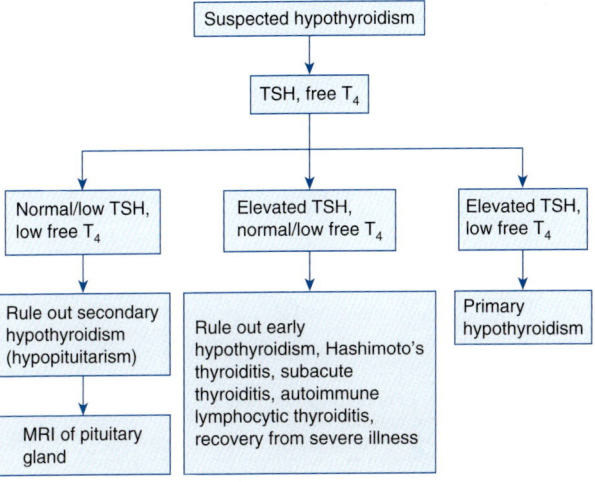

FIG. 3.112 Diagnostic algorithm.

76. Immunodeficiency (Fig. 3.113)

Diagnostic Imaging

Best Test(s)
- None

Ancillary Tests
- Chest x-ray

Laboratory Evaluation

Best Test(s)
- CBC with differential

Ancillary Tests
- Fig. 3.114

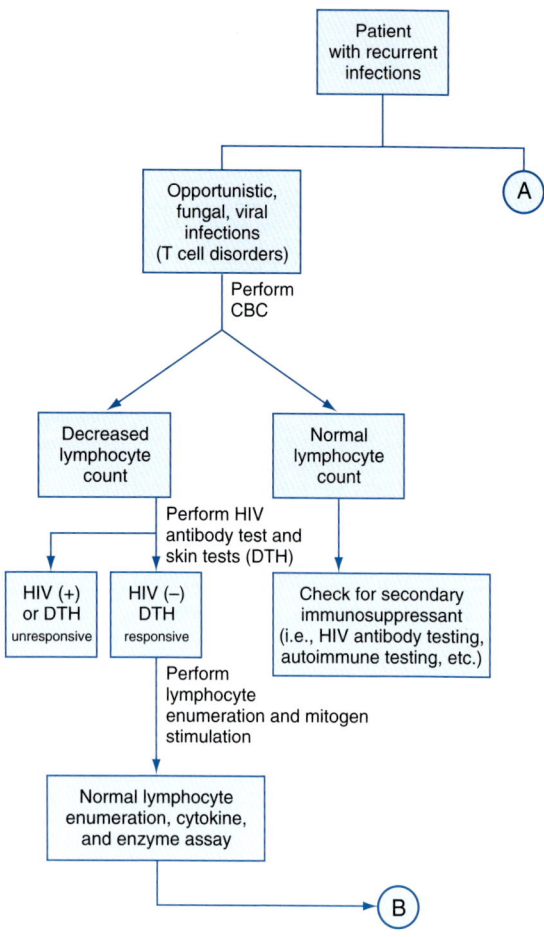

FIG. 3.113 Testing algorithm for immunodeficiency. *Ab,* antibody; *CBC,* complete blood count; *DTH,* delayed-type hypersensitivity skin test; *HIV,* human immunodeficiency virus; *Ig,* immunoglobulin; *NBT,* nitroblue tetrazolium slide test. (From McPherson RA, Pincus MR: *Henry's clinical diagnosis and management by laboratory methods,* ed 23, St. Louis, Elsevier, 2017.)

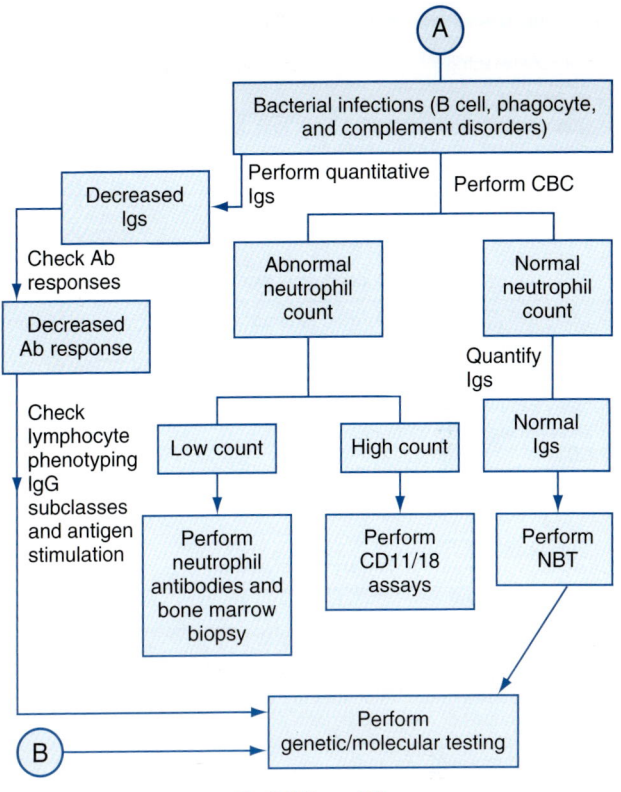

Primary Investigation

- History and physical examination
- CBC/differential
- Radiographs
- Quantitative immunoglobulins (IgG, IgM, IgA)
- Review of previous culture results
- Pulmonary function testing

Secondary Investigation

- Titers for vaccines administered (tetanus, diphtheria, pneumococcus)
- IgG subclass analysis
- Lymphocyte enumeration panel (CD3, CD4, CD8, CD19, CD16, CD56)
- Complement levels (CH50, C3, C4)
- Skin testing
- Mononuclear cell proliferation studies
- NBT

Tertiary Investigation

- Enzyme studies
  (adenosine deaminase, purine nucleotide phosphorylase)
- NK cell cytotoxic studies
- Phagocyte studies
- Histology and immunohistochemistry analysis (or flow cytometry)
  of lymph nodes or lymphoid organs
- Cytokine production
- Advanced complement studies
- Molecular biology

Fɪɢ. 3.114 Staged approach for laboratory investigation of primary immunodeficiency. *CBC,* complete blood count; *NBT,* nitroblue tetrazolium slide test; *NK cell,* natural killer cell. (From McPherson RA, Pincus MR: *Henry's clinical diagnosis and management by laboratory methods,* ed 23, St. Louis, Elsevier, 2017.)

77. Infertility (Fig. 3.115)

Diagnostic Imaging

Best Test(s)
- None

Ancillary Tests
- MRI of pituitary with contrast
- Hysterosalpingography (Fig. 3.116)
- Testicular ultrasound

Laboratory Evaluation

Best Test(s)
- Male: semen analysis
- Female: endometrial biopsy

Ancillary Tests
- FSH, LH
- Prolactin
- Serum testosterone (male)
- TSH
- CBC, ESR, FBS
- Urinalysis, VDRL, Mycoplasma culture, chlamydiae serology

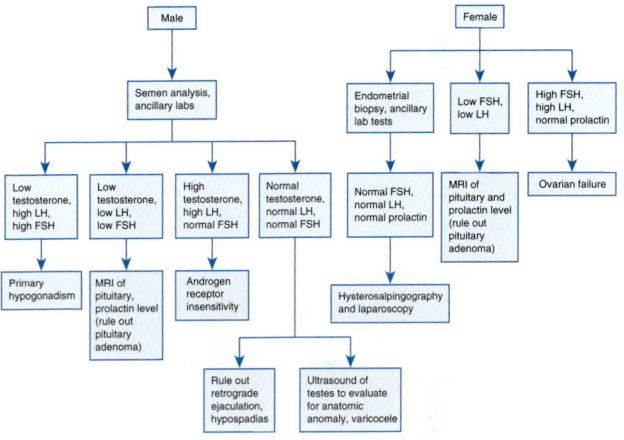

FIG. 3.115 Diagnostic algorithm.

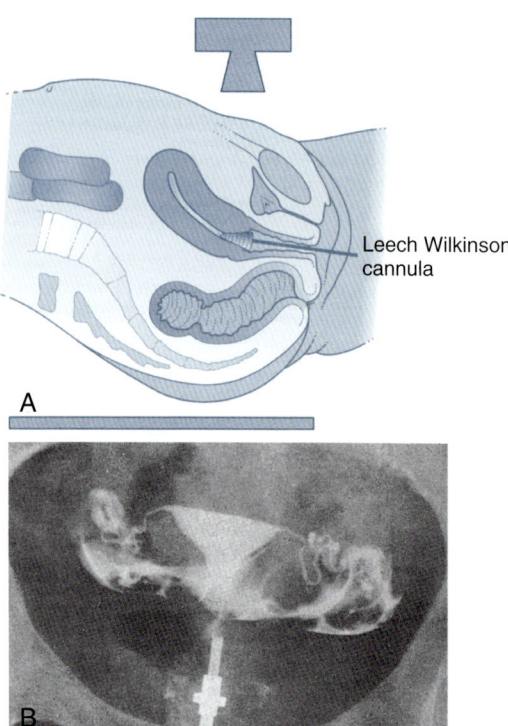

Fig. 3.116 A, Hysterosalpingography enables the site of tubal obstruction to be assessed and the uterine cavity to be examined for the presence of pathologic findings. B, The triangular outline of the uterine cavity and the spill of dye on both sides from the fimbrial ends of the fallopian tubes can be seen. The dye spreads over the adjacent bowel. (From Symonds EM, Symonds IM: *Essential obstetrics and gynecology*, ed 4, Edinburgh, Churchill Livingstone, 2004.)

78. Jaundice (Fig. 3.117)

Diagnostic Imaging

Best Test(s)
- CT of abdomen with contrast if painless jaundice
- Ultrasound of abdomen if painful jaundice

Ancillary Tests
- MRCP
- ERCP
- EUS
- Table 3.8 summarizes imaging studies for evaluation of jaundice

Laboratory Evaluation

Best Test(s)
- Bilirubin with fractionation
- Alkaline phosphatase, ALT, AST
- Viral hepatitis panel

Ancillary Test(s)
- Serum amylase, lipase, LDH
- CBC, reticulocyte count, antiglobulin test
- BUN, creatinine, electrolytes
- Liver biopsy

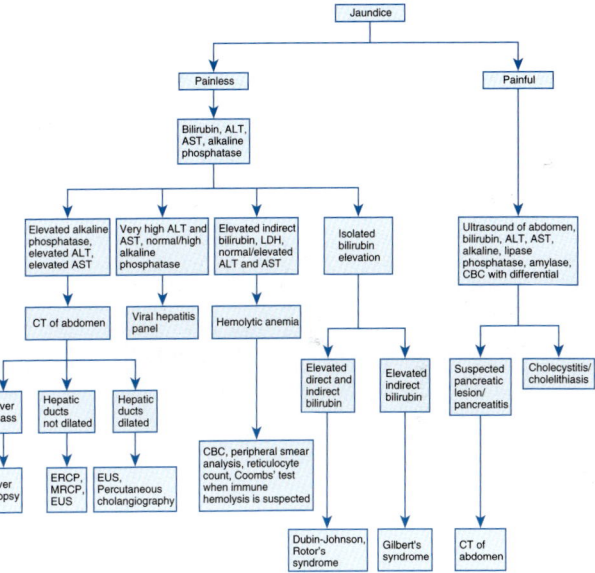

Fig. 3.117 Diagnostic algorithm.

TABLE 3.8	Imaging Studies for Evaluation of Jaundice				
TEST	SENSITIVITY (%)	SPECIFICITY (%)	MORBIDITY (%)	MORTALITY (%)	ADVANTAGES AND DISADVANTAGES
Abdominal US	55–91	82–95	0	0	*Advantages:* noninvasive, portable *Disadvantages:* bowel gas may obscure bile duct; difficult in obese persons, operator dependent
Abdominal CT	63–96	93–100	See disadvantages	0	*Advantages:* noninvasive, higher resolution than US, not operator dependent *Disadvantages:* potential for contrast-induced nephrotoxicity, anaphylaxis
MRCP	82–100	94–98	See disadvantages		*Advantages:* noninvasive, imaging of bile ducts superior to US and CT *Disadvantages:* requires breath holding, may miss small-caliber bile duct disease
ERCP	89–98	89–100	5	0.2	*Advantages:* provides direct imaging of bile ducts; permits direct visualization of periampullary region and acquisition of tissue distal to bifurcation of hepatic ducts; permits simultaneous therapeutic intervention, especially useful for lesions distal to the bifurcation of the hepatic ducts *Disadvantages:* requires sedation, cannot be performed if altered anatomy precludes endoscopic access to the ampulla (e.g., Roux-en-Y loop); may cause complications (e.g., pancreatitis)
Percutaneous THC	98–100	89–100	3.5	0.2	*Advantages:* provides direct imaging of the bile ducts, permits simultaneous therapeutic intervention, especially useful for lesions proximal to the common hepatic duct *Disadvantages:* more difficult with nondilated intrahepatic bile ducts; may cause complications
EUS	89–97	67–98	See disadvantages	0	*Advantages:* imaging of the bile ducts is superior to US and CT; permits needle aspiration of suspected neoplasms *Disadvantages:* requires sedation

79. Joint Effusion (Fig. 3.118)

Diagnostic Imaging

Best Test(s)
- Radiograph of affected joint if history of trauma

Ancillary Tests
- MRI of affected joint if ligament tear is suspected
- CT if suspected fracture not visible on plain films

Laboratory Evaluation

Best Test(s)
- Arthrocentesis with fluid analysis (Gram stain, C&S, cell count and differential, examination for crystals under polarized light)

Ancillary Tests
- Lyme disease serologic testing (Western blot) in endemic areas
- Thayer-Martin culture
- Anaerobic cultures
- TB and fungal cultures

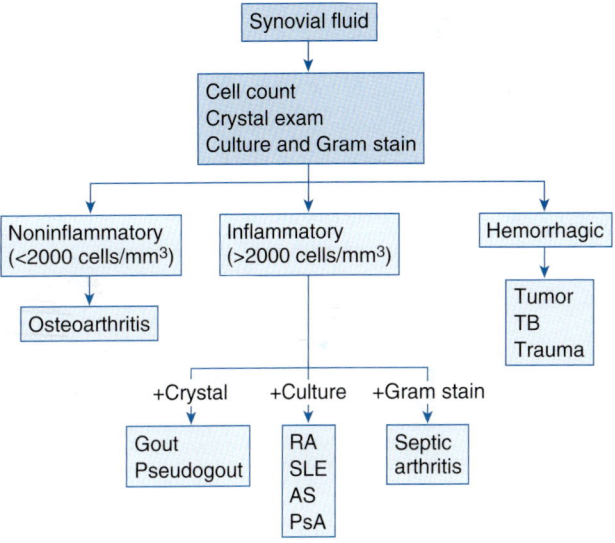

Fɪɢ. **3.118** Algorithm for analysis of joint fluid. Examples of inflammatory arthritis are indicated, although many conditions can produce these findings. *AS,* ankylosing spondylitis; *PsA,* psoriatic arthritis; *RA,* rheumatoid arthritis; *SLE,* systemic lupus erythematosus; *TB,* tuberculosis. (From Goldman L, Schafer AI: *Goldman's Cecil medicine,* ed 25, Philadelphia, Saunders, 2015.)

80. Liver Function Test Elevations (Fig. 3.119)

Diagnostic Imaging

Best Test(s)
- Ultrasound of liver is best initial test

Ancillary Tests
- CT of liver with contrast if ultrasound is inconclusive

Laboratory Evaluation

Best Test(s)
- Liver biopsy

Ancillary Tests
- ALT, AST, alkaline phosphatase
- Bilirubin
- GGTP
- Viral hepatitis panel
- Albumin level, PT
- ANA, transferrin saturation
- Serum protein IEP
- Ceruloplasmin level
- Serum copper, smooth muscle antibody, LKM-1 antibody
- Clues to the differential diagnosis of jaundice with laboratory studies are summarized in Table 3.9

TABLE 3.9	Clues to the Differential Diagnosis of Jaundice: Biliary Obstruction Versus Liver Disease	
PARAMETER	**BILIARY OBSTRUCTION**	**LIVER DISEASE**
History	Abdominal pain Fever, rigors Prior biliary surgery Older age	Anorexia, malaise, myalgias (viral prodrome) Known viral exposure History of blood product receipt or injection drug use Exposure to known hepatotoxin Family history of liver disease
Physical examination	Fever Abdominal tenderness Palpable abdominal mass Abdominal surgical scar	Spider telangiectasias Stigmata of portal hypertension (e.g., prominent abdominal veins, splenomegaly, ascites) Asterixis
Laboratory studies	Predominant elevation of serum alkaline phosphatase relative to aminotransferases* Prothrombin time (INR) normal or normalizes with vitamin K administration Leukocytosis Elevated serum amylase or lipase level	Predominant elevation of serum aminotransferase levels relative to alkaline phosphatase Prolonged prothrombin time that does not normalize with vitamin K administration Thrombocytopenia Serologies indicative of specific liver disease

*Except early after acute obstruction when the opposite pattern may be seen transiently.

From Feldman M, Friedman LS, Brandt LJ: *Sleisenger and Fordtran's gastrointestinal and liver disease*, ed 10, Philadelphia, Saunders, 2016.

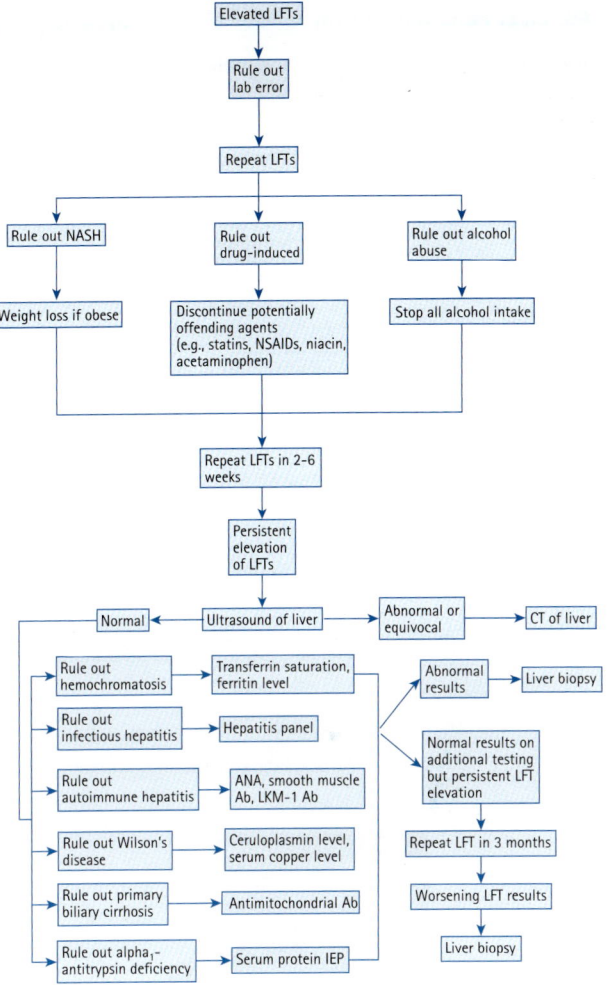

Fig. 3.119 Diagnostic algorithm.

81. Liver Mass (Fig. 3.120)

Diagnostic Imaging

Best Test(s)
- Ultrasound of liver is best initial test

Ancillary Tests
- CT of liver (Fig. 3.121)
- MRI of liver

Laboratory Evaluation

Best Test(s)
- None

Ancillary Tests
- ALT, AST, alkaline phosphatase
- CBC with differential
- PT (INR)
- Hepatitis screen
- Alpha-fetoprotein level

```
                        Ultrasound of liver

        Nonsolid lesion                          Solid lesion

Simple cyst  Complex cyst  Suspected    Suspected hemangioma   CT of liver,
                           abscess                             alpha-fetoprotein level

Reassure,   Serologic test                CT or MRI    Biopsy under CT or
no further  for E. histolytica,           of liver     ultrasound guidance
treatment   CT liver

            + serology for       - serology for E. histolytica,   Hemangioma confirmed,
            E. histolytica       pyogenic abscess suspected        reassure patient

            Stool testing for    Percutaneous drainage
            E. histolytica       under CT or ultrasound guidance
```

Fɪɢ. **3.120** Diagnostic algorithm.

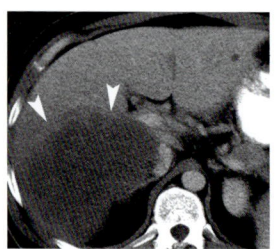

Fɪɢ. **3.121** Primary hepatic lymphoma. Portal venous phase computed tomography reveals a large, homogeneous, poorly enhancing mass (*arrowheads*) largely replacing the right lobe. Chest-abdomen-pelvis CT showed no additional evidence of lymphoma. A biopsy confirmed primary hepatic lymphoma. (From Webb WR, Brant WE, Major NM: *Fundamentals of body CT*, ed 4, Philadelphia, Saunders, 2015.)

82. Lymphadenopathy, Generalized (Fig. 3.122)

Diagnostic Imaging

Best Test(s)
- None

Ancillary Tests
- Chest radiograph (Fig. 3.123, *A*)
- CT of chest (Fig. 3.123, *B*)/abdomen/pelvis with IV contrast
- PET scan (Fig. 3.123, *C*)

Laboratory Evaluation

Best Test(s)
- Lymph node biopsy

Ancillary Tests
- CBC with differential
- HIV, viral titers (EB, CMV)
- Toxoplasma titers
- ALT, AST
- Blood culture × 2
- Bone marrow examination
- VDRL
- ESR (nonspecific)

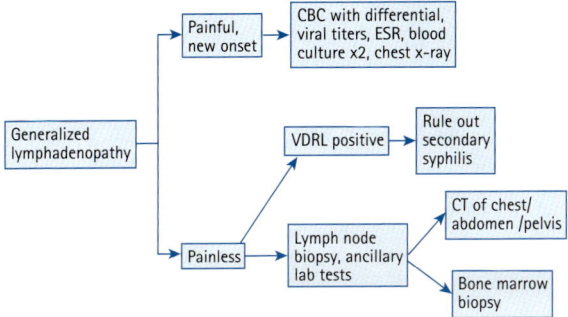

Fig. 3.122 Diagnostic algorithm.

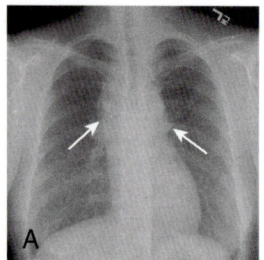

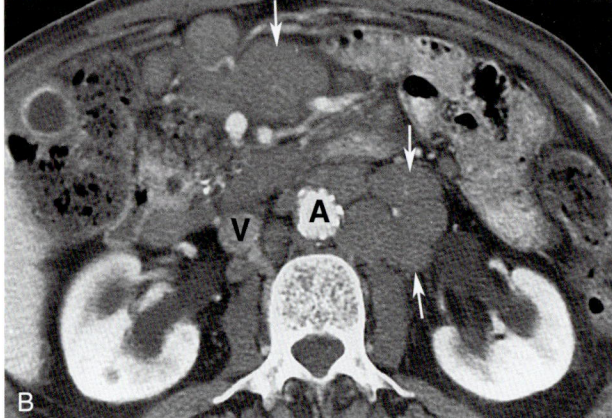

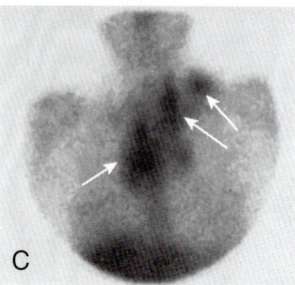

Fig. 3.123 A, Imaging of Hodgkin's lymphoma. Bulky Hodgkin's disease as seen on chest radiograph, computed tomography (CT) of the chest. B, Lymphoma. Enlarged lymph nodes (*arrows*) surround the aorta (*A*) and inferior vena cava (*V*) and are seen in the small-bowel mesentery. C, Imaging of Hodgkin's lymphoma. Bulky Hodgkin's disease as seen on chest radiograph and positron emission tomography. (A and C, From Goldman L, Schafer AI: *Goldman's Cecil medicine*, ed 24, Philadelphia, Saunders, 2012; B, From Webb WR, Brant WE, Major NM: *Fundamentals of body CT*, ed 4, Philadelphia, Saunders, 2015.)

83. Malabsorption, Suspected (Fig. 3.124)

Diagnostic Imaging

Best Test(s)
- Small-bowel series
- Abdominal sonogram

Ancillary Tests
- CT of pancreas with IV contrast
- Video capsule endoscopy
- EACP/MRCP

Laboratory Evaluation

Best Test(s)
- Biopsy of small bowel

Ancillary Tests
- Albumin, total protein
- ALT, AST, PT (INR)
- Serum electrolytes, BUN, creatinine
- Sudan III stain of stool for fecal leukocytes
- CBC, RBC folate, serum iron, serum carotene, cholesterol, serum calcium
- Hydrogen 14-C xylose breath test
- D-xylose test, secretin stimulation test
- Quantitative stool fat test
- Antigliadin antibody, IgA endomysial antibody
- Schilling test, 75SeHCAT test
- Quantitative culture of small-bowel intestinal aspirate
- Box 3.4 summarizes tests to establish the cause of malabsorption based on main symptoms

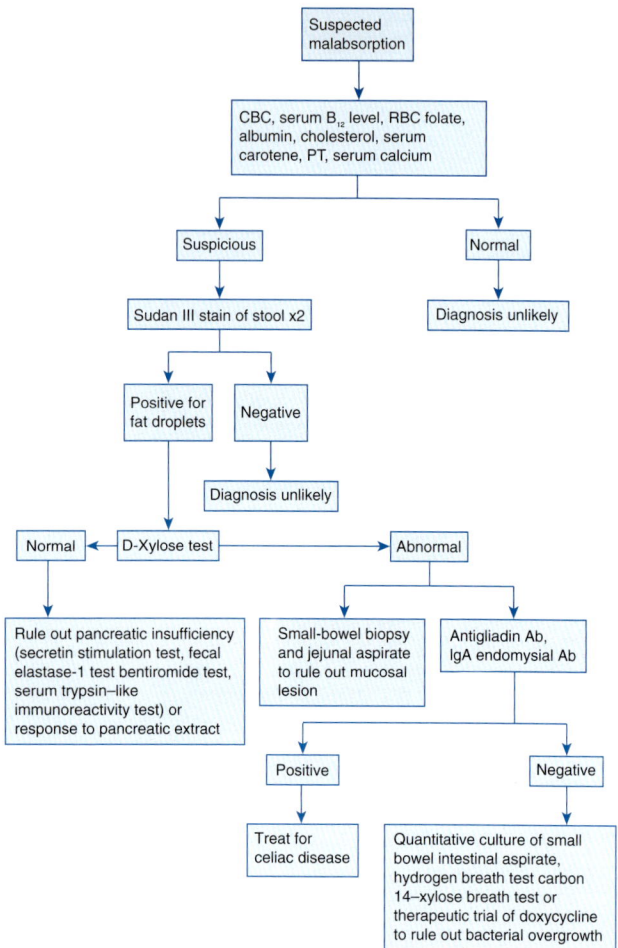

F<small>IG</small>. 3.124 Diagnostic algorithm.

BOX 3.4	Tests to Establish the Cause of Malabsorption Based on Main Symptoms

Weight Loss, Osteomalacia or Osteopenia, Diarrhea, Suspected Steatorrhea, or Deficiency of Fat-Soluble Vitamins

First-Line Tests
- Abdominal and small-intestinal US
- Chymotrypsin or elastase concentration in stool
- EGD with small-intestinal biopsies
- Endomysial or tissue transglutaminase antibodies
- Laboratory tests (CBC, white blood cell differential, cholesterol, TGs, electrolytes, calcium, magnesium, serum ALT, AST, AP, bilirubin levels, prothrombin time, serum albumin level, erythrocyte sedimentation rate and C-reactive protein, TSH)
- Ova, parasites, calprotectin and leukocytes in stool

Second-Line Tests
- Abdominal CT, MRI
- Endoscopic examination of the terminal ileum, including ileal biopsies
- Enteroscopy, including biopsies
- ERCP/MRCP
- More extensive laboratory investigation (immunoglobulins, HIV ELISA, antinuclear antibodies, ferritin, food allergen–specific IgE, ACTH cortisol, chromogranin A, gastrin, urinary 5-HIAA)
- Quantitative fecal fat
- Quantitative small-intestinal culture or breath tests for small intestinal bacterial overgrowth (SIBO)
- Small-bowel series/small-bowel MRI
- Special staining of small-intestinal biopsies (e.g., Congo red for amyloid, PAS for Whipple's disease)
- Therapeutic trial of pancreatic enzymes, antibiotics (tetracycline, metronidazole), cholestyramine, or a gluten-free diet
- Video capsule endoscopy

Tests in Unusually Difficult Cases (Third-Line Tests)
- Mesenteric angiography
- Antienterocyte antibodies
- EUS
- MRA
- PET
- Serum or plasma glucagon, somatostatin
- Somatostatin (octreotide) scan
- Special tests of intestinal biopsies (e.g., flow cytometry of intraepithelial lymphocytes for lymphoma and refractory celiac disease, PCR for *Tropheryma whipplei* or other infective organisms, chromogranin A stain for enteroendocrine cells)
- Spiral CT of the pancreas
- Tests for bile acid malabsorption
- Tube test for exocrine pancreatic secretion (secretin, CCK, or Lundh test)

Bloating, With or Without Diarrhea

First-Line Tests

- Fructose H_2 breath test
- Lactose H_2 breath test
- Lactose tolerance test

Continued

| BOX 3.4 | Tests to Establish the Cause of Malabsorption Based on Main Symptoms—cont'd |

Second-Line Tests
- Chymotrypsin or elastase concentration in stool
- EGD with duodenal biopsies
- Endomysial or tissue transglutaminase antibodies
- Genetic testing for hypolactasia
- Quantitative small-intestinal culture or breath tests for SIBO
- Stool pH (in patients with diarrhea)

Anemia and Suspected Malabsorption

Microcytic or Hypochromic Anemia (low MCV, MCH)
- EGD with duodenal biopsies
- Endomysial and tissue transglutaminase antibodies
- Iron, ferritin, and transferrin in serum
- Ova and parasites in stool
- Video capsule endoscopy
- Calprotectin
- FOBT

Macrocytic Anemia (high MCV, MCH)

First-Line Tests
- Folic acid in serum or red blood cells
- Vitamin B_{12} in serum

Second-Line Tests in Cases of Vitamin B_{12} Deficiency
- CT, small-bowel series, enteroclysis, video capsule endoscopy
- EGD with gastric and duodenal biopsies
- Endomysial and tissue transglutaminase antibodies
- Evaluation of ileum (e.g., colonoscopy to ileum with biopsy, balloon enteroscopy with biopsy)
- Ova and parasites in stool
- Quantitative small-intestinal culture or breath tests for SIBO
- Calprotectin
- Schilling test (with and without intrinsic factor)

Second-Line Tests in Cases of Folate Deficiency
- EGD with duodenal biopsies
- Endomysial or tissue transglutaminase antibodies

AP, alkaline phosphatase; *CBC,* complete blood cell count; *FOBT,* fecal occult blood test; *5-HIAA,* 5-hydroxyindoleacetic acid; *IgE,* immunoglobulin E; *MCH,* mean corpuscular hemoglobin; *MCV,* mean corpuscular volume; *PAS,* periodic acid–Schiff; *TG,* triglycerides; *TSH,* thyroid-stimulating hormone.
(From Feldman M, Friedman LS, Brandt LJ: *Sleisenger and Fordtran's gastrointestinal and liver disease*, ed 10, Philadelphia, Saunders, 2016.)

84. Meningioma (Fig. 3.125)

Diagnostic Imaging

Best Test(s)
- MRI or CT of brain with contrast (Fig. 3.126, *A* and *B*)

Ancillary Tests
- Cerebral angiography (Fig. 3.126, *C*)

Laboratory Evaluation

Best Test(s)
- None

Ancillary Tests
- None

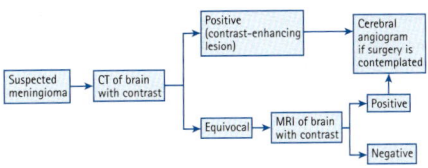

Fɪɢ. **3.125** Diagnostic algorithm.

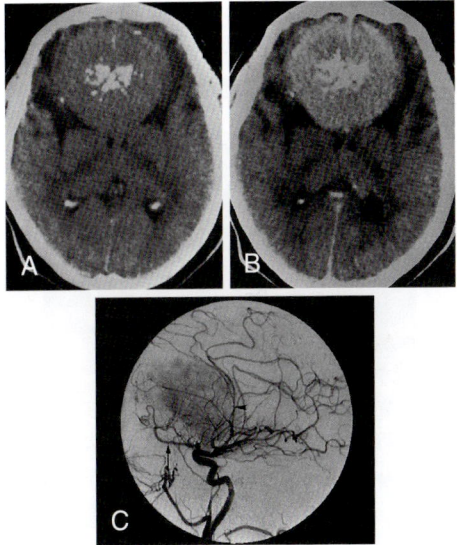

Fɪɢ. **3.126** Subfrontal meningioma. Computed tomography (CT) scan before (A) and after (B) intravenous contrast medium, and lateral projection of common carotid arteriogram (C). There is a large circumscribed mass in the anterior cranial fossa that is isodense to normal gray matter, contains foci of calcification centrally, and enhances homogeneously. There is edema in the white matter of both frontal lobes and posterior displacement and splaying of the frontal horns of the lateral ventricles. On the arteriogram (C), the mass is delineated by a tumor blush, and the anterior cerebral arteries are displaced posteriorly, mirroring the mass effect seen on CT. The ophthalmic artery is enlarged as its ethmoidal branches supply the tumor (*arrow*). (From Grainger RG, Allison DJ, Adam A, Dixon AK [eds]: *Grainger and Allison's diagnostic radiology*, ed 4, Philadelphia, Churchill Livingstone, 2001.)

85. Mesenteric Ischemia (Fig. 3.127)

Diagnostic Imaging

Best Test(s)
- CT of abdomen with IV contrast (Fig. 3.128): may reveal bowel wall thickening, venous dilation, venous thrombus

Ancillary Tests
- Angiography if CT is inconclusive
- Abdominal plain radiograph (generally nonspecific)

Laboratory Evaluation

Best Test(s)
- CBC with differential
- Serum electrolytes (metabolic acidosis is indicative of possible bowel infarction)
- Lactic acid

Ancillary Tests
- Serum amylase
- Hypercoagulopathy workup (e.g., PT, PTT, platelets, protein C, protein S, antithrombin III)
- D-dimer

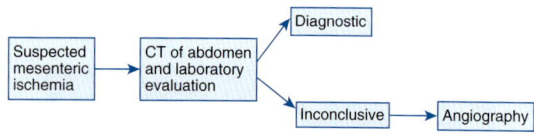

FIG. 3.127 Diagnostic algorithm.

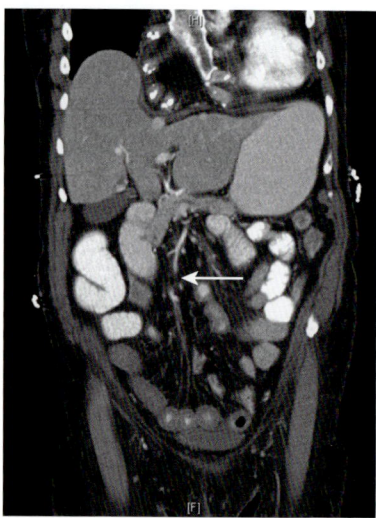

FIG. 3.128 Computed tomography scan after administration of oral and intravenous contrast in a patient with an embolism to the superior mesenteric artery and ischemia of the small bowel and right colon. The *arrow* points to the embolus in the superior mesenteric artery. (From Vincent JL, Abraham E, Moore FA, et al: *Textbook of critical care*, ed 6, Philadelphia, Saunders, 2011.)

86. Mesothelioma (Fig. 3.129)

Diagnostic Imaging

Best Test(s)
- CT of chest with IV contrast after initial abnormalities noted on chest x-ray (Fig. 3.130)

Ancillary Tests
- CT of abdomen with IV contrast and bone scan to assess extent of disease

Laboratory Evaluation

Best Test(s)
- Pleural biopsy

Ancillary Tests
- Cytologic examination from diagnostic thoracentesis is generally inadequate for diagnosis
- PFTs

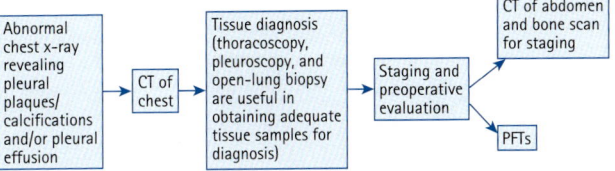

Fɪɢ. 3.129 Diagnostic algorithm.

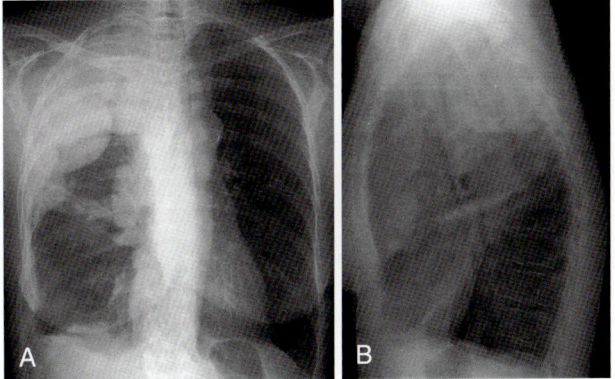

Fɪɢ. 3.130 Malignant pleural thickening. Posteroanterior (A) and lateral (B) chest radiographs. Malignant pleural thickening is characteristically lobulated and nodular. The extensive right-sided disease in this patient was caused by mesothelioma. Notice the extension of the tumor into the fissure. (From Grainger RG, Allison DJ, Adam A, Dixon AK [eds]: *Grainger and Allison's diagnostic radiology*, ed 4, Philadelphia, Churchill Livingstone, 2001.)

87. Metabolic Acidosis (Fig. 3.131)

Diagnostic Imaging   Laboratory Evaluation

Best Test(s)
- None

Best Test(s)
- Serum electrolytes
- ABGs

Ancillary Tests
- Chest radiograph

Ancillary Tests
- BUN, creatinine, glucose
- Determination of anion gap
- $AG = Na^+ - (Cl^- + HCO_3)$
- Lactate level
- Urine electrolytes
- Ethylene glycol, methanol, ethanol (if osmolar gap >12)

METABOLIC ACIDOSIS

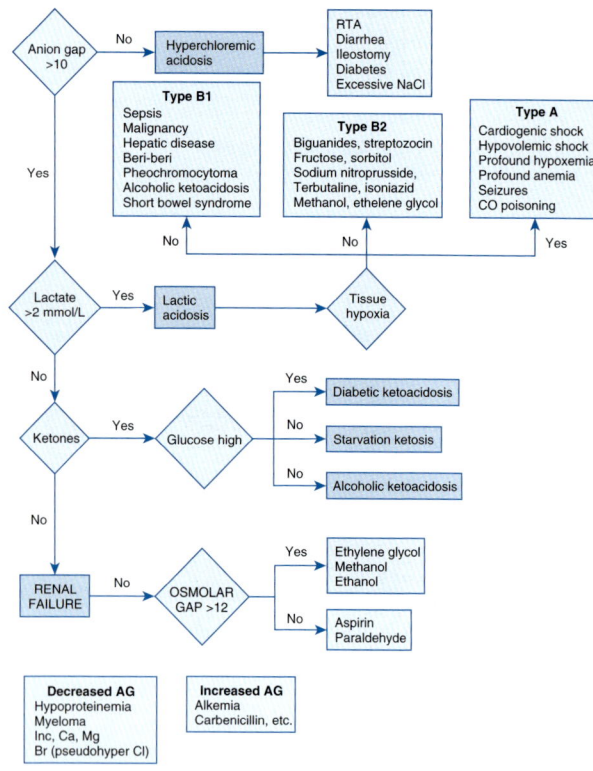

FIG. 3.131 Diagnostic approach to metabolic acidosis. (From Vincent JL, Abraham E, Moore FA, et al: *Textbook of critical care*, ed 6, Philadelphia, Saunders, 2011.)

88. Metabolic Alkalosis (Fig. 3.132)

Diagnostic Imaging

Best Test(s)
• None

Ancillary Tests
• Chest radiograph

Laboratory Evaluation

Best Test(s)
• ABGs
• Urine electrolytes, serum electrolytes

Ancillary Tests
• BUN, creatinine
• Urine potassium
• Plasma renin
• Plasma cortisol

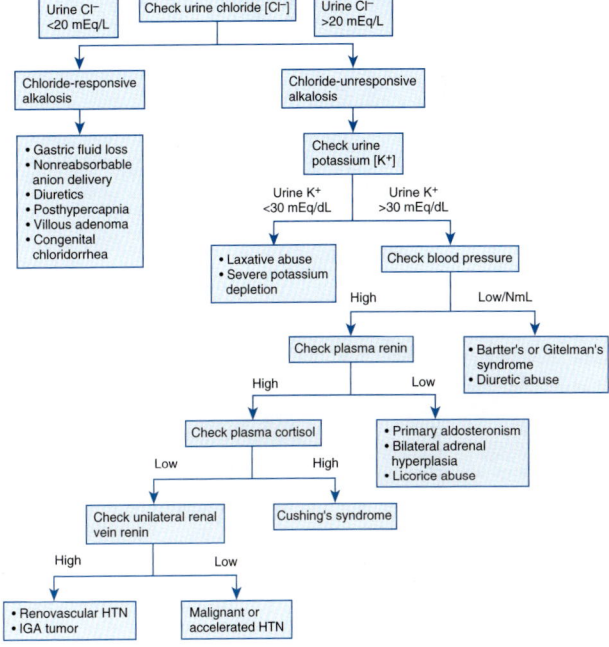

Fɪɢ. 3.132 Workup of metabolic alkalosis. (Data from DuBose TD Jr. Acid-base disorders. In Brenner BM, Rector FC: *Brenner and Rector's the kidney*, ed 10, Philadelphia, Saunders, 2016.)

89. Microcytosis (Fig. 3.133)

Diagnostic Imaging

Best Test(s)
- None

Ancillary Tests
- None

Laboratory Evaluation

Best Test(s)
- RDW

Ancillary Tests
- Reticulocyte count
- Ferritin level
- TIBC, serum iron
- Hemoglobin electrophoresis

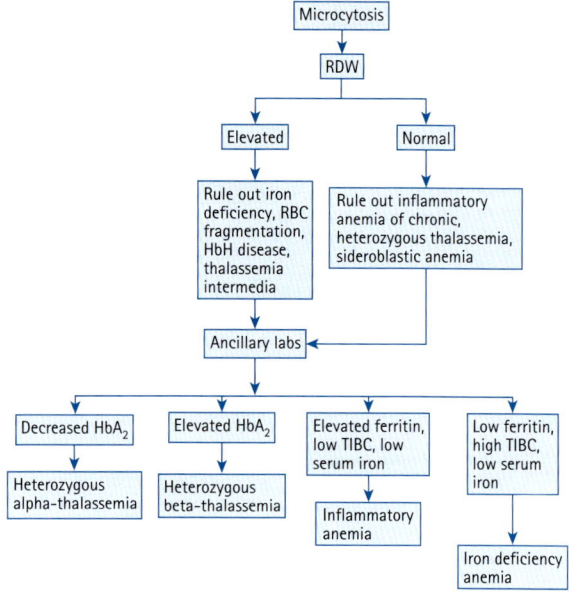

Fig. 3.133 Diagnostic algorithm.

90. Multiple Myeloma (Fig. 3.134)

Diagnostic Imaging

Best Test(s)
- None

Ancillary Tests
- Plain radiographs of painful areas may demonstrate punched-out lytic lesions
- Bone scan is not useful because lesions are not blastic
- Ultrasound of kidneys if renal insufficiency present

Laboratory Evaluation

Best Test(s)
- Serum and urine protein IEP (Fig. 3.135)
- Bone marrow examination

Ancillary Tests
- CBC and peripheral smear (normochromic, normocytic anemia, rouleaux formation, thrombocytopenia)
- Serum calcium (elevated)
- Urinalysis (proteinuria)
- BUN, creatinine, uric acid, LDH, total protein (all elevated)
- Serum electrolytes (decreased anion gap, hyponatremia)

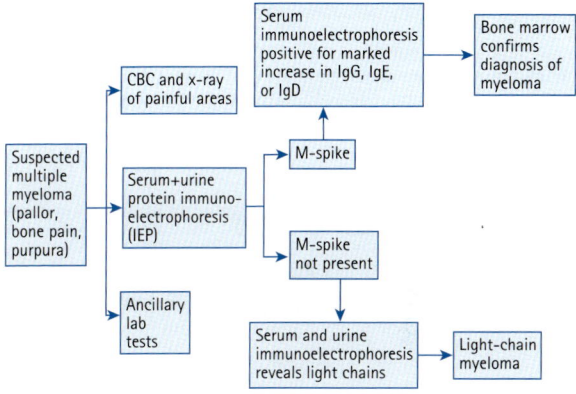

Fɪɢ. 3.134 Diagnostic algorithm.

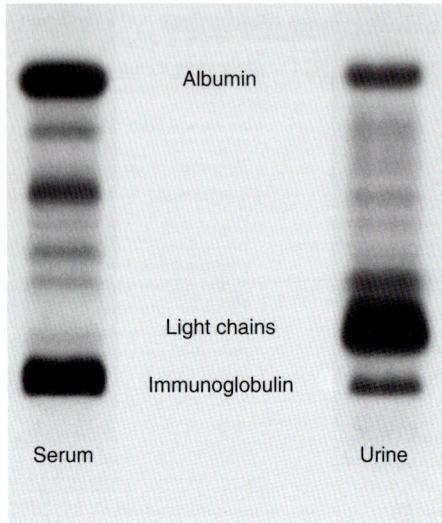

FIG. 3.135 Serum and urine protein electrophoretic patterns in a patient with multiple myeloma. (From McPherson RA, Pincus MR: *Henry's clinical diagnosis and management by laboratory methods*, ed 23, St. Louis, Elsevier, 2017.)

91. Multiple Sclerosis (Fig. 3.136)

Diagnostic Imaging

Best Test(s)
- MRI of brain with gadolinium (Figs. 3.137, 3.138)

Ancillary Tests
- MRI of C-spine with gadolinium in selected cases

Laboratory Evaluation

Best Test(s)
- LP with CSF analysis

Ancillary Tests
- Agarose electrophoresis of CSF (reveals discrete "oligoclonal" bands in the gamma region in approximately 90% of cases)
- Elevated CSF protein, mononuclear WBC, gamma globulin (mostly IgG)
- Presence of myelin basic protein in CSF
- CBC, ANA, B₁₂ level, ESR, LFTs

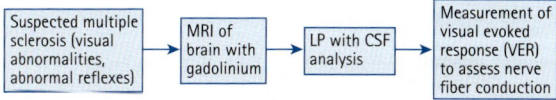

Fɪɢ. 3.136 Diagnostic algorithm.

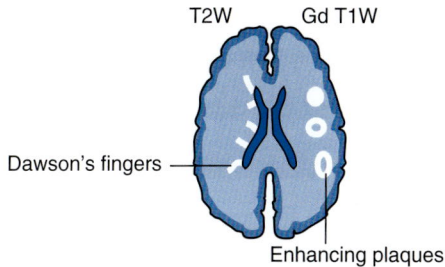

Fɪɢ. 3.137 Typical appearance of brain magnetic resonance imaging scan in multiple sclerosis. Dawson's fingers are perivenular extensions of elliptical structures deep into white matter. (From Weissleder R, Wittenberg J, Harisinghani MG, Chen JW: *Primer of diagnostic imaging*, ed 5, St. Louis, Mosby, 2011.)

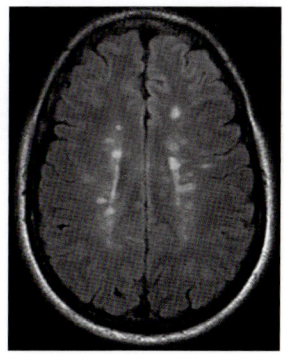

Fɪɢ. 3.138 Axial fluid attenuation inversion recovery image of the brain from a patient with multiple sclerosis, revealing classic multiple periventricular and deep white matter high-signal lesions. (From Goldman L, Schafer AI: *Goldman's Cecil medicine*, ed 24, Philadelphia, Saunders, 2012.)

92. Myalgias (Fig. 3.139)

Diagnostic Imaging	Laboratory Evaluation
Best Test(s)	*Best Test(s)*
• None	• CPK (with fractionation if elevated)
Ancillary Tests	*Ancillary Tests*
• None	• Serum aldolase
	• Serum electrolytes, BUN, creatinine, calcium, magnesium, phosphate, glucose
	• TSH, ANA, ESR, CRP
	• Muscle biopsy
	• CBC with differential
	• Lyme titer

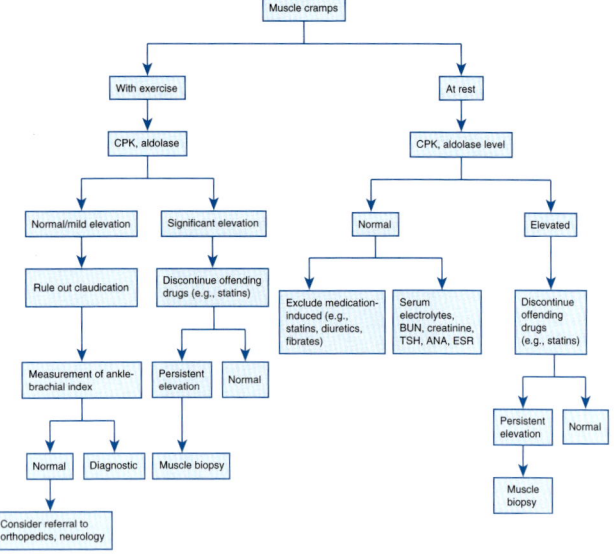

Fɪɢ. 3.139 Diagnostic algorithm.

93. Muscle Weakness (Fig. 3.140)

Diagnostic Imaging

Best Test(s)
- None

Ancillary Tests
- None

Laboratory Evaluation

Best Test(s)
- Muscle biopsy

Ancillary Tests
- CBC with differential
- CPK, aldolase
- ESR (nonspecific)
- TSH
- Serum electrolytes, BUN, creatinine, glucose, calcium
- B_{12} level, ANA

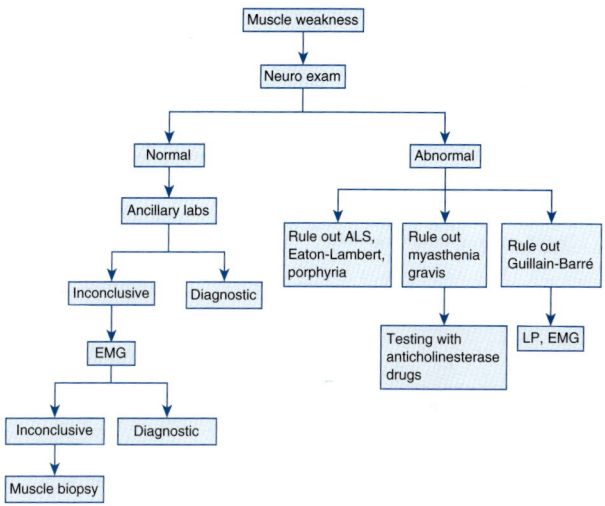

Fig. 3.140 Diagnostic algorithm.

94. Neck Mass (Fig. 3.141)

Diagnostic Imaging

Best Test(s)
- CT or MRI of neck

Ancillary Tests
- Ultrasound of soft tissues of neck
- Thyroid ultrasound
- Chest x-ray

Laboratory Evaluation

Best Test(s)
- None

Ancillary Tests
- CBC with differential
- TSH
- Serum calcium
- Monospot
- Throat C&S

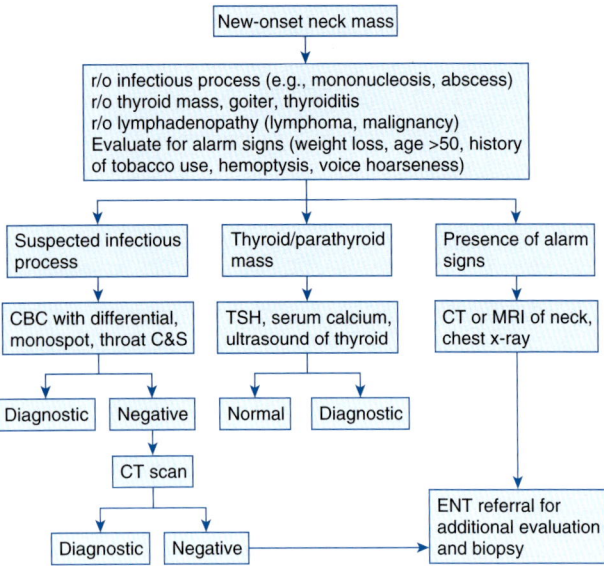

Fɪɢ. 3.141 Diagnostic algorithm.

95. Neuropathy (Fig. 3.142)

Diagnostic Imaging

Best Test(s)
- None

Ancillary Tests
- Chest radiograph (useful to r/o sarcoidosis, lung carcinoma)
- Plain bone films in suspected trauma

Laboratory Evaluation

Best Test(s)
- Electroneuromyography

Ancillary Tests
- CBC, glucose, electrolytes, ALT, AST, calcium, magnesium, phosphate
- Heavy metal screening in suspected toxic neuropathy
- Lyme disease serologic testing in endemic areas or in patients with suggestive history and physical examination
- TSH, B_{12}, folate level
- HIV in patients with risk factors

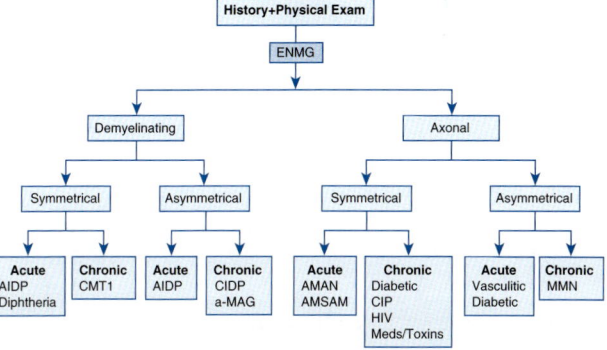

Fɪɢ. **3.142** A systematic approach to evaluating neuropathy. The diseases listed are examples of neuropathies associated with specific neurophysiologic and clinical findings. Diabetic distal, predominantly sensory neuropathies are manifested as chronic axonal neuropathies; acute asymmetric neuropathies can also occur with diabetes. Most neuropathies caused by toxins or as side effects of medication are chronic symmetric axonal neuropathies. Acute inflammatory demyelinating polyradiculoneuropathy (AIDP), acute motor axonal neuropathy (AMAN), and acute motor and sensory axonal neuropathy (AMSAN) are subtypes of Guillain-Barré syndrome. *CIDP,* chronic inflammatory polyradiculoneuropathy; *CIP,* chronic illness polyneuropathy; *CMT1,* Charcot-Marie-Tooth disease type 1, a genetic disorder; *ENMG,* electroneuromyography; *HIV,* human immunodeficiency virus–related neuropathy; *α-MAG,* antimyelin-associated glycoprotein; *MMN,* multifocal motor neuropathy. (Modified from Goldman L, Schafer AI: *Goldman's Cecil medicine,* ed 25, Philadelphia, Saunders, 2015.)

96. Neutropenia (Fig. 3.143)

Diagnostic Imaging

Best Test(s)
- None

Ancillary Tests
- CT of abdomen with IV contrast

Laboratory Evaluation

Best Test(s)
- Bone marrow examination

Ancillary Tests
- CBC with differential
- HIV
- Blood culture × 2, urine C&S
- B_{12} level, folate

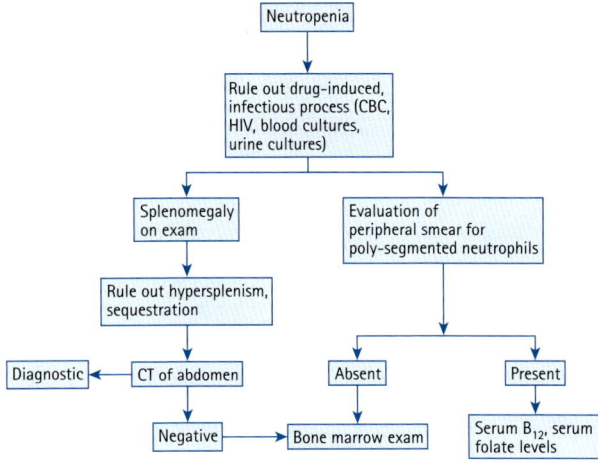

FIG. 3.143 Diagnostic algorithm.

97. Osteomyelitis (Fig. 3.144)

Diagnostic Imaging

Best Test(s)
- MRI of affected bone with contrast (Fig. 3.145)

Ancillary Tests
- Three-phase bone scan with Tc-99m MDP (useful if MRI is contraindicated or is not readily available)
- Doppler study of affected extremity (useful for determining vascular patency in selected patients [e.g., diabetics])
- Conventional radiograph of affected bone (initial imaging test)

Laboratory Evaluation

Best Test(s)
- Bone culture following surgical debridement

Ancillary Tests
- ESR, CRP
- CBC with differential
- Blood culture × 2

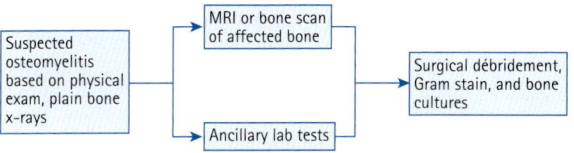

Fɪɢ. **3.144** Diagnostic algorithm.

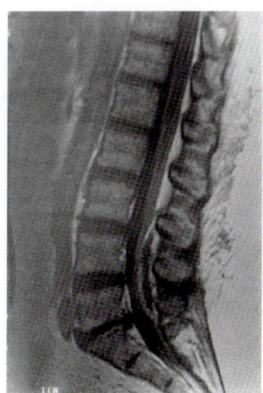

Fɪɢ. **3.145** Vertebral osteomyelitis. A sagittal contrast-enhanced conventional spin-echo MRI scan (T1-weighted) demonstrates a posteriorly located epidural abscess at the L4-L5 vertebral level with an enhancing rim and displacement of the nerve roots anteriorly. (Courtesy Dr. Joseph Mammone. From Cohen J, Powderly WG: *Infectious diseases*, ed 2, St. Louis, Mosby, 2004.)

98. Pancreatic Mass (Fig. 3.146)

Diagnostic Imaging

Best Test(s)
- High-resolution CT of pancreas with IV contrast (Fig. 3.147)

Ancillary Tests
- ERCP
- EUS
- MRCP
- MRI pancreas

Laboratory Evaluation

Best Test(s)
- Ultrasound-guided needle biopsy

Ancillary Tests
- Serum amylase, lipase
- Alkaline phosphatase, ALT
- Serum calcium

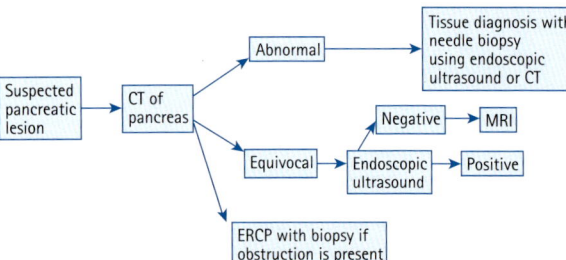

Fɪɢ. 3.146 Diagnostic algorithm.

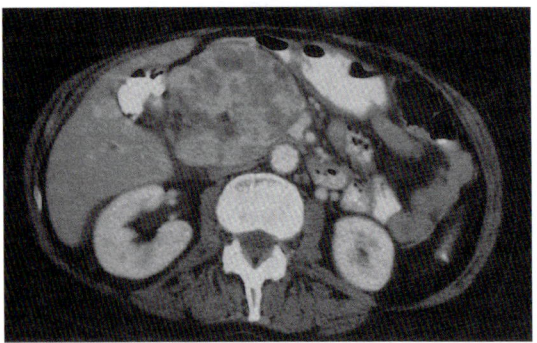

Fɪɢ. 3.147 Pancreatic carcinoma on computed tomography scan. (From Talley NJ, Martin CJ: *Clinical gastroenterology*, ed 2, Sidney, Churchill Livingstone, 2006.)

99. Pancreatitis, Acute (Fig. 3.148)

Diagnostic Imaging

Best Test(s)
- CT of abdomen with IV contrast (Fig. 3.149); CT can also be used to assess severity of acute pancreatitis (Table 3.10)

Ancillary Tests
- Ultrasound of abdomen when gallstone pancreatitis is suspected
- MRCP (>90% sensitivity for choledocholithiasis)
- Endoscopic ultrasound (useful to identify anatomic abnormalities and good sensitivity for small gallstones [≤5 mm])
- ERCP useful for biliary sphincterotomy and stone removal in retained bile duct stone
- Abdominal plain film (often performed as initial imaging test in acute setting)
- Chest radiograph

Laboratory Evaluation

Best Test(s)
- Serum amylase
- Serum lipase

Ancillary Tests
- Serum trypsin level and urinary trypsinogen level (sensitive tests but not readily available)
- ALT, AST, serum calcium, glucose, electrolytes, alkaline phosphatase but not readily
- CBC, urinalysis
- ABGs

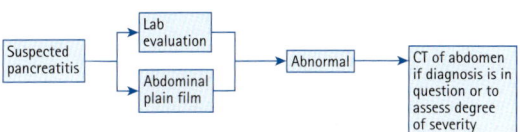

FIG. 3.148 Diagnostic algorithm.

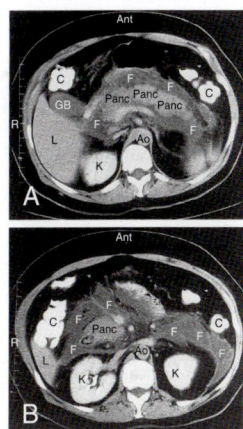

FIG. 3.149 Acute pancreatitis. A, A computed tomography scan in this young woman with hyperlipidemia shows the body of the pancreas (*Panc*) and surrounding fluid (*F*). The liver (*L*), kidney (*K*), gallbladder (*GB*), and colon (*C*) also are identified. Another image in the same patient obtained slightly more inferiorly (B) shows a marked amount of fluid (*F*) around the uncinate portion of the pancreas (*Panc*) and fluid extending around into the left paracolic gutter with associated thickening of Gerota's fascia. *Ao,* Aorta. (From Mettler FA: *Essentials of radiology,* ed 3, Philadelphia, Saunders, 2014.)

TABLE 3.10	Scoring Systems to Assess Severity of Acute Pancreatitis	
SYSTEM	CRITERIA	DEFINITION OF SEVERE PANCREATITIS

Ranson	At admission:
	    Age >55 years
	    WBC >16,000/μL
	    Glucose >200 mg/dL
	    LDH >350 IU/L
	    AST >250 IU/L
	
	Within next 48 hours:
	    Decrease in hematocrit by >10%
	    Estimated fluid sequestration of >6 L
	    Serum calcium <8 mg/dL
	    PaO_2 <60 mm Hg
	    BUN increase >5 mg/dL after hydration
	    Base deficit >4 mmol/L
		Total score ≥3

APACHE-II	Multiple clinical and laboratory factors. Calculator available at www.mdcalc.com/apache-ii-score-for-icu-mortality	Total score ≥8
BISAP	BUN >25 mg/dL Impaired mental status Presence of SIRS Age >60 years Pleural effusion	Total score >2
CT	A: Normal pancreas B: Focal or diffuse enlargement of pancreas C: Grade B plus pancreatic and/or peripancreatic inflammation D: Grade C plus a single fluid collection E: Grade C plus two or more fluid collections or gas in pancreas	Grade >C

CT severity index	CT grade
	    A = 0
	    B = 1
	    C = 2
	    D = 3
	    E = 4
	Plus necrosis grade
	    No necrosis = 0
	    <30% necrosis = 2
	    30%–50% necrosis = 4
	    >50% necrosis = 6
		Score >5

APACHE, Acute Physiology and Chronic Health Evaluation; *AST*, aspartate aminotransferase; *BISAP*, Bedside Index of Severity of Acute Pancreatitis; *BUN*, blood urea nitrogen; *CT*, computed tomography; *LDH*, lactate dehydrogenase; *PaO₂*, partial pressure of oxygen; *SIRS*, systemic inflammatory response syndrome; *WBC*, white blood count.
From Goldman L, Schafer AI: *Goldman's Cecil medicine*, ed 25, Philadelphia, Saunders, 2015.

100. Parapharyngeal Abscess (Fig. 3.150)

Diagnostic Imaging

Best Test(s)
- CT of soft tissues of neck with IV contrast

Ancillary Tests
- Lateral soft tissue radiograph of neck (often initial study)
- CT (Fig. 3.151) or MRI of neck

Laboratory Evaluation

Best Test(s)
- None

Ancillary Tests
- CBC with differential
- Throat C&S
- Blood cultures

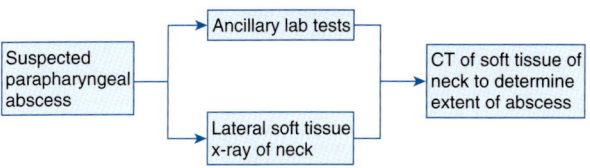

FIG. 3.150 Diagnostic algorithm.

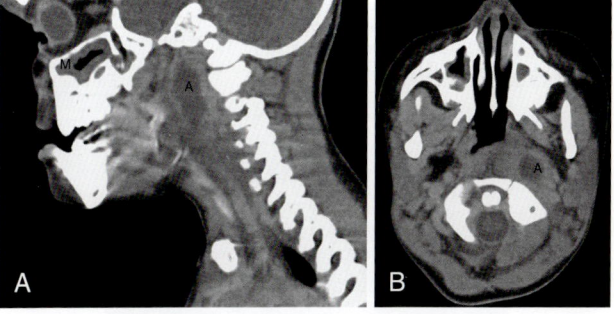

FIG. 3.151 Computed tomography scan of parapharyngeal abscess in a 3-year-old child. A, Sagittal section demonstrating parapharyngeal abscess (*A*) and mucosal swelling (*M*) in the maxillary sinus. B, Coronal section of parapharyngeal abscess (*A*). (Kliegman RM, Stanton B, St. Geme J, et al: *Nelson textbook of pediatrics*, ed 19, Philadelphia, Saunders, 2011.)

101. Pelvic Mass (Fig. 3.152)

Diagnostic Imaging

Best Test(s)
- Ultrasound of pelvis is best initial study (Fig. 3.153)

Ancillary Tests
- CT of pelvis with IV contrast
- MRI of pelvis

Laboratory Evaluation

Best Test(s)
- Biopsy of pelvic mass

Ancillary Tests
- CA 125
- CBC, creatinine
- Serum hCG

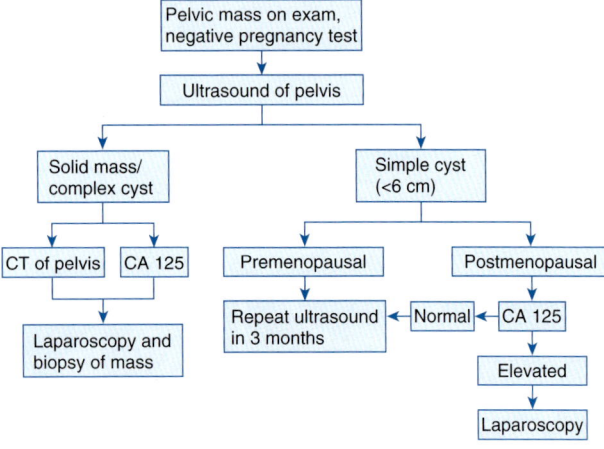

Fig. 3.152 Diagnostic algorithm.

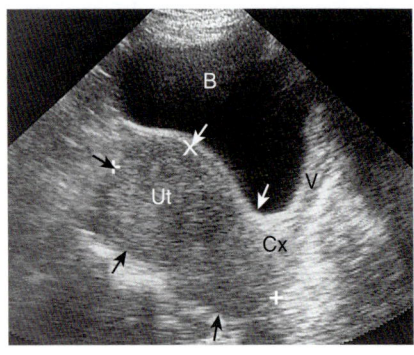

Fig. 3.153 Ultrasound of uterine fibroids. A longitudinal transabdominal view of the pelvis shows that the uterus (*Ut*) is enlarged and lumpy (*arrows*). The echo pattern within the uterus is inhomogeneous; this is the most common appearance of uterine fibroids. *B,* bladder; *Cx,* cervix; *V,* vagina. (From Mettler FA: *Essentials of radiology,* ed 3, Philadelphia, Saunders, 2014.)

102. Peripheral Arterial Disease (PAD) (Fig. 3.154)

Diagnostic Imaging

Best Test(s)
- Continuous-wave Doppler to measure systolic arterial pressure and ABI

Ancillary Tests
- Doppler ultrasound (can be used to locate occluded areas and assess patency of distal arterial system)
- MRA to locate occluded areas and to assess patency of distal arterial system or previous vein grafts
- Angiography (Fig. 3.155): performed if surgical reconstruction is being considered, for angioplasty, or for stent placement. Digital subtraction (invasive) angiography retains the gold standard, but computed tomographic angiography (CTA) and magnetic resonance angiography (MRA) have largely replaced catheter-based angiography in the evaluation of PAD. Invasive angiography, in most cases, is used only as part of an interventional procedure.

Laboratory Evaluation

Best Test(s)
- None

Ancillary Tests
- FBS
- Lipid panel
- Creatinine

Comments
PAD can be staged clinically using the Fontaine classification (Table 3.11).

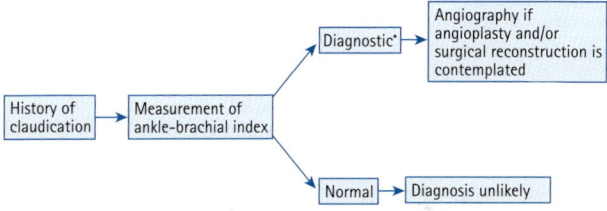

Fɪɢ. 3.154 Diagnostic algorithm. *ABI 0.91 to 0.99, borderline; 0.41 to 0.90, mild to moderate peripheral arterial disease (PAD); ≤0.40, severe PAD.

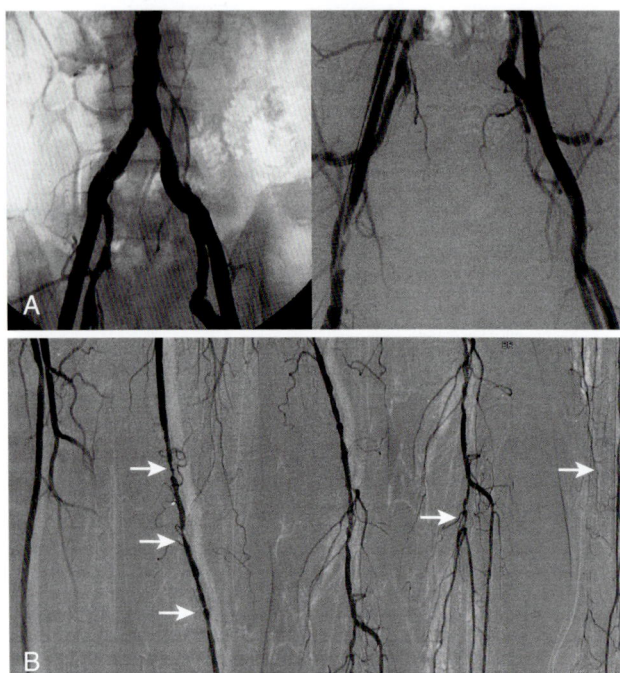

FIG. 3.155 Angiogram of a patient with disabling left calf claudication. A, The aorta and bilateral common iliac arteries are patent. B, The left superficial femoral artery has multiple stenotic lesions (*arrows*). There is a significant stenosis of the left tibioperoneal trunk and left posterior tibial artery (*arrows*). (From Zipes DP, Libby P, Bonow RO, Braunwald E: *Braunwald's heart disease*, ed 7, Philadelphia, Saunders, 2005.)

TABLE 3.11	Fontaine Classification of Peripheral Arterial Disease
STAGE	**SYMPTOMS**
I	Asymptomatic
II	Intermittent claudication
IIa	Pain-free, claudication walking >200 m
IIb	Pain-free, claudication walking <200 m
III	Rest and nocturnal pain
IV	Necrosis, gangrene

From Zipes DP, Libby P, Bonow RO, Braunwald E: *Braunwald's heart disease*, ed 7, Philadelphia, Saunders, 2005.

103. Pheochromocytoma (Fig. 3.156)

Diagnostic Imaging*

Best Test(s)
- MRI of adrenals without contrast or CT of adrenals with IV contrast

Ancillary Tests
- Scintigraphy with iodine-131-I MIBG (100% sensitivity); this norepinephrine analog localizes in adrenergic tissue and is particularly useful in locating extraadrenal pheochromocytomas
- 6[18F]-fluorodopamine PET (reserved for cases in which clinical symptoms and signs suggest pheochromocytoma and results of biochemical tests are positive but conventional imaging studies cannot locate tumor)

*Diagnostic imaging is indicated only after biochemical diagnosis.

Laboratory Evaluation

Best Test(s)
- Plasma-free metanephrines (plasma concentration of normetanephrines >2.5 pmol/mL or metanephrines >1.4 pmol/mL indicate pheochromocytoma with 100% specificity)

Ancillary Tests
- 24-hour urine collection for metanephrines (elevated)
- Clonidine suppression test

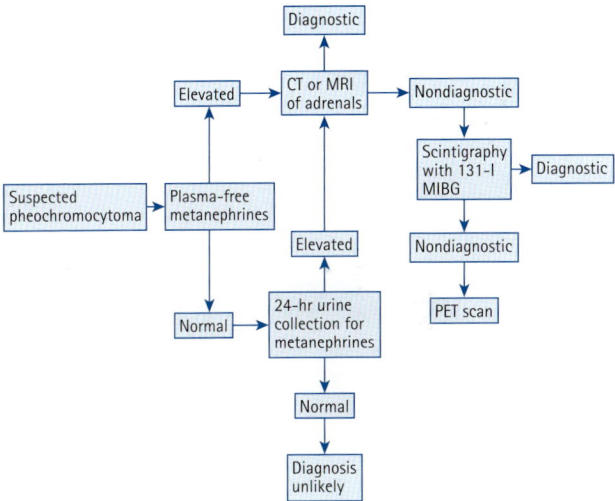

Fig. 3.156 Diagnostic algorithm.

104. Pituitary Adenoma (Fig. 3.157)

Diagnostic Imaging

Best Test(s)
- MRI of pituitary with gadolinium (Fig. 3.158)

Ancillary Tests
- CT of pituitary if MRI is contraindicated

Comments
Fig. 3.159 illustrates a radiologic classification of pituitary adenomas.

Laboratory Evaluation

Best Test(s)
- Tests for hormonal excess or deficit based on clinical presentation (e.g., serum prolactin level, IGF-I, free T_4, dexamethasone suppression test)

Ancillary Tests
- None

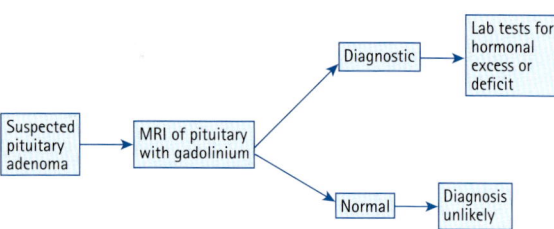

Fig. 3.157 Diagnostic algorithm.

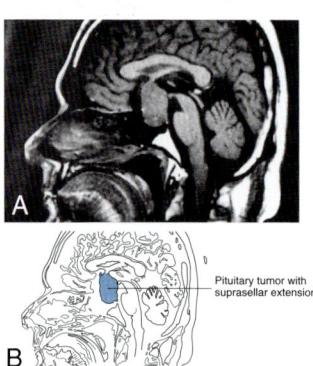

Fig. 3.158 Magnetic resonance imaging scan showing a large pituitary tumor with suprasellar extension causing acromegaly. (From Besser CM, Thorner MO: *Comprehensive clinical endocrinology*, ed 3, St. Louis, Mosby, 2002.)

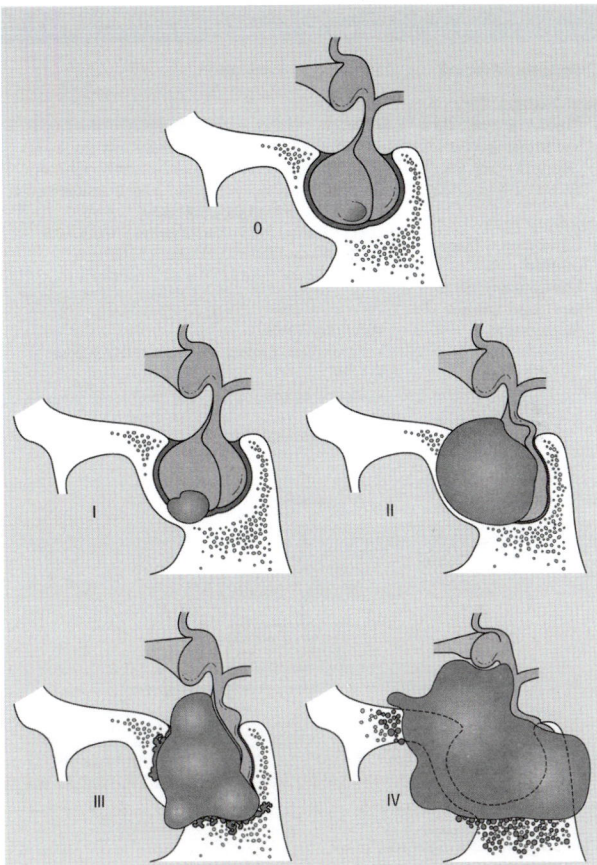

Fɪɢ. 3.159 Radiologic classification of pituitary adenomas. Pituitary tumors are commonly classified on the basis of their size, invasion status, and growth patterns, as proposed by Hardy and Vezina in 1979. Tumors less than or equal to 1 cm in diameter are designated microadenomas, whereas larger tumors are designated macroadenomas. Grade 0: Intra-pituitary microadenoma; normal sellar appearance. Grade I: Intrapituitary microadenoma; focal bulging of sellar wall. Grade II: Intrasellar macroadenoma; diffusely enlarged sella; no invasion. Grade III: Macroadenoma; localized sellar invasion and/or destruction. Grade IV: Macroadenoma; extensive sellar invasion and/or destruction. Tumors are further subclas-sified on the basis of their extrasellar extension, whether suprasellar or parasellar. (From Besser CM, Thorner MO: *Comprehensive clinical endocrinology*, ed 3, St. Louis, Mosby, 2002.)

105. Pleural Effusion (Fig. 3.160)

Diagnostic Imaging

Best Test(s)
- Chest radiograph PA and lateral upright (Fig. 3.161) and lateral decubitus (Fig. 3.162)

Ancillary Tests
- CT of chest without contrast
- Ultrasound (if necessary) for diagnostic thoracentesis

Laboratory Evaluation

Best Test(s)
- Diagnostic thoracentesis and pleural fluid analysis for pH, LDH, glucose, cell count, protein. Light's criteria are used to diagnose exudative vs. transudative effusion. An effusion is likely exudative if at least one of the following exists:
  (1) Pleural fluid protein/serum protein >0.5
  (2) Pleural fluid LDH/serum LDH >0.6
  (3) Pleural fluid LDH is >2/3 normal serum LDH

Ancillary Tests
- Serum total protein, albumin, LDH, glucose
- CBC
- Pleural biopsy
- ADA
- Cytologic analysis

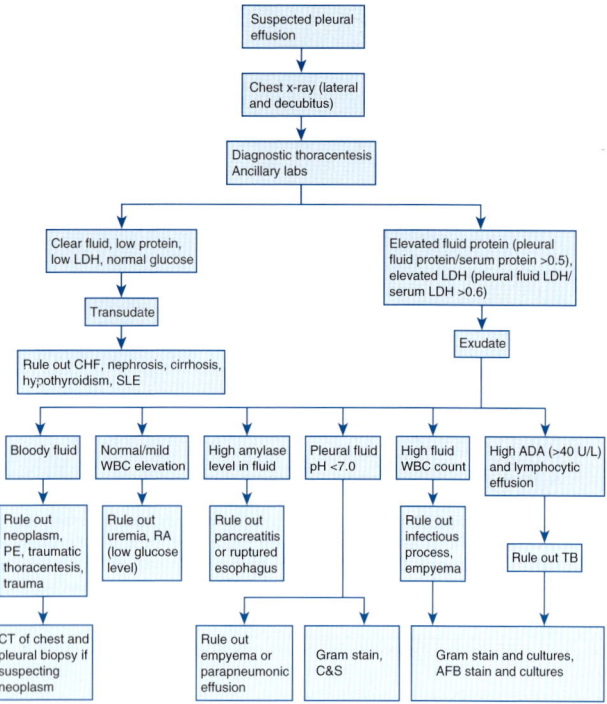

Fɪɢ. 3.160 Diagnostic algorithm.

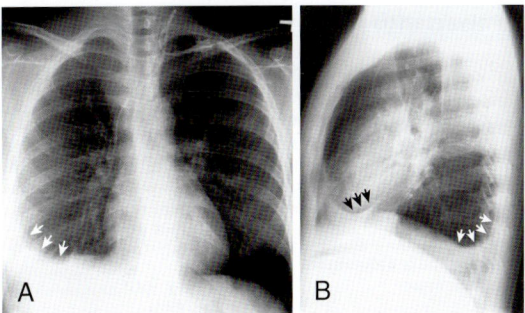

Fig. 3.161 Moderate-sized pleural effusion. On this upright posteroanterior chest x-ray (A), blunting of the right costophrenic angle (*arrows*) is due to pleural fluid. On the lateral view (B), fluid can be seen tracking up into the major fissure (*black arrows*), and blunting of the right posterior costophrenic angle is seen (*white arrows*). (From Mettler FA: *Essentials of radiology*, ed 3, Philadelphia, Saunders, 2014.)

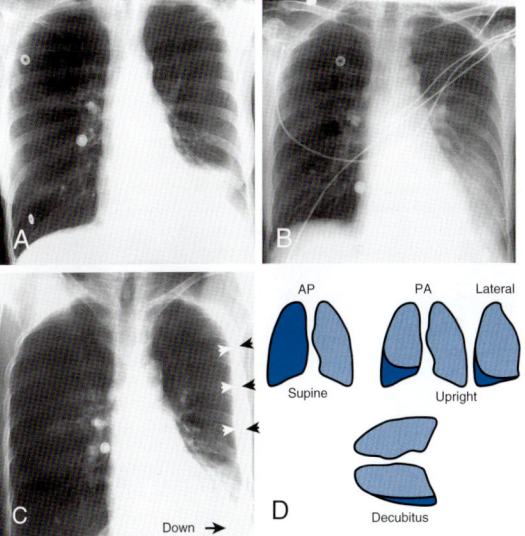

Fig. 3.162 The appearance of pleural effusions depending on patient position. On an upright posteroanterior chest x-ray (A), a large left pleural effusion obscures the left hemidiaphragm, the left costophrenic angle, and the left cardiac border. On a supine anteroposterior view (B), the fluid runs posteriorly, causing a diffuse opacity over the lower two-thirds of the left lung; the left hemidiaphragm remains obscured. This can easily mimic left lower lobe infiltrate or left lower lobe atelectasis. With a left lateral decubitus view (C), the left side of the patient is dependent, and the pleural effusion can be seen to be freely moving and layering (*arrows*) along the lateral chest wall. These findings are shown diagrammatically as well for a right pleural effusion (D). (From Mettler FA: *Essentials of radiology*, ed 3, Philadelphia, Saunders, 2014.)

106. Polyarteritis Nodosa (Fig. 3.163)

Diagnostic Imaging

Best Test(s)

- Abdominal arteriogram (Fig. 3.164) or CT angiography: reveals small or large aneurysms and arteries' focal constrictions between dilated segments, usually in renal, mesenteric, or hepatic arteries

Ancillary Tests

- CT or ultrasound of abdomen to exclude other causes

Laboratory Evaluation

Best Test(s)

- Biopsy of small- or medium-sized artery (biopsies of gastrocnemius muscle and sural nerve are commonly performed when potential involvement is confirmed by abnormal nerve conduction studies)

Ancillary Tests

- ESR, CBC
- HBsAg, ANA
- ALT, BUN, creatinine
- p-ANCA

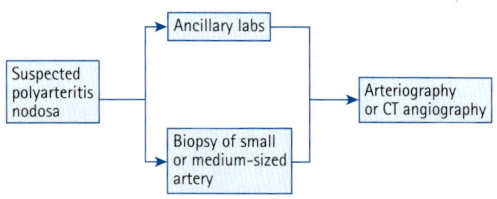

Fɪɢ. **3.163** Diagnostic algorithm.

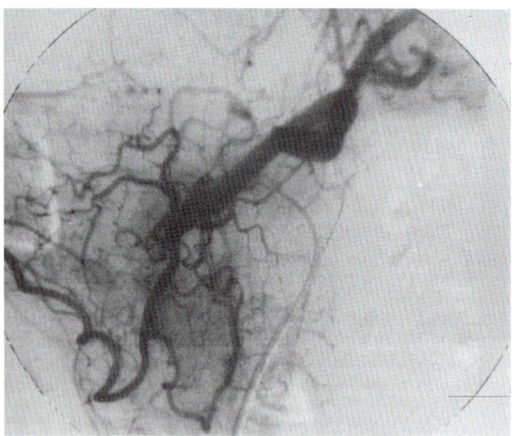

Fɪɢ. **3.164** Aneurysms on the mesenteric artery in polyarteritis. (From Hochberg MC, Silman AJ, Smolen JS, et al: *Rheumatology*, ed 3, St. Louis, Mosby, 2003.)

107. Polycythemia (Fig. 3.165)

Diagnostic Imaging

Best Test(s)
- None

Ancillary Tests
- None

Laboratory Evaluation

Best Test(s)
- Serum erythropoietin level (low)
- Peripheral blood *JAK2 V617* mutation screen

Ancillary Tests
- Leukocyte alkaline phosphatase (increased), B_{12} level (increased), uric acid level (increased)
- O_2 saturation

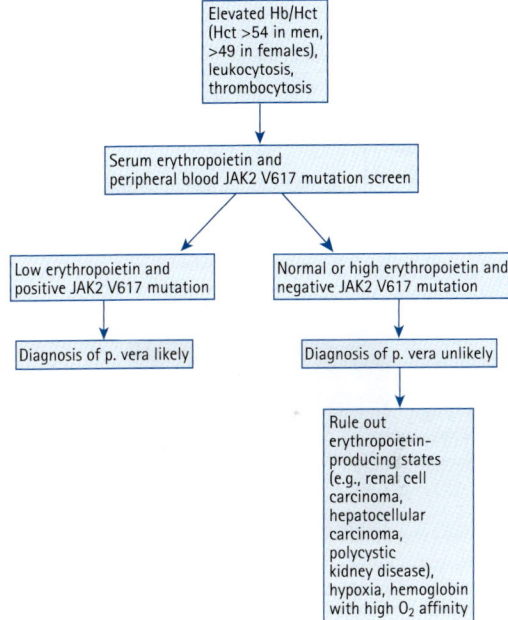

Fɪɢ. 3.165 Diagnostic algorithm.

108. Portal Vein Thrombosis (Fig. 3.166)

Diagnostic Imaging

Best Test(s)
- Doppler ultrasound of portal vein (Fig. 3.168)

Ancillary Tests
- Abdominal CT (Fig. 3.167) with IV contrast (r/o neoplasm)

Laboratory Evaluation

Best Test(s)
- None

Ancillary Tests
- CBC with differential
- Coagulopathy screen (PT, PTT, factor V Leiden protein C, protein S, antithrombin III, lupus anticoagulant)
- ALT, AST, creatinine, urinalysis
- ANA

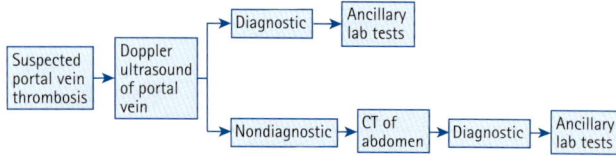

Fɪɢ. 3.166 Diagnostic algorithm.

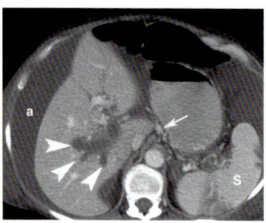

Fɪɢ. 3.167 Portal vein thrombosis. Postcontrast computed tomography shows a low-attenuation thrombus (*arrowheads*) filling and distending the right portal vein and branches. Collaterals have formed in the gastrohepatic ligament (*arrow*). The spleen (*S*) shows multiple infarctions. The splenic vein (not shown) was also thrombosed. Ascites (*a*) is present. (From Webb WR, Brant WE, Major NM: *Fundamentals of body CT*, ed 4, Philadelphia, Saunders, 2015.)

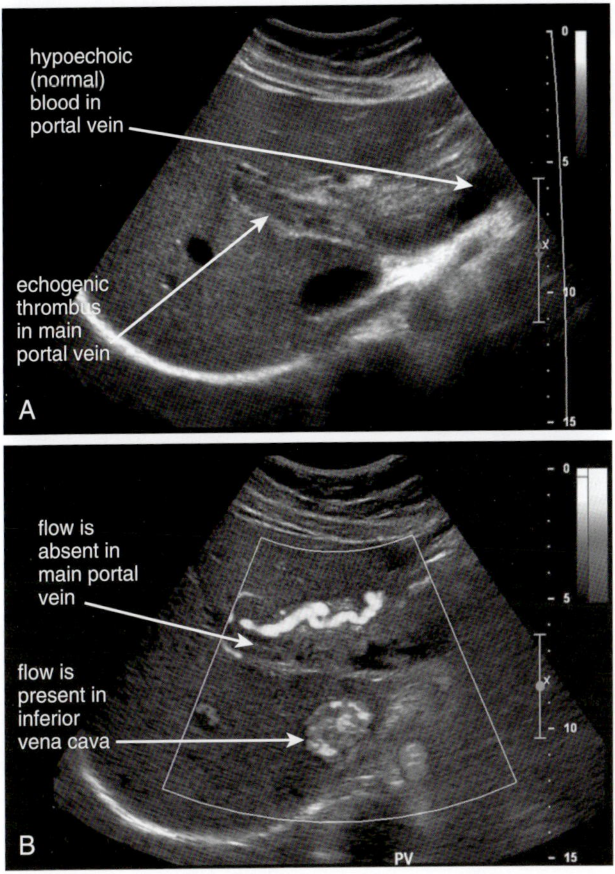

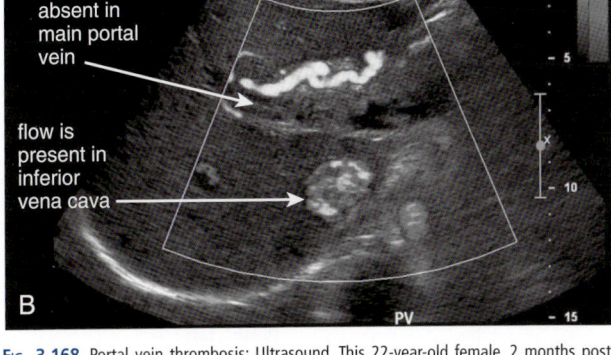

Fig. 3.168 Portal vein thrombosis: Ultrasound. This 22-year-old female, 2 months post-partum, presented with 1 week of right upper quadrant pain. Ultrasound was performed to evaluate for suspected cholecystitis or symptomatic cholelithiasis. Instead, portal vein thrombosis was discovered. The postpartum state is a risk factor for this condition. Hyper-coagulable states and inflammatory or neoplastic abdominal conditions, including pancreatitis and abdominal malignancies, also can result in portal vein thrombosis. A, Ultrasound grayscale image showing thrombus in the main portal vein. B, Doppler ultrasound showing no flow within the portal vein. (From Broder JS: *Diagnostic imaging for the emergency physician*, Philadelphia, Saunders, 2011.)

109. Precocious Puberty (Fig. 3.169)

Diagnostic Imaging

Best Test(s)
- MRI of hypothalamic-pituitary region with gadolinium

Ancillary Tests
- CT of hypothalamic-pituitary region if MRI is contraindicated
- MRI or CT of adrenal glands
- Ultrasound of testicle in patient with unilateral testicular swelling
- Pelvic ultrasound in females

Laboratory Evaluation

Best Test(s)
- LH, FSH
- GnRH

Ancillary Tests
- hCG
- Serum testosterone, estradiol
- TSH
- Adrenal androgens (DHEAS, D4 androstenedione)

Comments
Table 3.12 describes a protocol for investigation of precocious puberty.

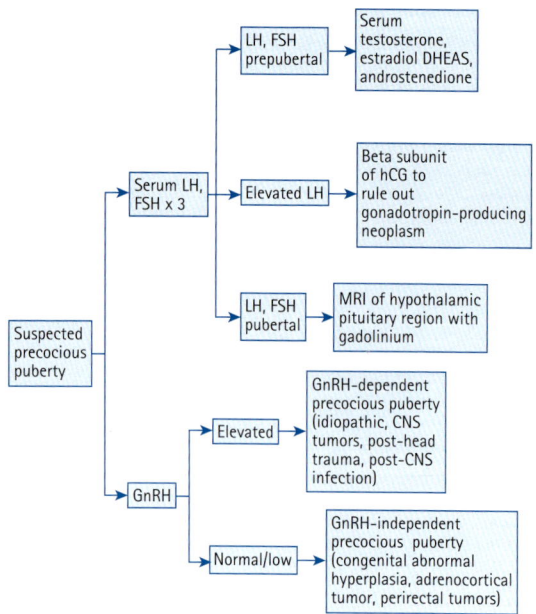

Fig. 3.169 Diagnostic algorithm.

TABLE 3.12	**Protocol for Investigation of Precocious Puberty**
History, family history	
Exposure to environmental sex steroids	
Physical examination	Cutaneous lesions (e.g., neurofibromatosis, McCune-Albright syndrome)
	CNS examination
Auxology	Pubertal staging (Tanner)
	Height, weight, height velocity
	Bone age
Hormone measurements	Testosterone/estradiol
	Luteinizing hormone/follicle-stimulating hormone
	Adrenal androgens (DHEA-S, androstenedione)
	17-hydroxyprogesterone + ACTH stimulation test hCG-β
	Gonadotropin-releasing hormone test
Radiology	Pelvic ultrasound
	Magnetic resonance imaging of the hypothalamic-pituitary region

DNA analysis for known genetic disorders (testotoxicosis, McCune-Albright syndrome)

ACTH, adrenocorticotropic hormone; *CNS,* central nervous system; *DHEA-S,* dehydroepiandrosterone sulfate; *hCG,* human chorionic gonadotropin.

From Besser CM, Thorner MO: *Comprehensive clinical endocrinology,* ed 3, St. Louis, Mosby, 2002.

110. Proteinuria (Fig. 3.170)

Diagnostic Imaging

Best Test(s)
• None

Ancillary Tests
• Ultrasound of kidneys

Laboratory Evaluation

Best Test(s)
• 24-hour urine protein collection

Ancillary Tests
• Serum glucose, BUN, creatinine
• Urine protein IEP
• ANA; serum C3, C4
• Serum c-ANCA, p-ANCA
• Anti-GB membrane Ab
• HBsAg, HCV, HIV
• CBC

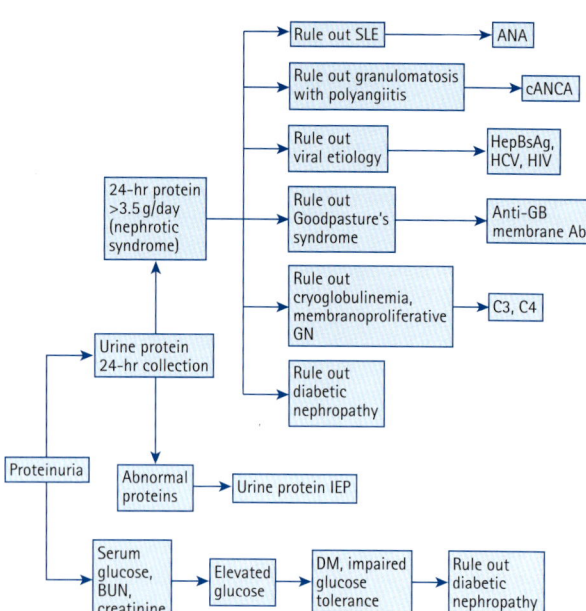

Fɪɢ. 3.170 Diagnostic algorithm.

111. Pruritus, Generalized (Fig. 3.171)

Diagnostic Imaging

Best Test(s)
• None

Ancillary Tests
• Chest radiograph
• CT of abdomen and chest with IV contrast

Laboratory Evaluation

Best Test(s)
• CBC with differential

Ancillary Tests
• ESR (nonspecific)
• FBS, BUN, creatinine
• ALT, AST, alkaline phosphatase, bilirubin
• TSH

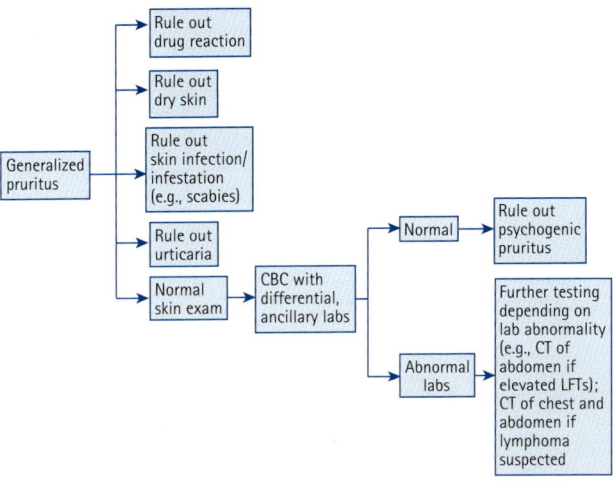

Fig. 3.171 Diagnostic algorithm.

112. Pulmonary Embolism (Fig. 3.172, Table 3.13)

Diagnostic Imaging

Best Test(s)
- CT with IV contrast (Fig. 3.173) is best screening test to r/o PE in patients with baseline chest radiograph abnormalities

Ancillary Tests
- V/Q scan (when readily available, is useful as initial test in patients without baseline chest x-ray abnormalities)
- Pulmonary angiography gold standard (but invasive and not readily available)
- Compressive duplex ultrasonography of lower extremities (useful in patients with inconclusive lung scan or spiral CT and high clinical suspicion)

Laboratory Evaluation

Best Test(s)
- D-dimer by ELISA (normal D-dimer assay in patients with low probability of PE excludes diagnosis; normal D-dimer in patients with nondiagnostic lung scan, negative Doppler ultrasound of extremities, and low clinical suspicion excludes DVT)

Ancillary Tests
- ABGs (decreased PaO₂, PaCO₂, elevated pH)
- A-a oxygen gradient; measure of the difference in oxygen concentration between alveoli and arterial blood; a normal A-a gradient

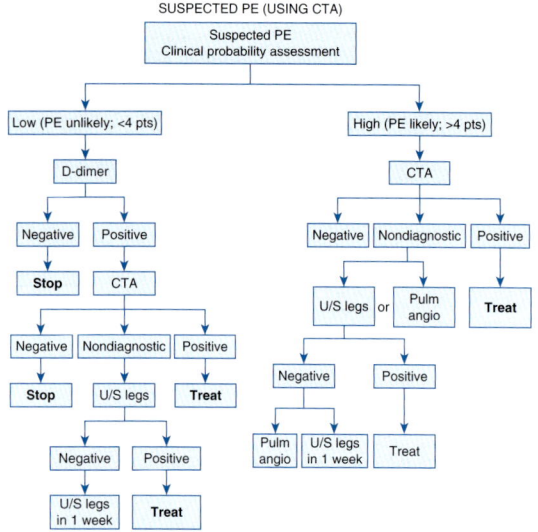

FIG. 3.172 Integrated strategy for diagnosis of pulmonary embolism (PE) using clinical probability assessment, measurement of D-dimer, and computed tomography angiography (CTA) as primary imaging test. Patients with low clinical probability (i.e., PE unlikely, negative D-dimer) need no further testing, but if D-dimer is positive, they should proceed to CTA, and if this is nondiagnostic, to ultrasonography of the legs. Then either treat or repeat ultrasound in 1 week. Patients with high clinical probability (i.e., PE likely) need not have D-dimer measured but should proceed directly to CTA. If CTA is not diagnostic, options are to perform ultrasonography of legs or proceed to pulmonary angiogram. If ultrasound of legs is negative, options are to repeat in 1 week or proceed to pulmonary angiography. (From Vincent JL, Abraham E, Moore FA, et al: *Textbook of critical care*, ed 6, Philadelphia, Saunders, 2011.)

TABLE 3.13	The Pulmonary Embolism Severity Index: Predictors of Low Prognostic Risk
PREDICTOR	**SCORE POINTS**
Age, per year	Age, in years
Male sex	10
History of cancer	30
History of heart failure	10
History of chronic lung disease	10
Pulse ≥110/min	20
Systolic blood pressure <100 mm Hg	30
Respiratory rate ≥30/min	20
Temperature <36° C	20
Altered mental status	60
Arterial oxygen saturation <90%	20

Low prognostic risk is defined as ≤85 points.
From Bonow RO, Mann DL, Zipes DP, et al: *Braunwald's heart disease*, ed 9, Philadelphia, Saunders, 2012.

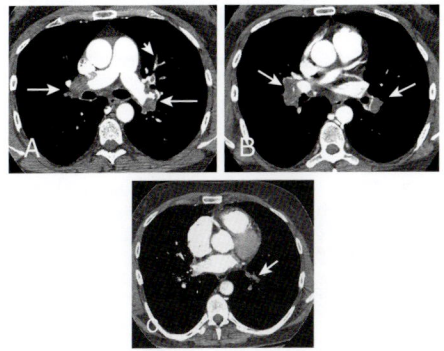

FIG. 3.173 A 46-year-old woman presented with acute shortness of breath and hypoxia. Chest radiograph is normal. A, Chest computed tomography shows a low-attenuation filling defect in left and right main pulmonary arteries (*large arrows*) and left upper lobe segmental artery (*small arrow*), representing massive pulmonary embolism. Emboli extend to left and right interlobar arteries (*arrows*) as well as to a left lower lobe segmental artery (*arrow*), seen in B and C, respectively. (From Vincent JL, Abraham E, Moore FA, et al: *Textbook of critical care*, ed 6, Philadelphia, Saunders, 2011.)

113. Pulmonary Hypertension (Fig. 3.174)

Diagnostic Imaging

Best Test(s)
- Cardiac catheterization with direct measurement of pulmonary arterial pressure (useful for ruling out cardiac causes of pulmonary hypertension)

Ancillary Tests
- Chest radiograph (may reveal dilation of central arteries with rapid tapering of the distal vessels) (Fig. 3.175)
- Echocardiogram (to assess ventricular function and exclude significant valvular pathologic conditions)
- CT of chest (Fig. 3.176) if pulmonary embolism is suspected

Laboratory Evaluation

Best Test(s)
- None

Ancillary Tests
- CBC (erythrocytosis)
- ABGs (decreased PaO_2, decreased oxygen saturation)
- PFTs (r/o obstructive or restrictive lung disease)
- ECG (right ventricular hypertrophy)
- ANA, BNP, ANCA

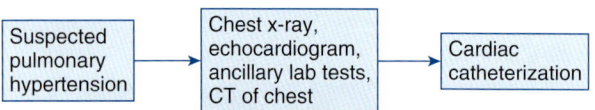

Fɪɢ. 3.174 Diagnostic algorithm.

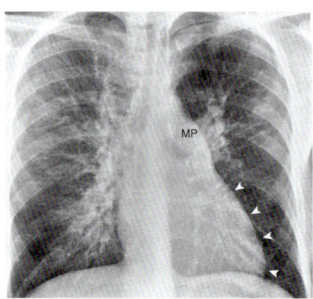

Fɪɢ. 3.175 Pulmonary hypertension in a 40-year-old man with mitral stenosis. The main pulmonary (*MP*) artery segment is enlarged, as are the central pulmonary artery branches. The peripheral arterial branches are not visible. The lower left heart border (*white arrowheads*) is displaced toward the left chest wall by the right heart changes, but overall heart size remains less than 50% of the chest dimension. (From Boxt LM, Abbara S: *The requisites: cardiac imaging*, ed 4, Philadelphia, Elsevier, 2016.)

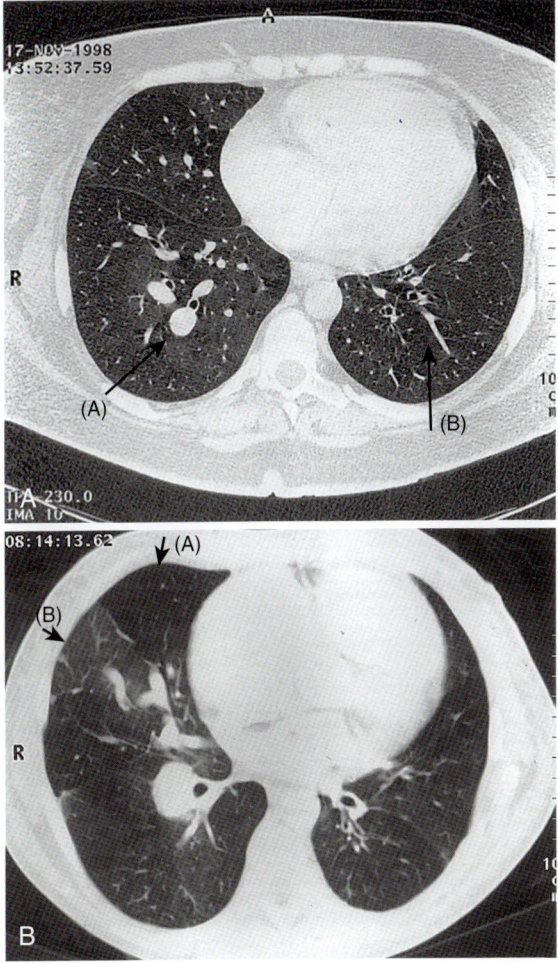

Fig. 3.176 Chest computed tomographic scans in a patient with chronic thromboembolic pulmonary hypertension. A, Helical scan with contrast medium enhancement of the pulmonary vasculature shows a marked disparity in vessel size between the involved vessels (*A*), which are enlarged from thrombus, and the uninvolved vessels (*B*). B, Noncontrast-enhanced high-resolution scan illustrates a marked mosaic pattern manifested by differences in density of regions of the lung parenchyma, reflecting the perfused areas (*B*) and the nonperfused areas (*A*), also consistent with underlying thromboembolic disease. (From Zipes DP, Libby P, Bonow RO, Braunwald E: *Braunwald's heart disease*, ed 7, Philadelphia, Elsevier, 2005.)

114. Pulmonary Nodule (Fig. 3.177)

Diagnostic Imaging

Best Test(s)
- CT of chest without contrast

Ancillary Tests
- PET scan

Laboratory Evaluation

Best Test(s)
- Needle biopsy of nodule

Ancillary Tests
- None

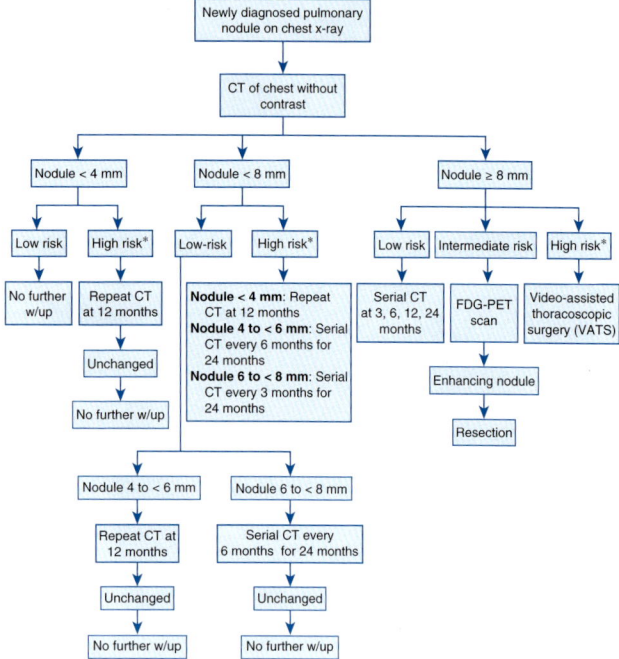

FIG. 3.177 Diagnostic algorithm. *High risk: history of smoking, history of malignancy, age >40.

115. Purpura (Fig. 3.178)

Diagnostic Imaging

Best Test(s)
- None

Ancillary Tests
- Chest radiograph when collagen vascular disease or granulomatosis with polyangiitis is suspected
- Echocardiogram when endocarditis is suspected

Laboratory Evaluation

Best Test(s)
- Platelet count

Ancillary Tests
- CBC, peripheral blood smear, PT, PTT
- ANA (r/o vasculitis, collagen vascular disease)
- ALT, AST, creatinine
- Serum cryoglobulins (r/o mixed cryoglobulinemia)
- Serum protein IEP (r/o hyperglobulinemia)
- Blood culture × 2 (r/o SBE)
- c-ANCA (r/o granulomatosis with polyangiitis), p-ANCA (r/o vasculitis)

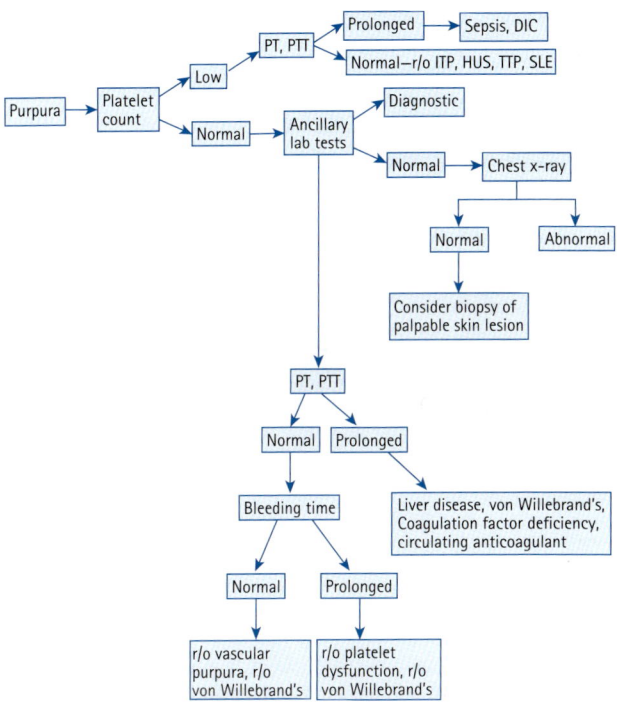

Fɪɢ. 3.178 Diagnostic algorithm.

116. Renal Artery Stenosis (Fig. 3.179)

Diagnostic Imaging

Best Test(s)
- MRA of renal arteries (Fig. 3.180)

Ancillary Tests
- CT of renal artery
- Renal Doppler: preferred as screening test in patients with reduced kidney function (<30 mL/min/1.73 m^2)
- Conventional kidney angiography (Fig. 3.181) is diagnostic and can also be used for intervention (angioplasty, stent placement)
- Ultrasound of kidneys
- Isotope renography (Fig. 3.182) if surgery is contemplated

Laboratory Evaluation

Best Test(s)
- Selective renal vein renin measurement (gold standard but is invasive and expensive)

Ancillary Tests
- Serum creatinine, BUN, electrolytes
- Urinalysis

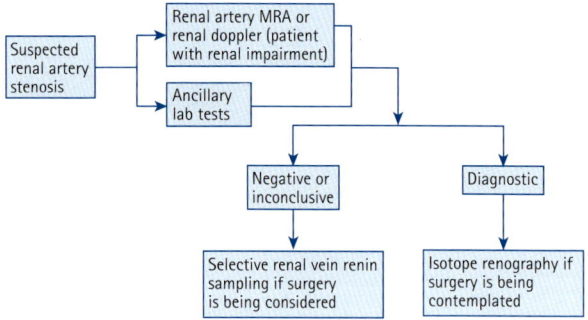

Fɪɢ. 3.179 Diagnostic algorithm.

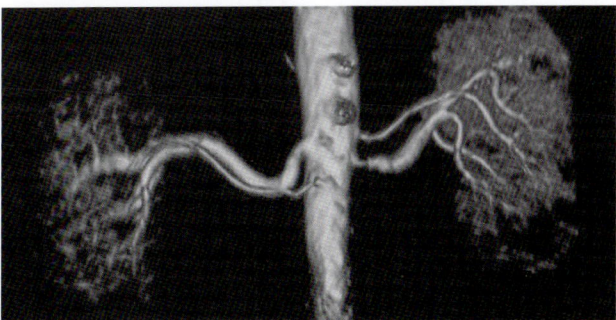

Fig. 3.180 Magnetic resonance (MR) angiogram reconstructed without gadolinium contrast. After warnings about the association of gadolinium with nephrogenic systemic fibrosis, contrast MR is less commonly used for atherosclerotic renovascular imaging than before. Newer techniques can allow excellent vascular imaging without contrast, as shown here. (From Skorecki K, Chertow GM, Marsden PA, et al: *Brenner and Rector's the kidney*, ed 10, Philadelphia, Elsevier, 2016.)

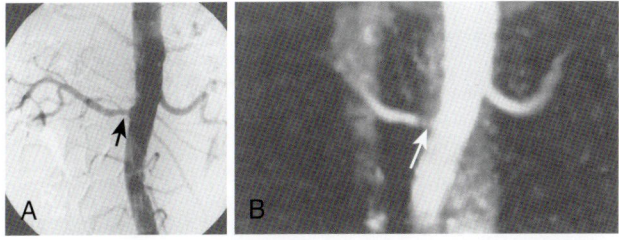

Fig. 3.181 A, Conventional renal digital subtraction arteriogram showing mild renal stenosis (*arrow*) on the right. B, Magnetic resonance angiogram of the same patient. The stenosed segment (*arrow*) is clearly seen in this coronal projection. (Photographs provided by Dr. W. Gedroyc. Souhami RL, Moxham J: *Textbook of medicine*, ed 4, London, Churchill Livingstone, 2002.)

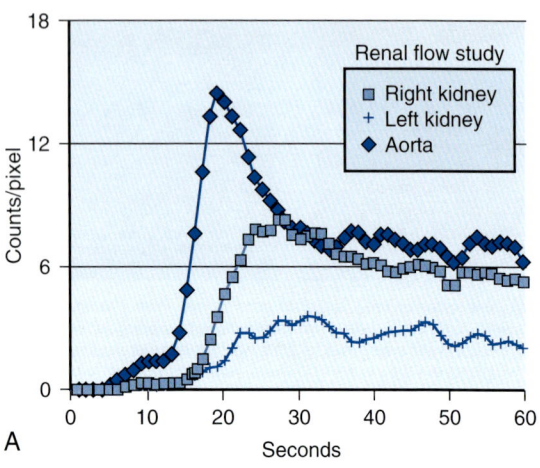

A

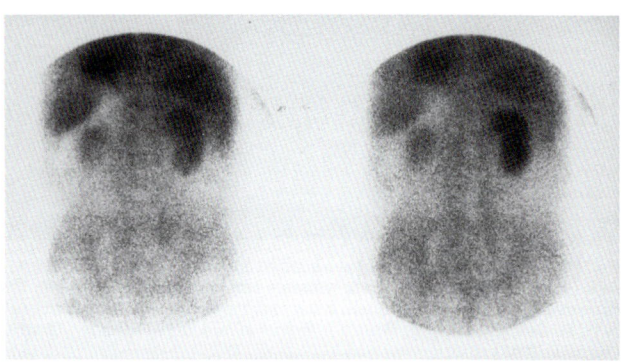

B

79 Y.O. Male
HIPPURAN SCAN

Fig. 3.182 Isotope renography in a patient with unilateral renal artery stenosis. A, Tc-99m–labeled diethylenetriaminepentaacetic acid (Tc-99–DTPA) scan demonstrates delayed circulation and excretion of isotope on the left. B, Hippuran scan (now replaced with Tc-99m–labeled mercaptoacetyltriglycine) provides a renogram demonstrating a small kidney with impaired renal function on the affected side. Radionuclide scans provide a comparative estimate of function from each kidney, which may facilitate selection of intervention, including the potential effect of nephrectomy (see text). (From Skorecki K, Chertow GM, Marsden PA, et al: *Brenner and Rector's the kidney*, ed 10, Philadelphia, Elsevier, 2016.)

117. Renal Mass (Fig. 3.183)

Diagnostic Imaging

Best Test(s)
- CT of kidneys with and without IV contrast (Fig. 3.184)

Ancillary Tests
- Renal ultrasound if cystic lesion
- MRI of kidneys when renal vein or caval thrombosis is suspected, or when CT is contraindicated

Laboratory Evaluation

Best Test(s)
- Urinalysis

Ancillary Tests
- Serum creatinine, calcium, albumin
- ESR (nonspecific)
- CBC

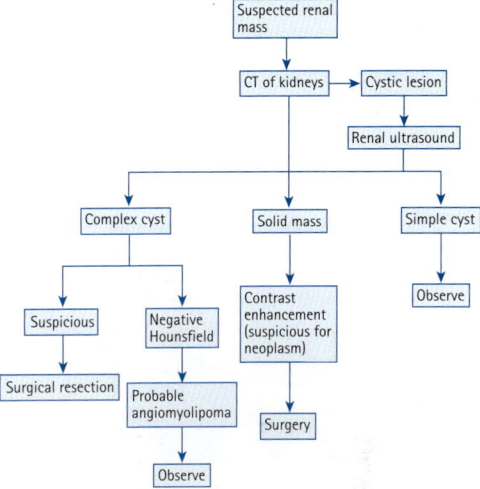

Fɪɢ. **3.183** Diagnostic algorithm.

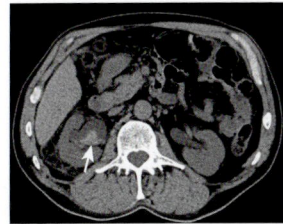

Fɪɢ. **3.184** Transitional cell carcinoma. An axial computed tomography image demonstrates a dilated right renal pelvis containing a nodular mass density (*arrow*), which was proven to be a transitional cell carcinoma. When these tumors are small or in the ureter, they are best visualized by cystoscopy and retrograde pyelogram. (From Mettler FA: *Essentials of radiology*, ed 3, Philadelphia, Saunders, 2014.)

118. Rotator Cuff Tear (Fig. 3.185)

Diagnostic Imaging

Best Test(s)
- MRI of shoulder without contrast (Figs. 3.186, 3.187)

Ancillary Tests
- Ultrasound of upper extremity may be used to diagnose suspected supraspinatus tear; major weakness in evaluating shoulder impingement is inability to visualize complete labrum; accuracy depends on skill of operator
- MRI arthrography (useful in shoulder instability for diagnosing labral tears)

Laboratory Evaluation

Best Test(s)
- None

Ancillary Tests
- None

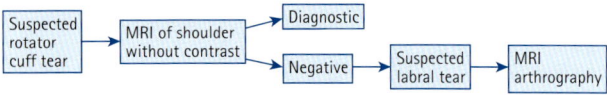

Fig. 3.185 Diagnostic algorithm.

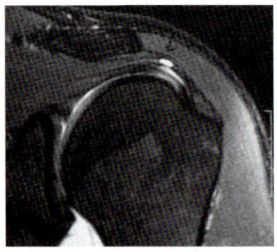

Fig. 3.186 Magnetic resonance imaging scan of complex rotator cuff pathologic condition. The image shows a partial articular surface rotator cuff tear along with a laminar defect and small cyst in the distal supraspinatus tendon. (From Hochberg MC, Silman AJ, Smolen JS, et al: *Rheumatology*, ed 3, St. Louis, Mosby, 2003.)

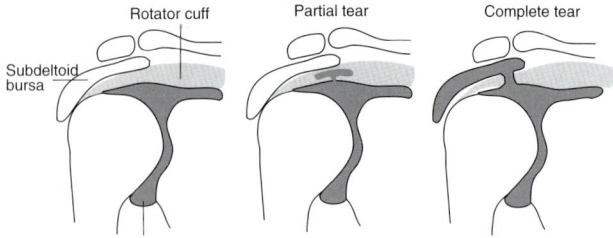

Fig. 3.187 Rotator cuff tear. (From Weissleder R, Wittenberg J, Harisinghani MG, Chen JW: *Primer of diagnostic imaging*, ed 4, St. Louis, Mosby, 2007.)

119. Sarcoidosis (Fig. 3.188)

Diagnostic Imaging

Best Test(s)
- Chest x-ray is best initial test (Fig. 3.189)

Ancillary Tests
- CT of chest for further investigation of abnormal chest x-ray

Laboratory Evaluation

Best Test(s)
- Biopsy of accessible tissue suspected of sarcoid involvement (e.g., lymph node, skin, conjunctiva) or bronchial and transbronchial biopsy (sensitivity 90%)

Ancillary Tests
- CBC, ESR
- ALT, AST, serum calcium
- Serum ACE level
- PFTs

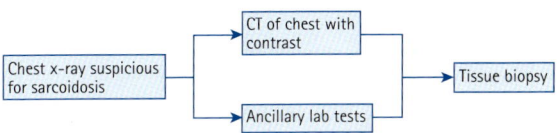

Fɪɢ. **3.188** Diagnostic algorithm.

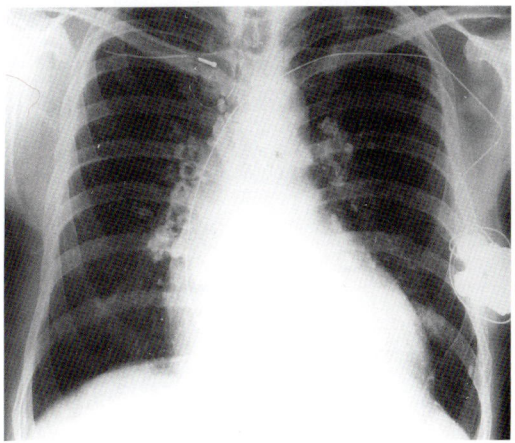

Fɪɢ. **3.189** "Eggshell" calcification in sarcoidosis. There is extensive calcification of the hilar and mediastinal lymph nodes; the calcification has a peripheral (eggshell) configuration, particularly in lymph nodes at the right hilum. Cardiac involvement necessitated the pacemaker. (From Grainger RG, Allison DJ, Adam A, Dixon AK [eds]: *Grainger and Allison's diagnostic radiology*, ed 4, Philadelphia, Churchill Livingstone, 2001.)

120. Scrotal Mass (Fig. 3.190)

Diagnostic Imaging

Best Test(s)
• Ultrasound of scrotum (Fig. 3.191)

Ancillary Tests
• MRI of pelvis with and without contrast for staging of neoplasm

Laboratory Evaluation

Best Test(s)
• None

Ancillary Tests
• Urinalysis
• CBC with differential
• Serum alpha-fetoprotein
• Serum hCG

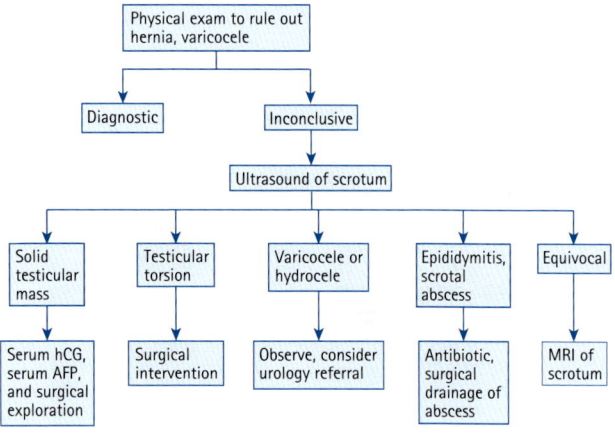

Fɪɢ. 3.190 Diagnostic algorithm.

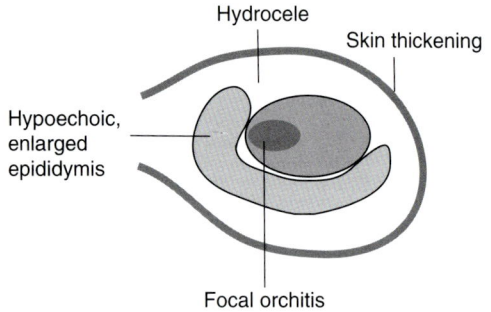

Fɪɢ. 3.191 Scrotal mass sonogram findings. (From Weissleder R, Wittenberg J, Harisinghani MG, Chen JW: *Primer of diagnostic imaging*, ed 4, St. Louis, Mosby, 2007.)

121. Small-Bowel Obstruction (Fig. 3.192)

Diagnostic Imaging

Best Test(s)
- Plain film of abdomen (Fig. 3.193); AP is best initial study

Ancillary Tests
- CT of abdomen and pelvis with IV contrast (Fig. 3.194): will confirm obstruction
- Small-bowel series (may be useful in selected cases for defining level of obstruction)

Laboratory Evaluation

Best Test(s)
- None

Ancillary Tests
- CBC with differential
- Serum electrolytes, BUN, creatinine
- ALT, AST, alkaline phosphatase
- Urinalysis, urine C&S
- Glucose, amylase

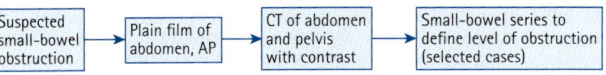

Suspected small-bowel obstruction → Plain film of abdomen, AP → CT of abdomen and pelvis with contrast → Small-bowel series to define level of obstruction (selected cases)

Fig. 3.192 Diagnostic algorithm.

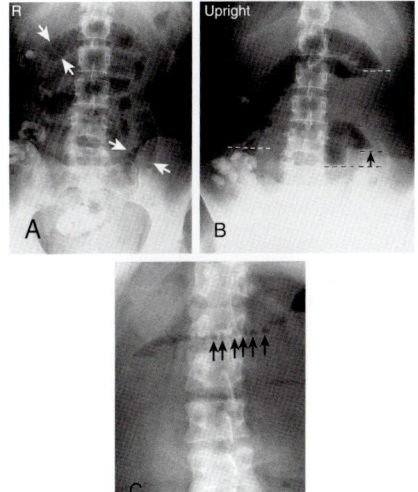

Fig. 3.193 Small-bowel obstruction. On a supine plain radiograph of the abdomen (A), a large amount of dilated small bowel is seen. This can be recognized as small bowel by the regular mucosal pattern of the valvulae extending across the lumen and resembling a set of thick, stacked coins (*arrows*). The small bowel normally should not exceed 3 cm in diameter. On an upright radiograph of the abdomen (B), the air and fluid levels within the same loop of bowel are at different heights. This indicates an obstruction rather than a paralytic ileus. A close-up view of the right midabdomen in another patient (C) shows the "string of pearls" air bubbles (*arrows*) that indicate a fluid-filled and obstructed small bowel. (From Mettler FA: *Essentials of radiology*, ed 3, Philadelphia, Saunders, 2014.)

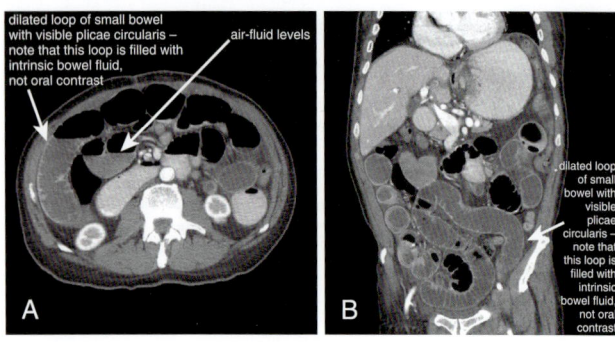

FIG. 3.194 Small-intestine obstruction. A, Section through the midabdomen shows dilated, mainly fluid-filled small intestine loops. The stretched valvula conniventes can be identified clearly in some segments. B, The obstruction was due to an incarcerated paraumbilical hernia, the edge of which can be identified (*arrow*). (From Broder JS: *Diagnostic imaging for the emergency physician*, Philadelphia, Saunders, 2011.)

122. Spinal Epidural Abscess (Fig. 3.195)

Diagnostic Imaging

Best Test(s)
- MRI with gadolinium of spinal cord (Fig. 3.196)

Ancillary Tests
- CT with myelography (more sensitive for cord compression)

Laboratory Evaluation

Best Test(s)
- Gram stain and C&S of abscess content

Ancillary Tests
- CBC with differential
- Blood culture × 2
- ESR

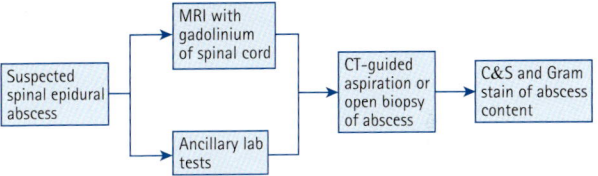

Fig. 3.195 Diagnostic algorithm.

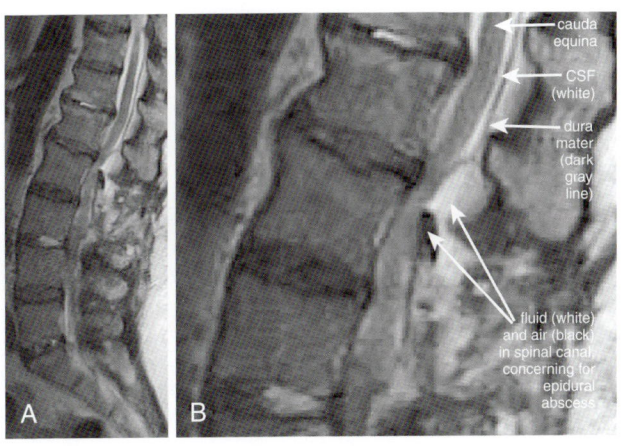

Fig. 3.196 Spinal epidural abscess. Noncontrast computed tomography showed air in the spinal canal, concerning for epidural abscess. Magnetic resonance imaging (MRI) of the lumbar spine without contrast was performed, as the patient was in acute renal failure. A, This T2-weighted sagittal MRI scan provides useful information even without gadolinium contrast. B, Close-up. On T2-weighted MRI sequences, fluid including cerebrospinal fluid appears white. Fat-containing tissues such as bone marrow and the spinal cord or cauda equina appear dark gray. Calcified bone appears nearly black because of an absence of resonating protons. Air appears completely black for the same reason. The midline sagittal image shows the cauda equina to be impinged on by an epidural fluid collection containing air—an epidural abscess. The dura mater is visible as a thin, dark gray line parallel to the spinal cord. It is indented in the region of the epidural abscess. (From Broder JS: *Diagnostic imaging for the emergency physician*, Philadelphia, Saunders, 2011.)

123. Splenomegaly (Fig. 3.197)

Diagnostic Imaging

Best Test(s)
- CT of abdomen (Fig. 3.198)

Ancillary Tests
- Ultrasound of abdomen (Fig. 3.199) if CT is contraindicated
- Echocardiography if CHF or SBE is suspected
- MRI of abdomen

Laboratory Evaluation

Best Test(s)
- CBC with differential

Ancillary Tests
- Viral titers (EB, CMV), Monospot, HIV
- ALT, AST, bilirubin, albumin, INR
- ESR, RF, ANA
- Blood culture × 2 if splenic abscess is suspected

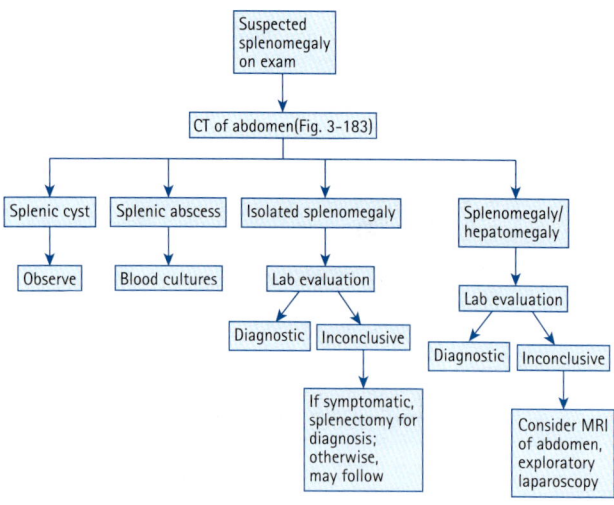

Fɪɢ. 3.197 Diagnostic algorithm for suspected splenomegaly on examination.

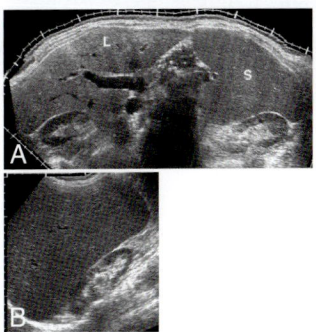

Fɪɢ. 3.198 Splenomegaly. A, Transverse, and B, coronal extended field of view (FOV) (Si-eScape); images demonstrate marked splenic (S) enlargement; L, liver. (From Rumack CM, Wilson SR, Charboneau JW, Levine D: *Diagnostic ultrasound*, ed 4, Philadelphia, Mosby, 2011.)

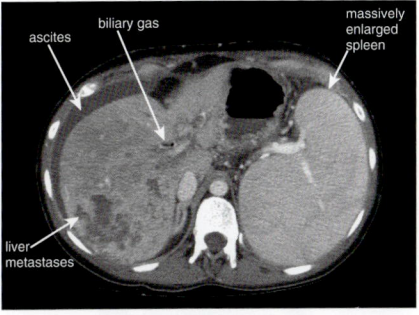

Fɪɢ. 3.199 Splenomegaly, computed tomography (CT) scan with intravenous contrast. Sple-nomegaly can occur from many causes. A normal spleen can be difficult to measure with ultrasound, because it lies sheltered by the left costal margin in many cases. An enlarged spleen can usually be seen with ultrasound. CT readily reveals splenomegaly, as does mag-netic resonance imaging (MRI). This patient has metastatic pancreatic cancer and related sple-nomegaly. The spleen is nearly as large as the liver in this cross section. This may be caused partly by thrombosis of the portal vein (not seen in this image but noted in other CT images and confirmed with MRI in this patient). Remember that the splenic vein drains ultimately to the portal vein, so thrombosis of the portal vein may lead to splenomegaly. (From Broder JS: *Diagnostic imaging for the emergency physician*, Philadelphia, Saunders, 2011.)

124. Stroke (Fig. 3.200)

Diagnostic Imaging

Best Test(s)
- CT of head without contrast initially, repeated subsequently with contrast (Figs. 3.201, 3.202)
- MRI of brain can be substituted for CT with contrast

Ancillary Tests
- Carotid Doppler ultrasound
- Echocardiogram

Laboratory Evaluation

Best Test(s)
- None

Ancillary Tests
- PT, PTT, platelet count
- Lipid panel, glucose
- ALT, electrolytes, BUN, creatinine
- CBC

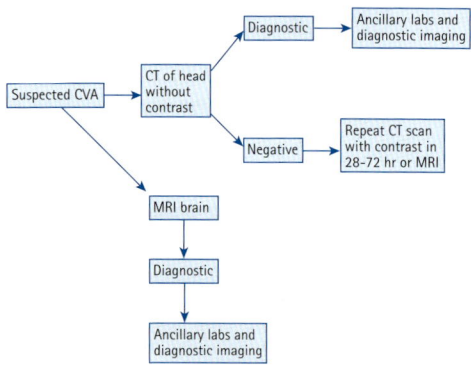

Fɪɢ. 3.200 Diagnostic algorithm.

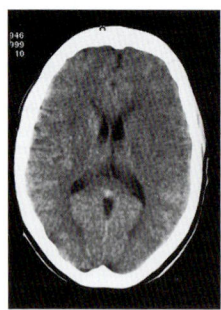

Fɪɢ. 3.201 Acute ischemic infarct in the left cerebral artery territory. Computed tomography scan less than 24 hours from onset of stroke shows loss of differentiation between gray and white matter in the left frontal region and obscuration of the caudate and lentiform nuclei. There is effacement of the left frontal sulci. (From Grainger RG, Allison DJ, Adam A, Dixon AK [eds]: *Grainger and Allison's diagnostic radiology*, ed 4, Philadelphia, Churchill Livingstone, 2001.)

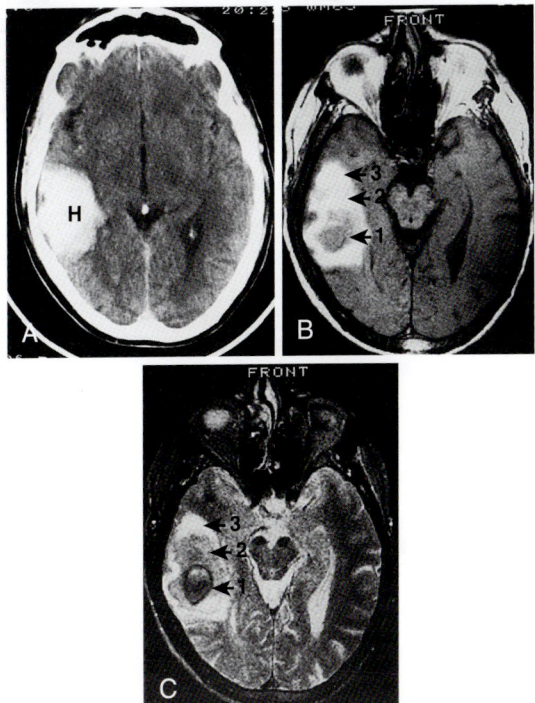

Fig. 3.202 Hemorrhage. Axial computed tomography image (A) demonstrates a large area of acute hemorrhage (*H*) in the right temporal lobe. T1-weighted (B) and T2-weighted (C) magnetic resonance imaging scans demonstrate the hemorrhage in various stages of breakdown. Center of lesion is dark on T1- and T2-weighted images, indicating oxyhemoglobin (*1*). Intermediate zone is bright on T1-weighted image and gray on T2-weighted image, indicating intracellular methemoglobin (*2*). Outer rim is bright on both T1- and T2-weighted images, indicating extracellular methemoglobin (*3*). (From Vincent JL, Abraham E, Moore FA, et al: *Textbook of critical care*, ed 6, Philadelphia, Saunders, 2011.)

125. Subarachnoid Hemorrhage (Fig. 3.203)

Diagnostic Imaging

Best Test(s)
- CT of brain without contrast (Figs. 3.204, 3.205)

Ancillary Tests
- CT angiogram or cerebral angiography to determine the origin of the SAH

Laboratory Evaluation

Best Test(s)
- None

Ancillary Tests
- PT, PTT, platelet count, serum troponin
- CBC, lytes, BUN, creatinine

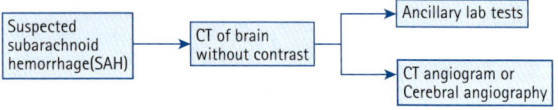

Fɪɢ. 3.203 Diagnostic algorithm.

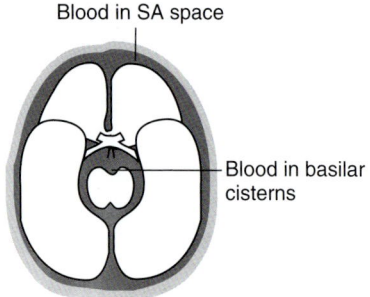

Fɪɢ. 3.204 Subarachnoid (SA) hemorrhage. (From Weissleder R, Wittenberg J, Harisinghani MG, Chen JW: *Primer of diagnostic imaging,* ed 4, St. Louis, Mosby, 2007.)

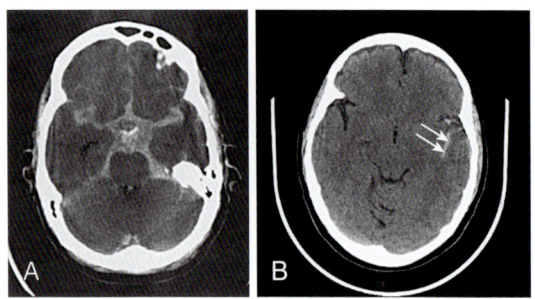

Fɪɢ. 3.205 Computed tomography scan appearance of subarachnoid hemorrhage. A, Blood filling the suprasellar cistern. B, Blood filling the sylvian cistern (*arrows*). (From Adams JG et al: *Emergency medicine: clinical essentials,* ed 2, Philadelphia, Elsevier, 2013.)

126. Subclavian Steal Syndrome (Fig. 3.206)

Diagnostic Imaging

Best Test(s)
- Arteriography of subclavian and vertebral innominate arteries (Fig. 3.207)

Ancillary Tests
- Doppler sonography of vertebral, subclavian, and innominate arteries (best initial screening test)

Laboratory Evaluation

Best Test(s)
- None

Ancillary Tests
- Lipid panel
- FBS
- ANA, ESR

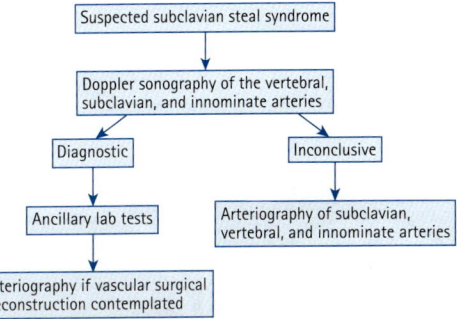

Fig. 3.206 Diagnostic algorithm.

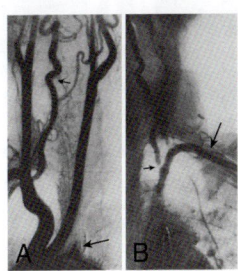

Fig. 3.207 Subclavian steal: arch aortogram. A, Left anterior oblique projection, arterial phase: proximal occlusion of the left subclavian artery (*large arrow*). Note irregularity and tortuosity of right vertebral artery related to degenerative changes in the cervical vertebrae (*small arrow*). B, Right anterior oblique projection late phase of aortogram; the distal segment of the left subclavian artery (*large arrow*) fills via retrograde flow in the left vertebral artery, despite this vessel being almost completely obstructed at its origin (*small arrow*). (From Grainger RG, Allison DJ, Adam A, Dixon AK [eds]: *Grainger and Allison's diagnostic radiology*, ed 4, Philadelphia, Churchill Livingstone, 2001.)

127. Subdural Hematoma (Fig. 3.208)

Diagnostic Imaging

Best Test(s)
• CT of brain without contrast (Fig. 3.209)

Ancillary Tests
• None

Laboratory Evaluation

Best Test(s)
• None

Ancillary Tests
• PT, PTT, platelet count
• CBC

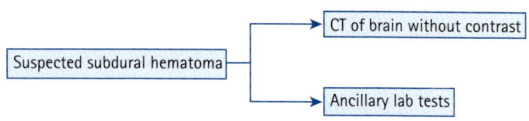

Fɪɢ. 3.208 Diagnostic algorithm.

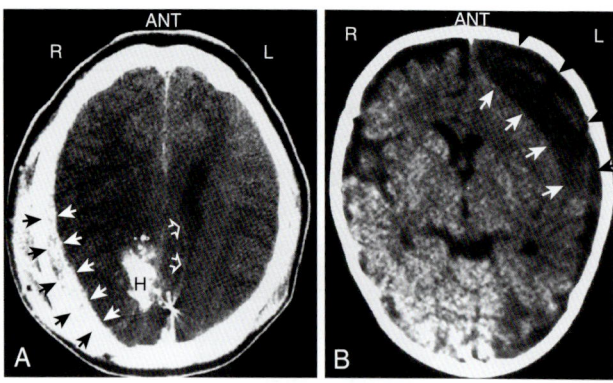

Fɪɢ. 3.209 Subdural hematomas. A noncontrasted computed tomography scan of an acute subdural hematoma (A) shows a crescentic area of increased density (*arrows*) in the right posterior parietal region between the brain and the skull. An area of intraparenchymal hemorrhage (H) also is seen; in addition, mass effect causes a midline shift to the left (*open arrows*). A chronic subdural hematoma is seen in a different patient (B). An area of decreased density appears in the left frontoparietal region, effacing the sulci, compressing the anterior horn of the left lateral ventricle, and shifting the midline somewhat to the right. (From Mettler FA: *Essentials of radiology*, ed 3, Philadelphia, Saunders, 2014.)

128. Superior Vena Cava Syndrome (Fig. 3.210)

Diagnostic Imaging

Best Test(s)
- Venography of superior vena cava: warranted only when an intervention (e.g., stent or surgery) is planned (Fig. 3.211)

Ancillary Tests
- Chest x-ray (Fig. 3.212) followed by CT of chest (Fig. 3.213)

Comments
Table 3.14 describes malignancies associated with superior vena cava syndrome.

Laboratory Evaluation

Best Test(s)
- None

Ancillary Tests
- Tissue biopsy of mediastinal mass (e.g., lymphoma, lung carcinoma) compressing superior vena cava

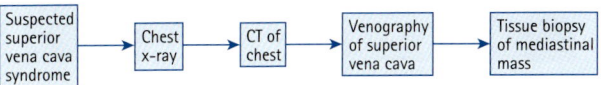

| Suspected superior vena cava syndrome | → | Chest x-ray | → | CT of chest | → | Venography of superior vena cava | → | Tissue biopsy of mediastinal mass |

FIG. 3.210 Diagnostic algorithm.

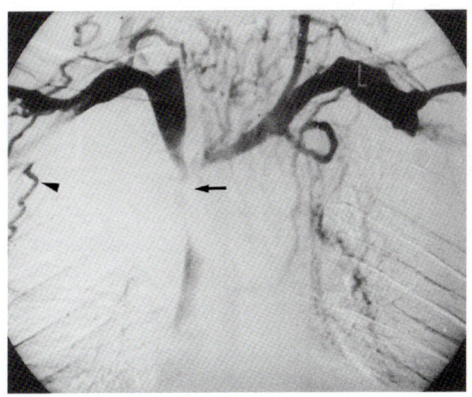

FIG. 3.211 Upper limb venography. Bilateral arm venography reveals obstruction of the superior vena cava (SVC) by tumor in the mediastinum (*arrow*). This was a case of SVC syndrome secondary to lung carcinoma metastatic disease. Note the collateral venous channels that have opened up (*arrowhead*). (From Grainger RG, Allison DJ, Adam A, Dixon AK [eds]: *Grainger and Allison's diagnostic radiology*, ed 4, Philadelphia, Churchill Livingstone, 2001.)

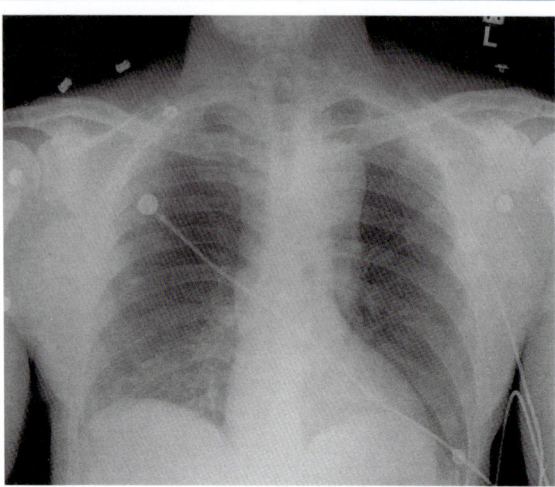

Fig. 3.212 Chest radiograph showing widened mediastinum. (From Adams JG et al: *Emergency medicine: clinical essentials*, ed 2, Philadelphia, Elsevier, 2013.)

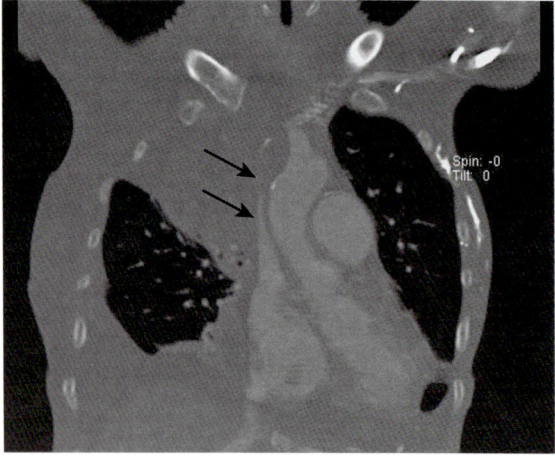

Fig. 3.213 Computed tomography scan showing blockage of the superior vena cava (*arrows*). (From Adams JG et al: *Emergency medicine: clinical essentials*, ed 2, Philadelphia, Elsevier, 2013.)

TABLE 3.14	Malignancies Associated With SVC Syndrome in Adults*	
NEOPLASTIC DIAGNOSIS	**PERCENTAGE OF SVC**	**PERCENTAGE OF DISEASE-ASSOCIATED SVC**
Lung cancer, stage 3B or 4:	48–81	15–45
Small-cell lung cancer		20–25
Squamous cell cancer		5–25
Adenocarcinoma		4–30
Large-cell carcinoma		
Lymphoma:	2–21	64
Diffuse large cell lymphoma		33
Lymphoblastic lymphoma		
Breast cancer	11	

SVC, superior vena cava.
*Include lung cancer, lymphomas, and metastases from other solid tumors. Of patients with SVC, 75% to 85% have neoplastic disease.
From Zipes DP, Libby P, Bonow RO, Braunwald E: *Braunwald's heart disease*, ed 7, Philadelphia, Saunders, 2005.

129. Syncope (Fig. 3.214)

Diagnostic Imaging

Best Test(s)
- None; diagnostic imaging should be guided by history and physical examination

Ancillary Tests
- Echocardiography (useful in patients with a heart murmur to r/o aortic stenosis, hypertrophic cardiomyopathy, or atrial myxoma)
- CT or MRI of brain and EEG if seizure is suspected
- Spiral CT of chest if PE is suspected
- Tilt table testing (Fig. 3.215)

Laboratory Evaluation

Best Test(s)
- None

Ancillary Tests
- Routine blood tests (rarely yield diagnostically useful information and should be done only when specifically suggested by history and physical examination)
- Serum pregnancy test (should be considered in women of childbearing age)
- CBC, electrolytes, BUN, creatinine
- Serum calcium, magnesium
- D-dimer
- ABGs
- ECG, 24-hour Holter monitor
- Cardiac troponins, isoenzymes if history of chest pain before syncope
- Toxicologic screen in selected patients
- Cardiac stress test
- EPS studies

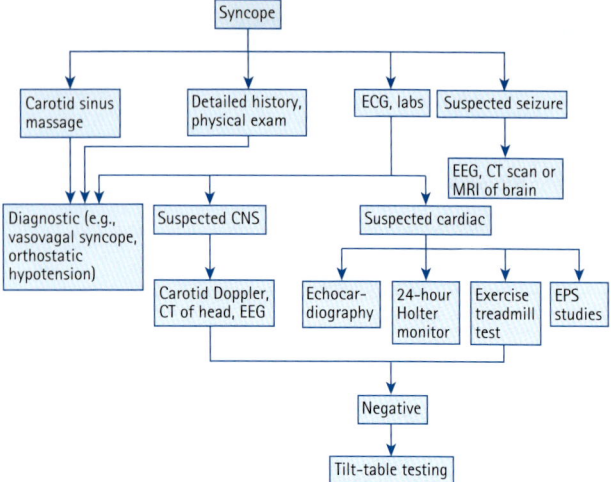

Fɪɢ. 3.214 Diagnostic algorithm.

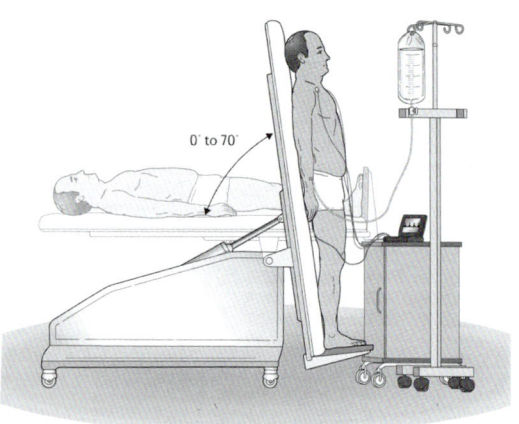

Fɪɢ. 3.215 A 70-degree tilt test using a motorized table with a footplate. Not illustrated are electrocardiogram leads for monitoring and a cuff or continuous blood pressure monitor. (From Crawford MH, DiMarco JP, Paulus WJ [eds]: *Cardiology*, ed 2, St. Louis, Mosby, 2004.)

130. Testicular Torsion (Fig. 3.216)

Diagnostic Imaging

Best Test(s)
- Doppler ultrasound of testicle (Fig. 3.217)

Ancillary Tests
- Tc-99m testicular scan (Fig. 3.218) also an excellent modality for evaluating blood flow to testicle (absence of blood flow centrally to testicle indicates torsion)

Laboratory Evaluation

Best Test(s)
- None

Ancillary Tests
- None

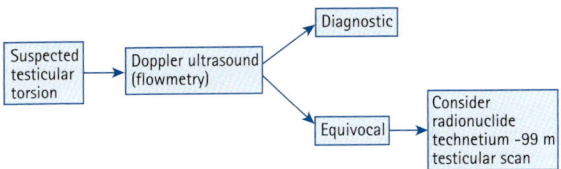

FIG. 3.216 Diagnostic algorithm.

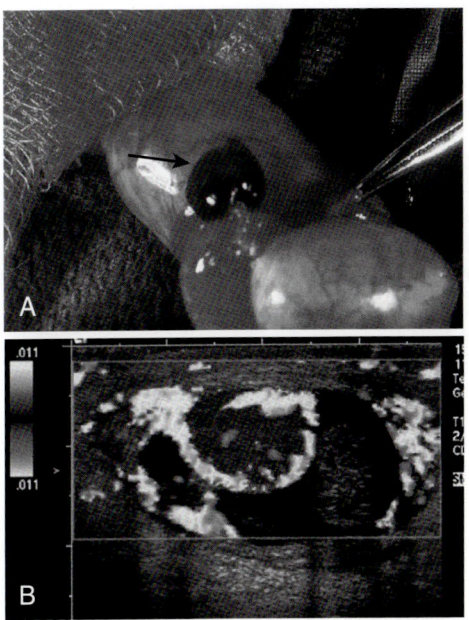

FIG. 3.217 A, Torsion of the appendix testis; the appendix testis is necrotic (*arrow*). B, Color Doppler scrotal sonogram showing hyperemia to the testis and absent flow to the appendix testis (right side). Symptoms resolved with medical therapy. (Kliegman RM, Stanton B, St. Geme J, et al: *Nelson textbook of pediatrics*, ed 19, Philadelphia, Saunders, 2011.)

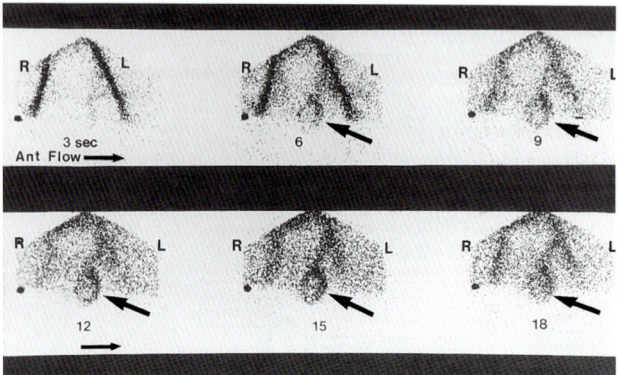

Fɪɢ. 3.218 Testicular torsion. Evaluation of blood flow to the testicle has been done by giving an intravenous bolus of radioactive material. The right and left iliac vessels are clearly identified, and sequential images are obtained every 3 seconds. Here increased flow is seen to the rim of the left testicle (*arrows*), and there is no blood flow centrally. This is the appearance of a testicular torsion in which the torsion has been present for more than approximately 24 hours. (From Mettler FA, Guiberteau MJ, Voss CM, Urbina CE: *Primary care radiology*, Philadelphia, WB Saunders, 2000.)

131. Thoracic Outlet Syndrome (Fig. 3.219)

Diagnostic Imaging

Best Test(s)
- Arteriography or venography when vascular pathology is suspected clinically

Ancillary Tests
- Chest radiograph (r/o cervical rib)
- X-ray of cervical spine (r/o cervical disk disease)
- CT of chest (r/o lung neoplasm)

Laboratory Evaluation

Best Test(s)
- None

Ancillary Tests
- EMG, nerve conduction studies

Comments
Fig. 3.220 illustrates the thoracic outlet structures.

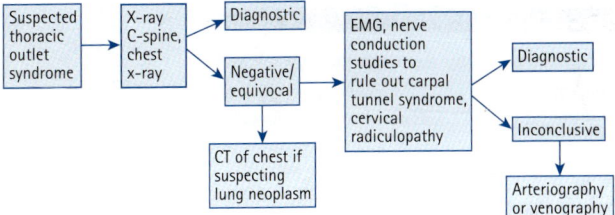

Fig. 3.219 Diagnostic algorithm.

THE THORACIC OUTLET

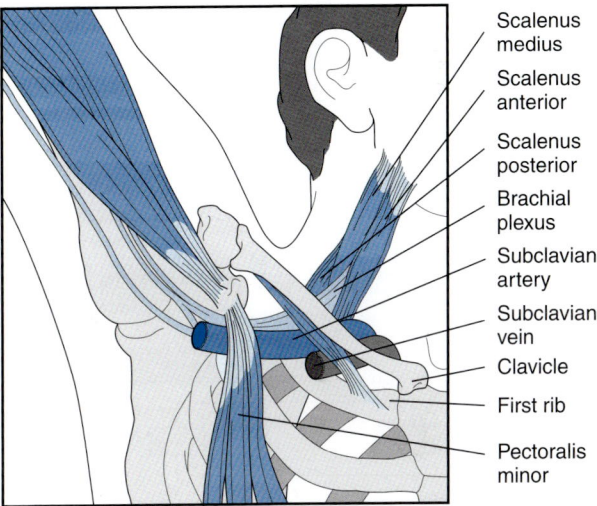

Scalenus medius

Scalenus anterior

Scalenus posterior

Brachial plexus

Subclavian artery

Subclavian vein

Clavicle

First rib

Pectoralis minor

Fig. 3.220 The thoracic outlet. Three narrow channels of the outlet include the scalene triangle, the costoclavicular passage, and the pectoralis minor attachment at the coracoid process. (From Hochberg MC et al: *Rheumatology*, ed 5, St. Louis, Mosby, 2011.)

132. Thrombocytopenia (Fig. 3.221)

Diagnostic Imaging

Best Test(s)
- None

Ancillary Tests
- CT of abdomen

Laboratory Evaluation

Best Test(s)
- Bone marrow examination

Ancillary Tests
- CBC, PT, PTT
- LDH, ALT, AST, TSH
- HIV, ANA, EB virus, viral hepatitis screen, CMV, rickettsia
- Antiplatelet Ab
- D-dimer, fibrinogen level
- Antiglobulin test, B$_{12}$, folate
- Heparin-induced platelet aggregation assay, 14-c-serotonin release assay in suspected heparin-induced thrombocytopenia
- *ADAMTS13* activity assay when suspecting TTP

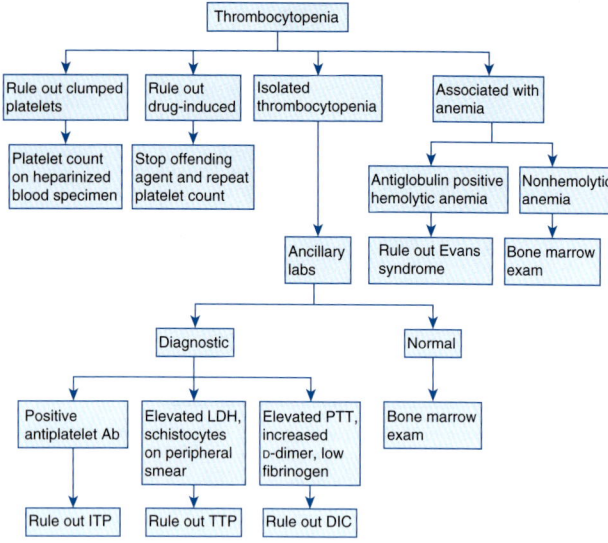

Fig. 3.221 Diagnostic algorithm.

133. Thrombocytosis (Fig. 3.222)

Diagnostic Imaging

Best Test(s)
• None

Ancillary Tests
• CT of chest and abdomen

Laboratory Evaluation

Best Test(s)
• Bone marrow examination

Ancillary Tests
• CBC
• Reticulocyte count
• Stool for occult blood × 3
• Serum ferritin, TIBC, iron

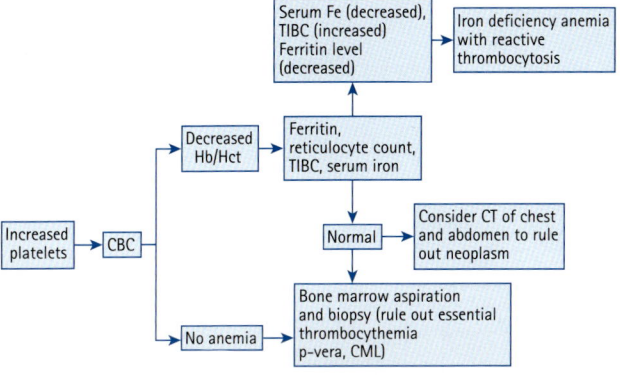

Fɪɢ. 3.222 Diagnostic algorithm.

134. Thyroid Nodule (Fig. 3.223)

Diagnostic Imaging

Best Test(s)
- Thyroid ultrasound to evaluate size and composition of nodule (solid vs. cystic)

Ancillary Tests
- Thyroid scan with Tc-99m classifies nodule as hyperfunctioning (hot), normally functioning (warm), or nonfunctioning (cold); thyroid scan can also be performed with iodine

Laboratory Evaluation

Best Test(s)
- FNAB

Ancillary Tests
- TSH, free T$_4$
- Antimicrosomal Ab
- Serum calcium
- Serum thyroglobulin level in patients with confirmed thyroid carcinoma

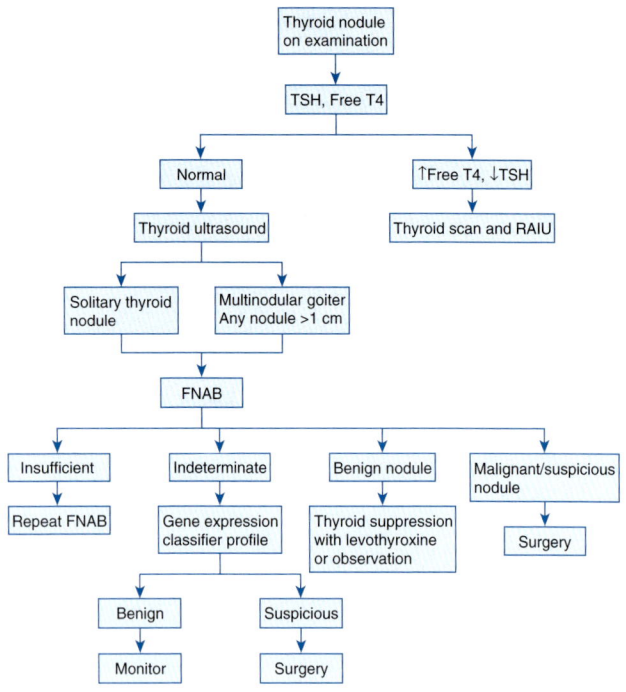

Fig. 3.223 Diagnostic algorithm.

135. Thyroiditis (Fig. 3.224)

Diagnostic Imaging

Best Test(s)
- 24-hour RAIU scan (Fig. 3.225): Useful in distinguishing Graves' disease (increased RAIU) from thyroiditis (normal/low RAIU)

Ancillary Tests
- Thyroid ultrasound

Laboratory Evaluation

Best Test(s)
- None

Ancillary Tests
- TSH, free T_4
- CBC with differential (leukocytosis with left shift occurs with subacute and suppurative thyroiditis)
- Antimicrosomal Ab (detected in >90% of patients with Hashimoto's thyroiditis)
- Serum thyroglobulin level (increased in patients with autoimmune lymphocytic thyroiditis)

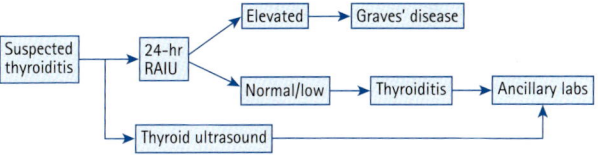

Fig. 3.224 Diagnostic algorithm.

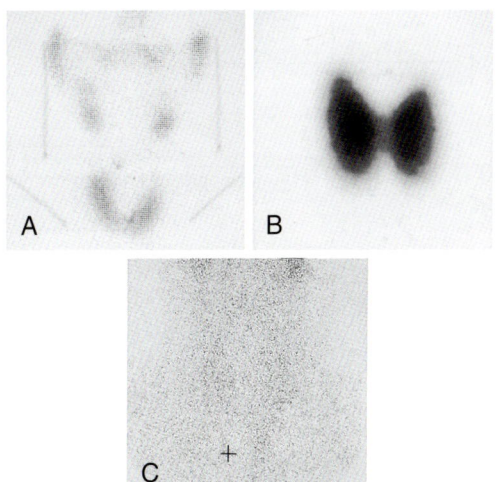

Fig. 3.225 Technetium-99m thyroid scans. A, Normal. B, Graves disease, showing enlarged thyroid with increased uptake. C, Low patchy uptake in thyroiditis. (From Besser CM, Thorner MO: *Comprehensive clinical endocrinology*, ed 3, St. Louis, Mosby, 2002.)

136. Tinnitus (Fig. 3.226)

Diagnostic Imaging

Best Test(s)
- Audiometry/tympanometry

Ancillary Tests
- Carotid Doppler ultrasound
- MRI of brain and auditory canals
- Brain MRA

Laboratory Evaluation

Best Test(s)
- None

Ancillary Tests
- CBC
- Lipid panel

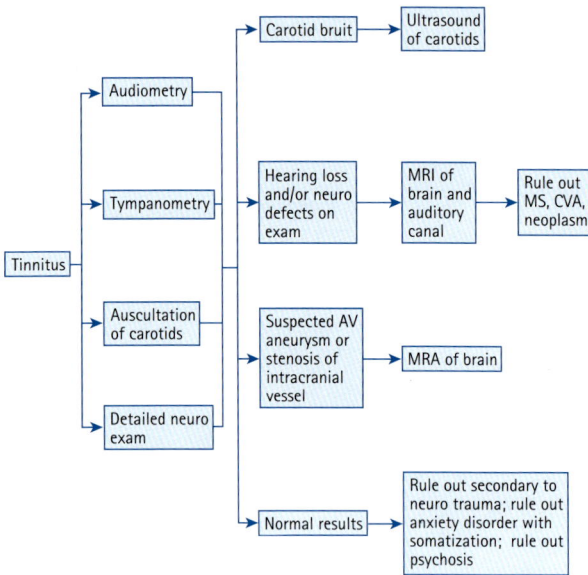

Fɪɢ. 3.226 Diagnostic algorithm.

137. Transient Ischemic Attack (TIA) (Fig. 3.227)

Diagnostic Imaging

Best Test(s)
- Carotid Doppler (to identify carotid stenosis)

Ancillary Tests
- CT of head without contrast (r/o hemorrhage or subdural hematoma)
- Echocardiography (if cardiac source is suspected)
- Brain MRA if posterior circulation TIA is suspected

Laboratory Evaluation

Best Test(s)
- None

Ancillary Tests
- PT, PTT, platelet count
- Lipid panel, FBS
- ESR

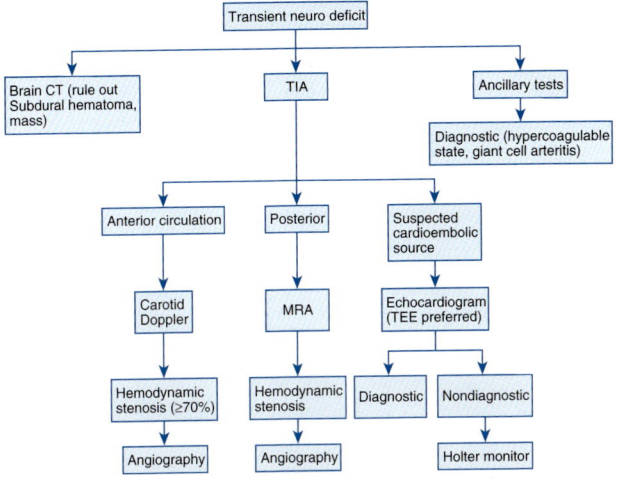

Fig. 3.227 Diagnostic algorithm.

138. Urethral Discharge (Fig. 3.228)

Diagnostic Imaging

Best Test(s)
• None

Ancillary Tests
• None

Laboratory Evaluation

Best Test(s)
• Urine gonorrhea PCR assay
• Urine *C. trachomatis* PCR assay
• Gram stain of exudate

Ancillary Tests
• Wet mount for *Trichomonas*
• HIV, VDRL

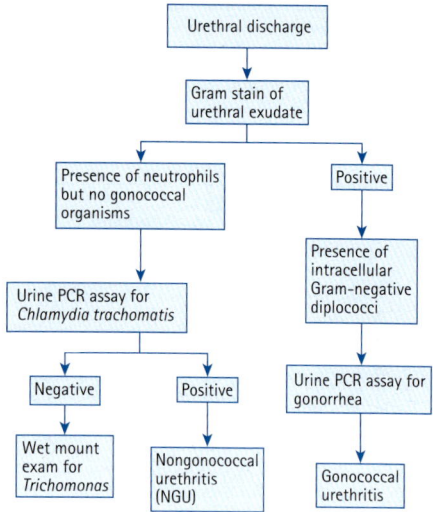

Fɪɢ. 3.228 Diagnostic algorithm.

139. Urolithiasis (Fig. 3.229)

Diagnostic Imaging

Best Test(s)
- Noncontrast CT of urinary tract (Fig. 3.230)

Ancillary Tests
- Renal sonogram
- IVP (demonstrates size and location of calculus)
- Plain film of abdomen (can identify radiopaque stones [e.g., calcium])

Laboratory Evaluation

Best Test(s)
- Urinalysis

Ancillary Tests
- Chemical analysis of recovered stone
- Serum calcium, phosphate, uric acid, BUN, creatinine
- Urine C&S
- 24-hour urine for calcium in patients with calcium stones

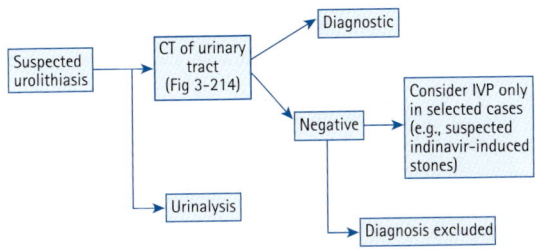

Fɪɢ. **3.229** Diagnostic algorithm.

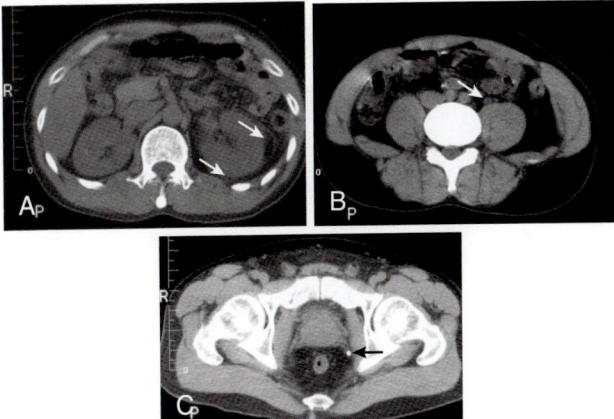

Fɪɢ. **3.230** Obstructing left ureteral calculus. On this noncontrasted computed tomography scan, at the level of the kidneys (A) there is stranding (*arrows*) around the left kidney but not around the right kidney. At a lower level (B) a dilated left ureter is seen (*arrow*), and at the level of the bladder (C), a calculus is seen at the left ureterovesicular junction (*arrow*). The small rim of soft tissue around the calculus helps distinguish it from a phlebolith. (From Mettler FA: *Essentials of radiology*, ed 3, Philadelphia, Saunders, 2014.)

140. Urticaria (Fig. 3.231)

Diagnostic Imaging

Best Test(s)
• None

Ancillary Tests
• Chest radiograph

Laboratory Evaluation

Best Test(s)
• None

Ancillary Tests
• CBC with differential
• ANA
• Viral hepatitis screen
• ALT, Monospot
• Thyroid antibodies
• Stool for ova and parasites

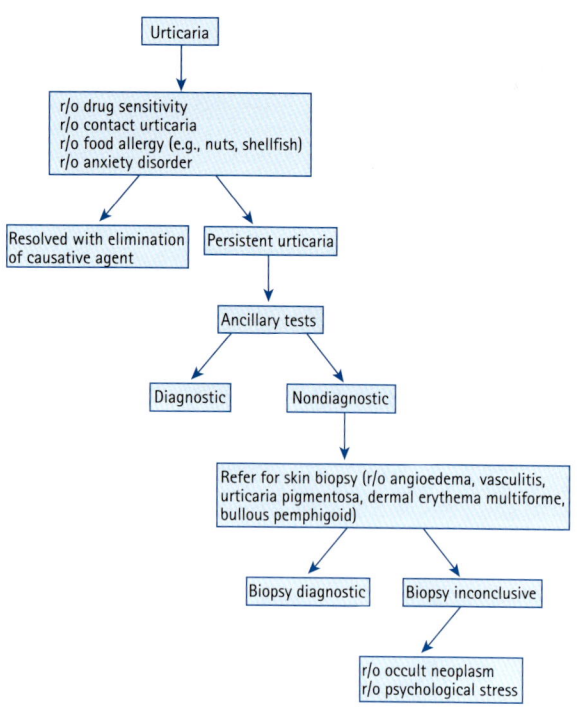

Fɪɢ. 3.231 Diagnostic algorithm.

141. Vaginal Discharge (Fig. 3.232)

Diagnostic Imaging

Best Test(s)
- None

Ancillary Tests
- None

Laboratory Evaluation

Best Test(s)
- Wet mount and KOH preparation

Ancillary Tests
- PCR assay or gonococcal culture on Thayer-Martin medium
- PCR assay or culture for *C. trachomatis*
- HIV
- Characteristics of normal vaginal secretions are described in Table 3.15

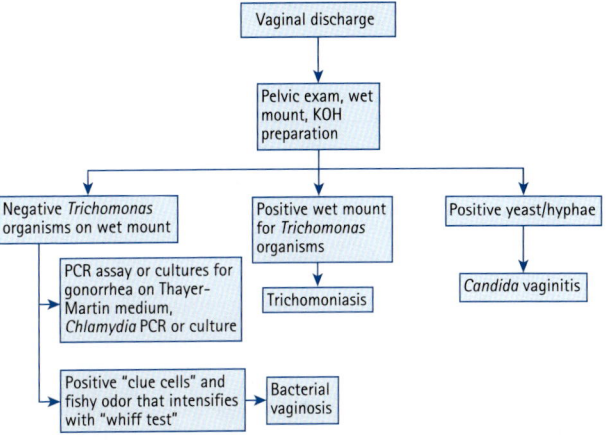

Fig. 3.232 Diagnostic algorithm.

TABLE 3.15	**Characteristics of Normal Vaginal Secretions**
Heterogeneous suspension of desquamated vaginal epithelial cells in water	
Lactobacilli dominate the microbiota	
pH of 3.5–4.6	
Odorless	
Do not cause itching or irritation	
Usually do not soil underclothing	

From Bennett JE, Dolin R, Blaser MJ: *Mandell, Douglas, and Bennett's principles and practice of infectious diseases*, ed 8, Philadelphia, Saunders, 2015.

142. Vertigo (Fig. 3.233)

Diagnostic Imaging

Best Test(s)
- MRI of brain

Ancillary Tests
- MRA of posterior circulation (CT of cerebellopontine region if MRI is contraindicated)

Laboratory Evaluation

Best Test(s)
- None

Ancillary Tests
- CBC with differential
- Serum glucose, creatinine, BUN, ALT, electrolytes

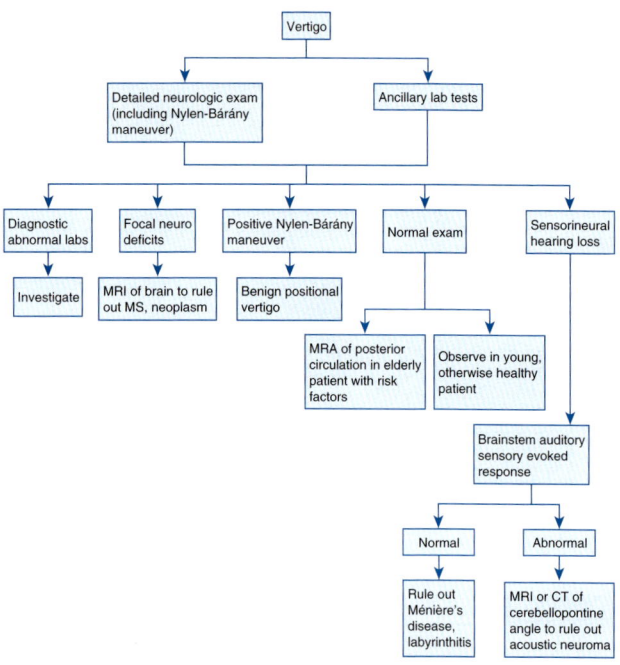

Fig. 3.233 Diagnostic algorithm.

143. Viral Hepatitis (Fig. 3.234)

Diagnostic Imaging

Best Test(s)
- None

Ancillary Tests
- None

Laboratory Evaluation

Best Test(s)
- Hepatitis panel (should include HBsAg, anti-HBc IgM, anti-HAV IgM, anti-HCV)

Ancillary Tests
- ALT, AST
- Alkaline phosphatase, bilirubin, PT
- ANA, ASMA, AMA if autoimmune hepatitis is suspected

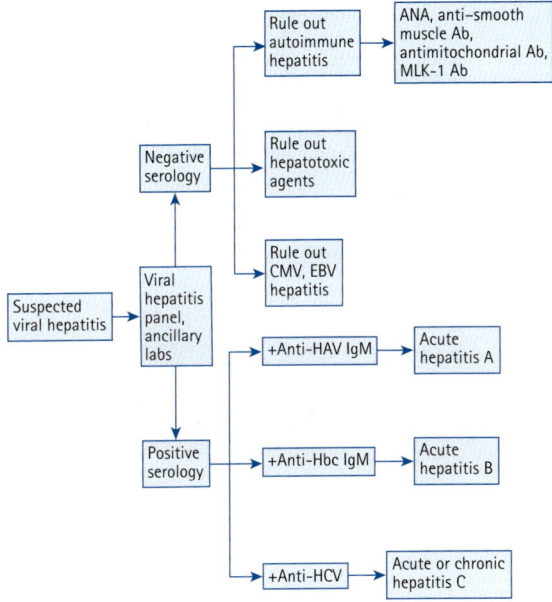

Fɪɢ. 3.234 Diagnostic algorithm.

144. Weight Gain (Fig. 3.235)

Diagnostic Imaging

Best Test(s)
- None

Ancillary Tests
- Ultrasound of abdomen if cirrhosis or nephrosis suspected

Laboratory Evaluation

Best Test(s)
- TSH

Ancillary Tests
- FBS, BUN, creatinine, ALT, AST, albumin
- Urinalysis
- Serum dehydroepiandrosterone level

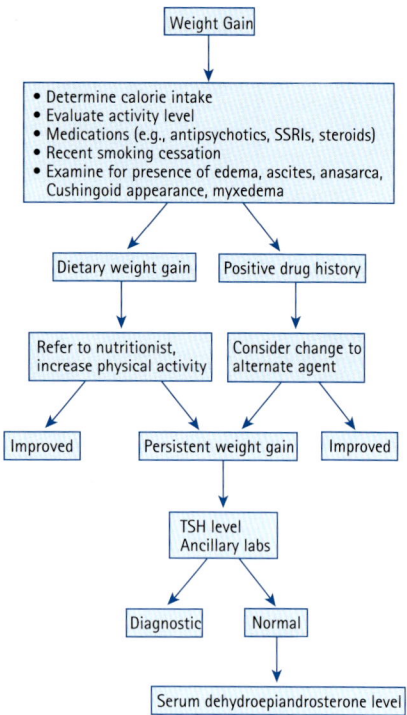

Fig. 3.235 Diagnostic algorithm.

145. Weight Loss, Involuntary (Fig. 3.236)

Diagnostic Imaging

Best Test(s)
- None

Ancillary Tests
- Chest radiograph

Laboratory Evaluation

Best Test(s)
- TSH, free T_4

Ancillary Tests
- CBC, glucose
- ESR (nonspecific)
- ALT, creatinine, serum albumin
- Urinalysis

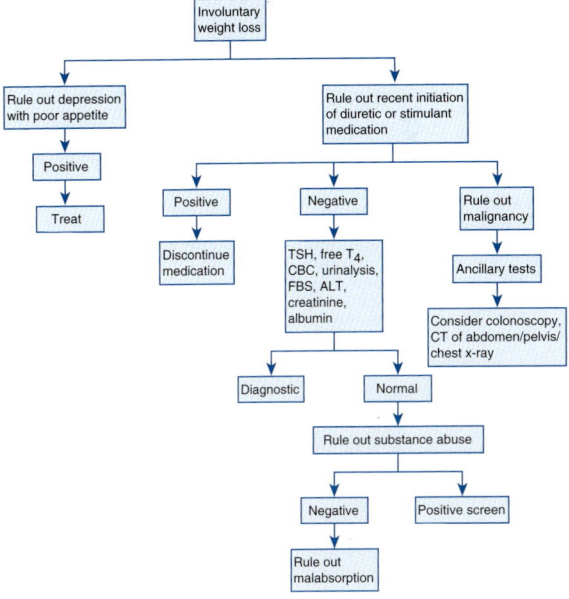

Fig. 3.236 Diagnostic algorithm.

References

Ballinger A. Kumar & Clark's essentials of clinical medicine. 5th ed. Edinburgh: Saunders; 2012.

Besser CM, Thorner MO. Comprehensive clinical endocrinology. 3rd ed. St. Louis: Mosby; 2002.

Broder JS. Diagnostic imaging for the emergency physician. Philadelphia: Saunders; 2011.

Cameron JL, Cameron AM. Current surgical therapy. 12th ed. Philadelphia: Saunders; 2017.

Cohen J, Powderly WG. Infectious diseases. 2nd ed. St. Louis: Mosby; 2004.

DeLee JC, Drez D Jr. DeLee and Drez's orthopaedic sports medicine. 2nd ed. Philadelphia: Saunders; 2003.

DuBose TD Jr. Acid-base disorders. In: Brenner BM, Rector FC, editors. Brenner and Rector's the kidney. 10th ed. Philadelphia: Saunders; 2016.

Ferri F. Ferri's best test. 3rd ed. St. Louis: Mosby; 2015.

Ferri F. Practical guide to the care of the medical patient. 9th ed. St. Louis: Mosby; 2013.

Fielding JR, Brown DL, Thurmond AS. Gynecologic imaging. Philadelphia: Saunders; 2011.

Goldman L, Schafer AI. Goldman's Cecil medicine. 24th ed. Philadelphia: Saunders; 2012.

Greer IA, Cameron IT, Kitchener HC, Prentice A. Mosby's color atlas and text of obstetrics and gynecology. London: Harcourt; 2000.

Crawford MH, DiMarco JP, Paulus WJ, editors. Cardiology. 2nd ed. St. Louis: Mosby; 2004.

Grainger RG, Allison DJ, Adam A, Dixon AK, editors. Grainger and Allison's diagnostic radiology. 4th ed. Philadelphia: Churchill Livingstone; 2001.

Hochberg MC, Silman AJ, Smolen JS, et al. Rheumatology. 3rd ed. St. Louis: Mosby; 2003.

Townsend CM, Beauchamp RD, Evers BM, Mattox KL, editors. Sabiston textbook of surgery. 17th ed. Philadelphia: Saunders; 2004.

Johnson RJ, Feehally J. Comprehensive clinical nephrology. 2nd ed. St. Louis: Mosby; 2000.

Kliegman RM, Stanton B, St. Geme J, et al. Nelson textbook of pediatrics. 19th ed. Philadelphia: Saunders; 2011.

Kuhn JP, Slovis TL, Haller JO. 10th ed. Caffey's pediatric diagnostic imaging, vol. 2. Philadelphia: Mosby; 2004.

Mettler FA, Guiberteau MJ, Voss CM, Urbina CE. Primary care radiology. Philadelphia: WB Saunders; 2000.

Souhami RL, Moxham J. Textbook of medicine. 4th ed. London: Churchill Livingstone; 2002.

Symonds EM, Symonds IM. Essential obstetrics and gynecology. 4th ed. Edinburgh: Churchill Livingstone; 2004.

Talley NJ, Martin CJ. Clinical gastroenterology. 2nd ed. Sidney: Churchill Livingstone; 2006.

Vincent JL, Abraham E, Moore FA, et al. Textbook of critical care. 6th ed. Philadelphia: Saunders; 2011.

Weissleder R, Wittenberg J, Harisinghani MG, Chen JW. Primer of diagnostic imaging. 5th ed. St. Louis: Mosby; 2011.

Zipes DP, Libby P, Bonow RO, Braunwald E. Braunwald's heart disease. 7th ed. Philadelphia: Saunders; 2005.

General Guidelines for Ordering Computed Tomography[1]

	WITHOUT IV CONTRAST	WITH IV CONTRAST	WITH AND WITHOUT IV CONTRAST
Head	(Procedure code 70450) • Trauma • Bleed	(Procedure code 70460) • Soft-tissue mass • Known brain lesion • Metastases	
Sinus/Face	(Procedure code 70486) • Trauma • Fracture • Sinusitis	(Procedure code 70487) • Soft-tissue mass	
Temporal Bones/ Orbits	(Procedure code 70480) • Hearing loss • Orbit fracture • Mastoiditis • Cholesteatoma	(Procedure code 70481) • Soft-tissue mass	
Neck	(Procedure code 70490) • Stones	(Procedure code 70491) • Mass • Difficulty swallowing	(Procedure code 70492)
Spine	Cervical (Procedure code 72125) • Stenosis • Pain • Fracture Thoracic (Procedure code 72128) • Stenosis • Pain • Fracture Lumbar (Procedure code 72131) • Stenosis • Pain • Fracture	• Infection Cervical (Procedure code 72126) Thoracic (Procedure code 72129) Lumbar (Procedure code 72132)	
Extremities	Upper (Procedure code 73200) • Fracture • Evaluation joint • Hip fracture Lower (Procedure code 73700) • Fracture • Evaluation joint • Hip fracture	Upper (Procedure code 73201) • Soft-tissue mass • Osteomyelitis/infection Lower (Procedure code 73701) • Soft-tissue mass • Osteomyelitis/infection	Upper (Procedure code 73202) Upper (Procedure code 73702)

Continued

	WITHOUT IV CONTRAST	WITH IV CONTRAST	WITH AND WITHOUT IV CONTRAST
Chest	(Procedure code 71250) • Nodule • Abnormal chest x-ray • Interstitial lung disease (high resolution) • Cough	(Procedure code 71260) • Lymph nodes (adenopathy) • Hilar mass	(Procedure code 71270)
Abdomen and Pelvis	(Procedure code 74176) • Kidney stones • Evaluation for aortic aneurysm	(Procedure code 74177) • Generalized abdominal pain (diverticulitis, appendicitis, pancreatitis) • Cancer staging • Crohn's (enterography)	(Procedure code 74178) • Urogram for hematuria
Abdomen			(Procedure code 74170) • 3-phase examinations (liver, pancreas mass) • Kidney mass
Bony pelvis	(Procedure code 72192) • Pelvic fracture		

Note: NPO 2 hours for contrast examinations. Please obtain serum creatinine and eGFR on all patients receiving IV contrast who have diabetes or renal disease or are over the age of 75 years.

General Guidelines for Ordering Computed Tomography Angiography (CTA)[1]

Brain	(Procedure code 70496) • Brain aneurysm
Neck	(Procedure code 70498) • Carotids
Chest	(Procedure code 71275) • Pulmonary embolism • Aortic dissection
Abdomen/Pelvis	(Procedure code 74174) • Evaluation of aneurysm graft
Runoff	(Procedure code 75635) • Claudication

Note: NPO 2 hours for contrast examinations. Please obtain serum creatinine and eGFR on all patients receiving IV contrast who have diabetes or renal disease or are over the age of 75 years.

General Guidelines for Ordering Magnetic Resonance Imaging (MRI)[1]

	NEUROLOGIC	MUSCULOSKELETAL (MSK)	BODY
Without IV Contrast	(Procedure code 70551) Brain	Upper-Extremity Pain (Procedure code 73218)	Kidneys (Procedure code 74181)
	• Headaches	Nonjoint	• Renal insufficiency
	• Concern for infarct	• Arm	
	• Aneurysm	• Clavicle	
	• AVM	• Scapula	
	• Syncope	• Hand	
	Congenital Epilepsy	(Procedure code 73221)	
	• Seizure disorder	Joint	
	(Procedure code 70551)	• Shoulder	
	MRA Brain	• Wrist	
	(Procedure code 70544)	• Elbow	
	Face	• Finger	
	(Procedure code 70540)	Lower-Extremity Pain (Procedure code 73718)	
	• TMJ	Nonjoint	
		• Femur	
		• Tibia/Fibula	
		• Foot	
		(Procedure code 73721)	
		Joint	
		• Hip	
		• Knee	
		• Ankle	
		• Toe	
		(Procedure code 72195)	
		Bony Pelvis	
		• Pain	
		• Sports hernia	
		• Fracture/stress fracture	
		(Procedure code 72141)	
		Cervical Spine	
		• Pain	
		• Radiculopathy	
		• Chronic myelopathy	
		(Procedure code 72146)	
		Thoracic Spine	
		• Pain	
		• Radiculopathy	
		• Chronic myelopathy	
		(Procedure code 72148)	
		Lumbar Spine	
		• Pain	
		• Radiculopathy	
		• Chronic myelopathy	

Continued

	NEUROLOGIC	MUSCULOSKELETAL (MSK)	BODY
With/ Without IV Contrast	(Procedure code 70543) Brain/Orbits • Optic nerve tumors • Infection • Focal orbital disease • Visual loss • Optic neuritis (Procedure code 70553) Brain • Mass • Infection/inflammation • Metastatic disease • Multiple sclerosis, known • Pituitary disorders • Hypothalamic parasellar pathology • Internal auditory canal • Hearing loss/tinnitus • Vestibular disease • Trigeminal neuralgia • Cranial nerve • Focal pain with all cranial nerves • Lesions (Procedure code 70543) Neck • Vascular anatomy • Mass assessment (Procedure code 70549) MRA Neck	Upper-Extremity Pain Mass, osteomyelitis/ infection, osteoarthritis (Procedure code 73220) Nonjoint • Arm • Clavicle • Scapula • Hand (Procedure code 73223) Joint • Shoulder • Wrist • Elbow • Finger Lower-Extremity Pain Mass, osteomyelitis/ infection, Morton's neuroma (Procedure code 73720) Nonjoint • Femur • Tibia/fibula • Foot (Procedure code 73723) Joint • Hip • Knee • Ankle • Toe (Procedure code 72197) Bony Pelvis • Mass • Osteomyelitis/ infection • Sacroiliitis (Procedure code 72197) Cervical Spine • Tumor/mass • Acute myelopathy • Known MS • Osteomyelitis/discitis/ infection (Procedure code 72157) Thoracic Spine • Tumor/mass • Acute myelopathy • Known MS • Osteomyelitis/discitis/ infection (Procedure code 72158) Lumbar Spine • Tumor/mass • Acute myelopathy • Prior back surgery • Osteomyelitis/discitis/ infection	Brachial Plexus (Procedure code 77058) Breast (Procedure code 71552) Chest MRA (Procedure code 71555) • Chest • (Procedure code 74185)Renal Enterography (Procedure code 74183) Abdomen • Crohn's disease (Procedure code 72197) PELVIS • Crohn's disease (Procedure code 74183) Abdomen • Overview • Concern for mass (Procedure code 74183) Pancreas • Bile ducts • Gallbladder • Pancreas (Procedure code 72197) Female Pelvis (Procedure code 72197) Prostate (Procedure code 72197) Adnexal Mass

Note: NPO 2 hours for contrast examinations. Please obtain serum creatinine and eGFR on all patients receiving IV contrast who have diabetes or renal disease or are over the age of 75 years.
Courtesy of Rhode Island Medical Imaging (www.rimirad.com).

Note: Page numbers followed by *b*, *t*, and *f* indicate boxes, tables, and figures, respectively.